BRANDEIS UNIVERSITY
SUMMER INSTITUTE IN THEORETICAL PHYSICS, 1969

Atomic Physics and Astrophysics

Volume 2

Brandeis University
Summer Institute in Theoretical Physics

The Brandeis University Summer Institute in Theoretical Physics was first held in 1957, then annually since 1959. The lectures of all the Institutes have been published in book form. Recent volumes include the following topics:

1963
Volume 1 *Strong and Electromagnetic Interactions*
Volume 2 *Astrophysics and Weak Interactions*
(Gordon and Breach)

1964
Volume 1 *General Relativity*
Volume 2 *Particle and Field Theory*
(Prentice Hall)

1965
Volume 1 *Axiomatic Field Theory*
Volume 2 *Particle Symmetries*
(Gordon and Breach)

1966
Two Volumes *Statistical Physics, Phase Transitions and Superfluidity*
(Gordon and Breach)

1967
Two Volumes *Elementary Particle Physics and Scattering Theory*
(Gordon and Breach)

1968
Two Volumes *Astrophysics and General Relativity*
(Gordon and Breach)

1969
Two Volumes *Atomic Physics and Astrophysics*
(Gordon and Breach)

BRANDEIS UNIVERSITY
SUMMER INSTITUTE IN THEORETICAL PHYSICS, 1969

Atomic Physics and Astrophysics

Volume 2

Edited by

M. Chrétien and E. Lipworth

GORDON AND BREACH SCIENCE PUBLISHERS
New York · London · Paris

Gordon and Breach, Science Publishers, Inc.
440 Park Avenue South
New York, N.Y. 10016

Editorial office for the United Kingdom

Gordon and Breach, Science Publishers Ltd.
42 William IV Street
London W.C.1

Editorial office for France

Gordon & Breach
7–9 rue Emile Dubois
Paris 14^e

Library of Congress catalog card number 79–141420. ISBN 0 677 14910 7 (*cloth*) 0 677 14915 8 (*paper*). Printed in east Germany

LECTURERS

W. R. BENNETT, *Dunham Laboratory, Yale University*

S. J. BRODSKY, *Stanford Linear Accelerator Center, Stanford University*

A. DALGARNO, *Harvard College Observatory and Smithsonian Astrophysical Observatory, Cambridge*

D. KLEPPNER, *Massachusetts Institute of Technology*

R. NOVICK, *Columbia University*

P. G. H. SANDARS, *Clarendon Laboratory, Oxford University*

FOREWORD

The Brandeis Summer Institutes in Theoretical Physics are seeking to fill a particular role in the advanced training of physicists, by occupying a middle ground between the standard courses in a graduate curriculum and the specialized research reports of conferences and meetings. Although the subject matter of the Institute is of contemporary research interest, it is presented primarily in a few well-organized lecture courses rather than in a series of diverse reports. These lectures are intended to appeal both to the intermediate and advanced graduate students and to postdoctoral students, and are centered around one or two special topics which change from year to year. The 12th Institute, which was held from June 16 to July 25, 1969, was organized about the topic of Atomic Physics and Astrophysics. The interplay between these fields has been very pronounced in recent years, particularly on the experimental side, and this fact was recognized by the composition of the lecturing staff. Six of the eight lecturers were experimentalists and two, Novick and Barrett, now prominent in astrophysics, started their careers as experimental atomic physicists.

Dr. Barrett discussed the theory and observations of the hydrogen and helium recombination spectra, the hydrogen 21 cm line, the Λ-doublet spectrum of the hydroxyl radical, and the recently discovered lines of water, ammonia, and formaldehyde molecules.

Dr. Kleppner discussed the fine structure and hyperfine structure of atomic hydrogen and the experiments designed to obtain an accurate measurement of the hyperfine splitting and Lamb shift. He also described the hydrogen maser and its applications.

Dr. Brodsky developed the exact quantum electrodynamic theory for the hydrogenic atom. In particular he derived expressions for the Lamb shift and hyperfine splitting and compared these with the experimental results discussed by Dr. Kleppner.

Dr. Novick gave a review of the experimental techniques used in x-ray astronomy, emphasizing the very recent development of x-ray telescopes and polarimeters. He discussed the observations and theoretical models of Sco X-1, the Crab Nebula, and the diffuse x-ray background.

Dr. Marrus discussed some of the recent refinements in atomic-beam techniques. He showed how atomic-beam experiments can be used to study the hyperfine structure of atoms and measure various nuclear parameters.

Dr. Dalgarno first discussed the one- and two-photon decay models of the metastable states of atoms which are of astrophysical importance. He then considered the quantum mechanics of simple radiative and collision processes.

Dr. Sanders introduced the diagram method for solving problems in angular-momentum theory and used it to prove the Yutsis, Levinson, and Vanagas theorems. He discussed the significance of the $3nj$ symbols and gave many physical examples of the use of the method.

Dr. Bennett first discussed the properties of self-reproducing optical modes of a cavity laser. He then considered the interaction of laser radiation with excited atomic states and the methods used to obtain population inversions.

In addition to the eight lecture courses, the following colloquia were given during the course of the Summer Institute.

Dr. Hale Bradt, Massachusetts Institute of Technology: *X-Ray Observations of Pulsar NP 0532 in the Crab Nebula.*

Dr. Howard Grotch, Pennsylvania State University: *Effective Potentials and Nuclear Corrections to Hydrogen Levels.*

Dr. Robert Novick, Columbia University: *Two-Photon Decay of Metastable Hydrogenic Atoms and Muonium.*

Dr. James C. Baird, Brown University: *A Measurement of the Fine-Structure Constant by Level-Crossing Spectroscopy on Atomic Hydrogen.*

Dr. Stanley Kaufman, Yale University: *Precision Measurements of Fine-Structure Intervals in Hydrogen.*

Dr. H. Gursky, American Science and Engineering: *The Distribution of Galactic X-Ray Sources.*

Dr. Donald Yennie, Cornell University: *Vector Mesons and Nuclear Optics.*

Dr. Stuart Crampton, Harvard University: *Spin Exchange Collisions in the Hydrogen Maser.*

Dr. Ben Zuckerman, University of Maryland: *Observation of Interstellar Formaldehyde.*

Dr. M. M. Litvak, Lincoln Laboratory: *Non-Equilibrium Processes Related to Interstellar Molecules.*

Dr. D. L. Lin, State University of New York at Buffalo: *Coherence of Spontaneous Radiation and Photon Trapping.*

Dr. Alan Lurio, IBM Watson Labs: *Lifetime and Isotope Shift Measurement in the Singlet Spectrum of Helium.*

Dr. Hugh Kelly, University of Virginia: *Some Recent Many-Body Calculations in Atomic Physics: Hyperfine Structure, Polarizability, and Long-Range Interactions between Atoms.*

Dr. Leo Goldberg, Harvard University: *Microwave Recombination Lines.*

The published notes of our Summer Institutes are supposed to fill the gap between text books and review articles on one side and publications in scientific journals and conference reports on the other side. They therefore do not contain the seminar lectures mentioned above. For various reasons it was not possible to include the lectures of Dr. Barrett and Dr. Marrus in the present volumes.

The editors are deeply indebted to the lecturers and notetakers in their efforts in preparing the notes and getting them into their final form. In addition we wish to thank Mrs. J. Prentice, the Institute secretary, for her assistance in preparing the manuscripts, the publishers, in particular Mrs. A. Wick, for their conscientious work, and the National Science Foundation for the continued support and encouragement which allowed us to hold the Institute.

MAX CHRÉTIEN
EDGAR LIPWORTH
Codirectors

CONTENTS

Volume 2

CONTENTS

Volume 1

Some Aspects of the Physics of Gas Lasers

W. R. BENNETT, JR.

Dunham Laboratory, Yale University

Contents

1. PROPERTIES OF OPTICAL CAVITY MODES

1.1 Background Information

Properties of the optical cavity modes have played a key role in the development of the laser. This role is seldom shown adequate appreciation. The properties of these modes were very poorly understood before the development of the first cw lasers and, for example, one finds references to entities such as "walk-off modes" in the early literature which have since been shown to be devoid of physical meaning as applied to the oscillator. In order to appreciate the importance of the problem, it is helpful to review some very basic considerations.

The first such consideration which enters involves the degree of mode isolation required for the construction of a laser. Consider an allowed

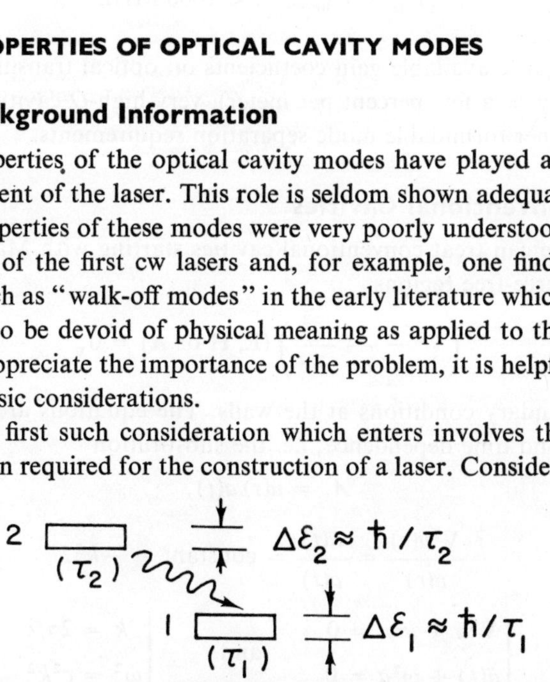

transition between a pair of excited states of an atom in which the phase interruption rates for the two levels are given by

$$R_1 = 1/\tau_1 \quad \text{and} \quad R_2 = 1/\tau_2$$

in terms of their respective lifetimes (τ_1, τ_2). From the uncertainty principle ($\Delta\varepsilon\, \Delta\tau \approx h/2\pi$) one expects the breadth of the frequency region of interaction of the atom with light to be

$$\Delta\nu \approx \frac{\Delta\varepsilon_1 + \Delta\varepsilon_2}{h} \approx \frac{R_1 + R_2}{2\pi}. \tag{1.1}$$

If R_1, R_2 are entirely determined by spontaneous radiative decay [with typical rates $\approx (10^7$ to $10^9)$/sec], then $\Delta\nu \approx 1$ to 100 MHz.

As a bare minimum requirement, it is clearly necessary to have optical modes in a laser cavity which are separated by $\gg \Delta\nu$ (1 to 1000) MHz. Otherwise many different modes would compete for the same atomic excitation and the output spectrum would be extremely noisy.

More stringent requirements on mode isolation arise if we are to insist on a single resonant frequency within the entire Doppler linewidth of an atom. Due to the atoms thermal motion, the resonant frequencies of the atoms are typically Doppler-shifted by amounts large compared to the natural width.

$$\Delta\nu_D \approx V_{\text{thermal}}/\lambda \approx 1000 \text{ MHz}$$

in the visible.

Because available gain coefficients on optical transitions are quite low (typically $\approx$ a few percent per meter), very-high-Q cavities are needed and with rather formidable mode separation requirements.

1.2 Conventional Cavities

One can treat conventional cavities starting with Maxwell's equations for source-free regions,

$$\left(\nabla^2 - \frac{1}{c^2}\frac{\partial^2}{\partial t^2}\right)(\mathbf{E}, \mathbf{H} \text{ or } \mathbf{A}) = 0, \tag{1.2}$$

and the boundary conditions at the walls. The equations are separable in their space and time dependence; i.e. the substitution

$$A_x = u(\mathbf{r})\, a(t), \tag{1.3}$$

yields

$$\frac{\nabla^2 u(r)}{u(r)} = \frac{\ddot{a}(t)}{a(t)} = \text{constant} \equiv -k^2.$$

Hence

$$\begin{cases} \nabla^2 u + k^2 u = 0 \\ \ddot{a}(t) + \omega^2 a = 0 \end{cases} \quad \text{and} \quad \begin{cases} k = 2\pi/\lambda \\ \omega^2 = c^2 k^2 \end{cases}.$$

For a cubic box with conducting walls we want spatial solutions that vanish on the boundaries. Hence,

$$u \sim (\sin k_1 x)(\sin k_2 y)(\sin k_3 z), \tag{1.4}$$

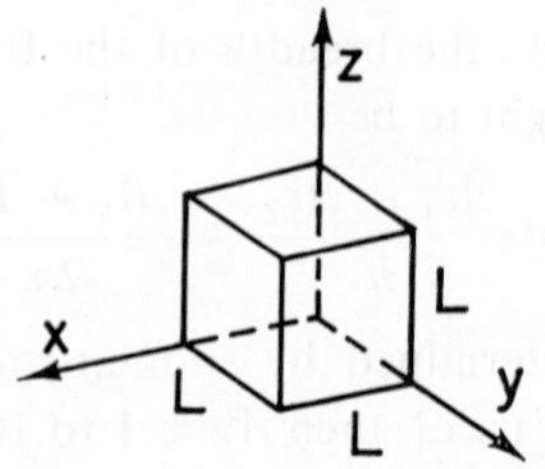

where $k_1 = n_1\pi/L$, $k_2 = n_2\pi/L$, etc. $n_1, n_2, n_3 = 1, 2, 3 \ldots$ or

$$\mathbf{k} = \frac{\pi}{L}\mathbf{n},$$

$$\mathbf{k}^2 = \left(\frac{\pi}{L}\right)^2 n^2 = \left(\frac{\omega}{c}\right)^2. \tag{1.5}$$

For $L \gg \lambda$, there are a large number of modes having the same value of n and hence approximately the same resonant frequency, ω. The different modes of the cubical box having the same value of n will describe the octant of a spherical surface of radius n constructed in a space bounded by orthogonal axes on which the positive integers n_1, n_2, n_3 are plotted. The number of modes, dN, between n and $n + dn$ (corresponding to a change in resonant frequency from ω to $\omega + d\omega$) will just be the volume of the octant of the corresponding spherical shell of thickness dn.

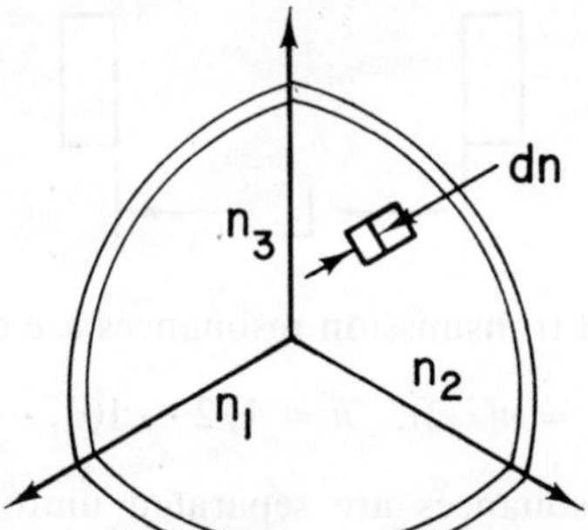

Hence

$$dN \approx 2 \times \frac{4\pi n^2\, dn}{8} = \pi\left(\frac{L}{\pi}\right)^3 \frac{\omega^2\, d\omega}{c^3}, \tag{1.6}$$

where the factor of 2 allows for the two orthogonal planes of polarization and the factor of 8 enters in the denominator because we are considering the area in the octant of a sphere. Hence

$$\frac{dN/d\nu}{L^3} = \frac{8\pi}{c^3}\nu^2 \approx 0.08 \text{ cm}^3 \text{ sec/cycle at 1 micron.} \tag{1.7}$$

Or equivalently, the mode spacing $= [(dN/d\nu)/L^3]^{-1} \approx 12$ cps-cm^3 at a wavelength of one micron.

Therefore, for $\Delta\nu_{\text{nat}} \approx 10^7$ cps, there would be $\approx 10^6$ modes within the natural width of a typical visible atomic transition with a cubic cavity of 1 cm^3 volume.

Even worse, consider a cavity made with 99 percent reflecting mirrors separated by 1 meter (suitable for a typical laser having a gain in excess of 1 percent/meter). With closed side-walls there would be $\approx 10^8$ modes within a typical Lorentz width. The probability of one mode dominating over the others would be negligible and the output would be just noise. If there were no other solution to the problem, there would be little reason to continue the present discussion beyond this point.

1.3 Mode Discrimination with Fabry–Perot Cavities

The possible use of Fabry–Perot interferometers for mode isolation was suggested by a number of people (Dicke, 1958; Prokorov, 1958; Schawlow and Townes, 1958; Gould, 1958). The classical Fabry–Perot interferometer consists of a pair of flat mirrors separated by a distance

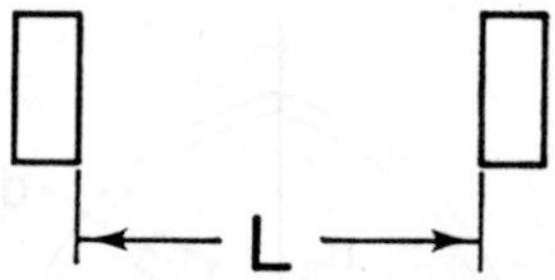

$L \gg \lambda$ and longitudinal transmission resonances are expected whenever

$$L = n(\lambda/2), \quad n = 1, 2 \cdots 10^6, \cdots.$$

These transmission resonances are separated uniformly in frequency by $\delta\nu = c/2L \approx 150$ MHz for $L = 1$ meter and are in about the general range required. It is intuitively reasonable that there would be normal modes for at least the longitudinal propagation resonances. For a standing wave to occur we need a closed loop for which the phase shift in one round trip satisfies

$$\Delta\phi = n\,2\pi = (2\pi\nu_n)\,(2L/c),$$

where $2L/c$ is the time delay around the loop, or

$$\nu_n = n(c/2L),$$

which is just the normal transmission resonance condition.

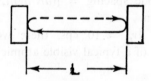

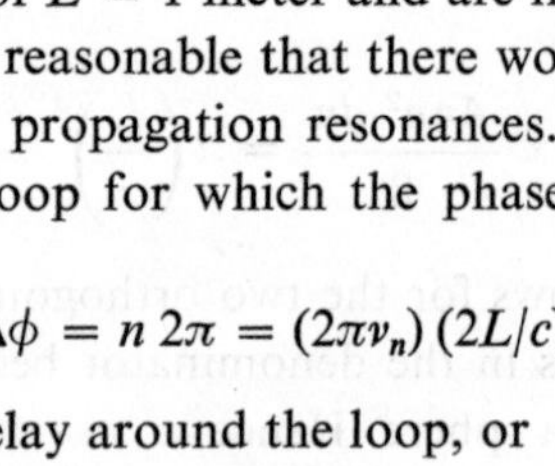

There is an important difference between passive transmission resonances in a driven device and these internal resonances which was not appreciated early in the field. With the "driven" interferometer, transmission resonances will occur whenever there are successive multiple paths through the Fabry–Perot which differ by an integral multiple of the wavelength. Thus the Heidinger fringes in a plane parallel plate or circular rings in the Fabry–Perot are observed when the device is driven with a monochromatic plane wave.

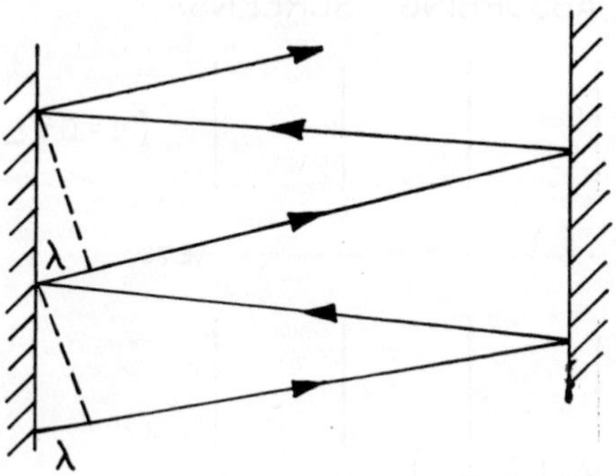

Schawlow and Townes also suggested using such "walk-off" modes in a laser cavity.

Although such passive transmission resonances are well-known, they clearly do not represent situations in which a localized field distribution launched at one point in space propagates reiteratively in a closed loop, thereby returning periodically to the same point in space from which it started. Hence, although such "walk-off" modes of propagation can lead to a greatly increased optical path through the amplifying medium, it seems clear that they can have no simple analogue in the actual steady-state resonant modes which develop in a laser.

1.4 The Fox and Li Modes

Fox and Li (1961, 1963) were the first people to give a rigorous mathematical treatment of the internal resonant mode problem in the plane parallel Fabry–Perot interferometer. Before their work on this problem, it was far from obvious that any internal resonant modes would even exist. Because of the open side walls, the boundary conditions in this type of cavity are drastically different from those previously studied in detail in the theory of microwave resonant structures.

Fox and Li took a model of the Fabry–Perot in which the successive passes through the interferometer were unfolded mathematically as shown in Fig. 1-1. They then asked the question: "Would there be self-reproducing field distributions on successive transists which would arise purely from diffraction propagation through the successive apertures?" If so, such self-reproducing field distributions would then be identified as normal resonances of the system.

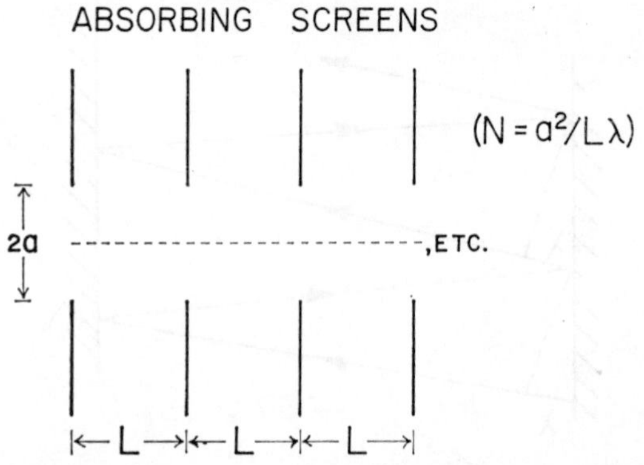

FIGURE 1-1. Model of the Fabry–Perot used by Fox and Li.

1.5 Diffraction Integrals Pertinent to the Fox and Li Modes

From Maxwell's equations, the spatial components of **E**, **H** satisfy

$$\nabla^2 u + k^2 u = 0 \tag{1.8}$$

in source-free regions. It is helpful to use a standard Green's function argument to simplify the problem (Fig. 1-2).

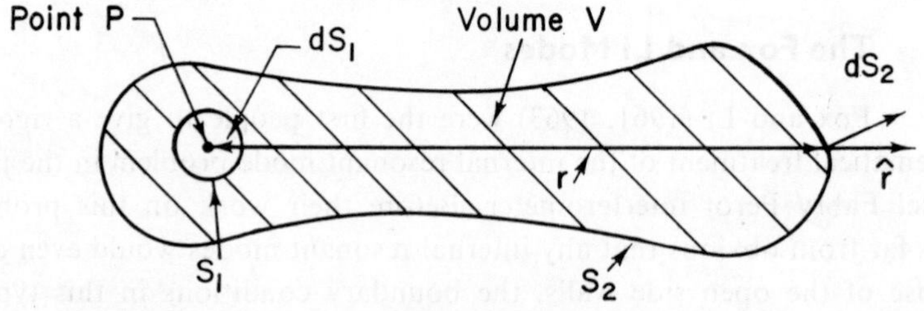

FIGURE 1-2. Bounded volume used in the Green's function argument to determine the field at point P from sources on the outer surface, S_2.

For any well-behaved functions F, G it follows from Gauss' theorem that

$$\int_V \nabla\cdot(F\nabla G - G\nabla F)\,dV = \int_V (F\nabla^2 G - G\nabla^2 F)\,dV$$
$$= \int_{S_1} (F\nabla G - G\nabla F)\cdot d\mathbf{S}_1 + \int_{S_2} (F\nabla G - G\nabla F)\cdot d\mathbf{S}_2. \tag{1.9}$$

If we let $F = e^{-ik}/r$ and $G = U$, then the integrand of the volume integral vanishes. Next, let S_1 shrink to zero about point P. Since ∇u is finite, the only nonzero term from the integral over S_1 is

$$\int -U\nabla\left(\frac{e^{-ikr}}{r}\right) dS_1 \to \frac{+Ue^{-ikr}}{r^2}4\pi r^2 \tag{1.10}$$

and

$$\lim_{r\to 0}\int_{S_1} \to U_p. \tag{1.11}$$

Therefore

$$U_p = \frac{1}{4\pi}\int_{S_2}\left[\frac{e^{-ikr}}{r}\nabla U - U\nabla\left(\frac{e^{-ikr}}{r}\right)\right]\cdot d\mathbf{S}, \tag{1.12}$$

a result which is exact so far. In the limit $r \gg \lambda$,

$$\nabla\left(\frac{e^{-ikr}}{r}\right) = -\left(ik + \frac{1}{r}\right)\frac{e^{-ikr}}{r}\hat{r} \approx \frac{-ik}{r}e^{-ikr}\,\hat{r}. \tag{1.13}$$

Note that $k = 2\pi/\lambda \approx 10^3$ and is very large compared to $1/r$ ($\approx 10^{-2}$). Therefore this approximation is good to about 10 ppm typically. Next let $\cos\theta = \hat{r}\cdot d\mathbf{S}$. Therefore,

$$U_p \approx \frac{1}{4\pi}\int_{S_2}\frac{e^{-ikr}}{r}[\nabla U\cdot\hat{n} + ik\cos\theta]\,dS, \tag{1.14}$$

where $\hat{n}$ is the outward unit normal to dS.

Next consider the case where there is only a source over a rectangular aperture on surface S_2 and that this aperture falls in the x–y plane.

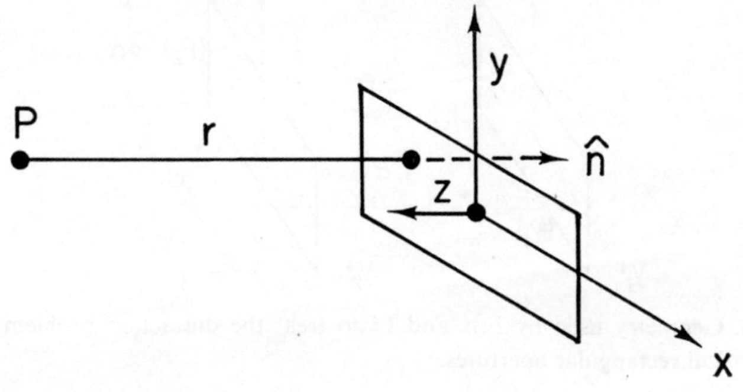

Letting

$$U(r) = A(r)\,e^{-i\phi(r)},$$

then

$$\nabla U \cdot d\mathbf{S} = U\left(\frac{\nabla A}{A} - \nabla\phi\right) \cdot d\mathbf{S}, \tag{1.15}$$

where

$$d\mathbf{S} \equiv -\hat{z}\,dS_{x,y}.$$

Assuming transvers polarization,

$$\nabla U \cdot d\mathbf{S} = U\left[-\frac{1}{A}\frac{\partial A}{\partial z} + i\frac{\partial\phi}{\partial z}\right] dS_{x,y}, \tag{1.16}$$

where the first term represents the fractional change in amplitude per unit distance and $\partial\phi/\partial z \approx k = 2\pi/\lambda$. Since $(1/A)\,(\partial A/\partial z) \ll 2\pi/\lambda$,

$$U_p \approx \frac{ik}{4\pi}\int U\frac{e^{-ikr}}{r}[1 + \cos\theta]\,dS_{x,y}, \tag{1.17}$$

which is the form of Huygen's principle with which Fox and Li started analysis of the self-reproducing modes. Note that the result is an approximation valid in the limit that the light is polarized at right angles to the laser axis and that $r \gg \lambda$.

1.6 Form for Rectangular Apertures

Consider two rectangular apertures (Fig. 1-3). From Huygen's principle,

$$U_2(x_2, y_2) = \frac{i}{2\lambda}\int_{-c}^{c}\int_{-a}^{a} U_1(x_1, y_1)\frac{e^{-ikR}}{R}\left(1 + \frac{L}{R}\right)dx_1\,dy_1, \tag{1.18}$$

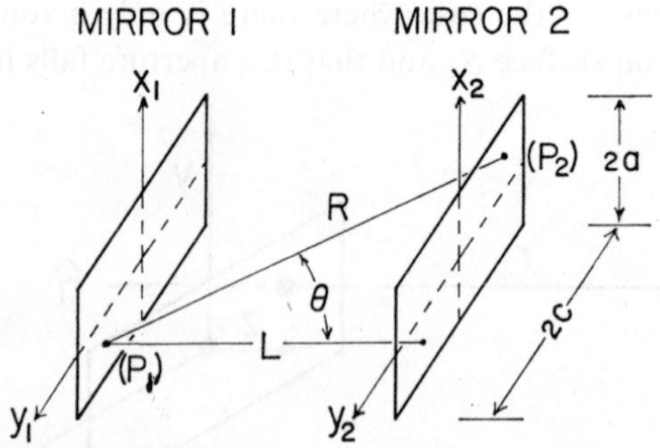

FIGURE 1-3. Geometry used by Fox and Li to treat the diffraction problem between a pair of identical rectangular apertures.

where

$$R = \sqrt{L^2 + (x_2 - x_1)^2 + (y_1 - y_1)^2}, \qquad k = 2\pi/\lambda.$$

In the limit

$$(a^2/L\lambda) \ll (L/a)^2 \quad \text{and} \quad (c^2/L\lambda) \ll (L/c)^2, \tag{1.19}$$

which frequently holds for lasers, the problem is separable into products of integrals of the type

$$U_2(x_2) = \frac{e^{i(\pi/4 - kL/2)}}{\sqrt{\lambda L}} \int_{-a}^{+a} U_1(x_1)\, e^{-i\pi N_x (x_2 - x_1)^2/a^2}\, dx_1 \tag{1.20}$$

Requirement (1.19) is, in fact, much more stringent than necessary from seaprability. In many instances the region in which approximation (1.19) fails is also a region in which the field U is negligibly small. In the latter situation, separability is still a good approximation. Also note that the problem scales with the Fresnel numbers

$$N_x = a^2/L\lambda \quad \text{and} \quad N_y = c^2/L\lambda.$$

Or, is we define x_1, x_2 in units of a (which is equivalent to normalizing the aperture to the region $-1 \leqq x \leqq 1$),

$$U_2(x_2) = e^{i(\pi/4 - \pi L/\lambda)} \int_{-1}^{+1} U_1(x_1)\, e^{-i\pi N (x_2 - x_1)^2}\, dx_1. \tag{1.21}$$

The term $e^{-i\pi L/\lambda}$ represents the geometrical phase shift and the total field is then made up of products of the type $U_2(x_2)\, U_2(y_2)$. Linear combination of such products with different polarizations can also be used to generate a variety of transverse electromagnetic waves with different polarization characteristics across the aperture.

By reiterative numerical solutions to Eq. (1.21) for the infinite-strip problem, Fox and Li showed that the field would eventually settle down to a set of normal self-reproducing modes across the aperture such that

$$U_n^{t+1}(x) = \gamma_n U_n^t(x) \tag{1.22}$$

in going from the tth to the $(t + 1)$th aperture. γ_n is a complex multiplier whose real part gives the amplitude loss per transit and hence the Q of the mode.‡ The imaginary part gives the phase shift of the mode per transit (leading in respect to the geometrical phase shift) and thereby the resonant frequency of the mode.

‡ As used here γ_n is the reciprocal of the eigenvalue used by Fox and Li.

The integral form of the eigenvalue problem for infinite-strip plane parallel mirrors corresponding to Eq. (1.22) is then

$$U_n(x_2) = (1/\gamma_n) \int_{-1}^{+1} K(x_2, x_1)\, U_n(x_1)\, dx_1, \tag{1.23}$$

where the kernel $K(x_2, x_1)$ is given by

$$K(x_2, x_1) = \sqrt{N}\, e^{i\pi/4}\, e^{-i\pi N(x_2 - x_1)^2}$$

and the geometrical phase shift is not included. Equation (1.23) is a homogeneous integral equation of the second kind. Because the kernel is continuous and symmetric (in respect to the interchange of x_1, x_2), its eigenfunctions U_n corresponding to distinct eigenvalues $1/\gamma_n$ are orthogonal over the aperture. Thus, through appropriate choice of normalization,

$$\int_{-1}^{+1} U_m(x)\, U_n(x)\, dx = \delta_{m,n}. \tag{1.24}$$

The eigenfunctions are, in general, complex and Eq. (1.24) represents a non-Hermitian type of orthogonality relation which arises because of the diffraction loss in the system.

Although analytic solutions have since been obtained to the equivalent form of Eq. (1.23) for a number of limiting cases involving curved mirror surfaces, solutions for plane parallel mirrors have only been obtained through reiterative numerical solution to Eqs. (1.21) and (1.22). This numerical process is, in fact, very much like the physical process which actually goes on in the laser cavity as the field propagates back and forth between the mirrors. One starts with some arbitrary field distribution across the aperture and lets the computer successively launch the wave back and forth between the mirrors. In this work one is assuming a rectangular aperture transmission function which is symmetric about the optical axis of the cavity. The solutions to the corresponding eigenvalue problem therefore possess definite symmetry.

In practice, it is difficult to obtain more than the first two dominant modes of opposite symmetry from the numerical solutions. That is, if the initial distribution has even symmetry about the axis, the computer solution will settle down to the dominant mode with even symmetry. If the initial distribution has odd symmetry, the dominant odd-symmetry mode will be obtained. However, in the process of settling down to a self-reproducing mode, there are clearly observable beating effects and fluctuations in the energy loss per transit which are indicative of a whole family of higher-order modes. How important or pronounced these effects are, of course, depends on how close the arbitrarily chosen distribution is to a self-reproducing mode for the particular Fresnel number.

The latter consideration may be put on a more formal basis by utilizing the completeness and orthonormality relation for the modes (Bennett, 1970a). For example, we may expand the initial field distribution, $E_y(x)$, across the aperture in terms of the set of self-reproducing modes

$$E_y^0(x) = \sum_k A_k^0 U_k(x). \qquad (1.25)$$

After t-reflections, the kth mode distribution becomes

$$U_k^t(x) = \gamma_k^t U_k(x) = (\gamma_k^0)^t\, U_k(x)\, e^{i\phi_k t}, \qquad (1.26)$$

where we have written γ_k in polar form,

$$\gamma_k \equiv \gamma_k^0\, e^{i\phi_k}.$$

Consequently, after t transits the initial field becomes

$$E_y(x, t) = \sum_k A_k^t U_k(x)\, e^{i\phi_k t}, \qquad (1.27)$$

where

$$A_k^t = A_k^0(\gamma_k^0)^t.$$

Letting

$$E_y^*(x, t) = \sum_j (A_i^t)^*\, U_j^*(x)\, e^{-i\phi_j t}$$

and expanding

$$U_j^*(x) = \sum_r C_{jr} U_r(x),$$

the intensity is given as a function of position across the aperture by

$$|E_y(x, t)|^2 = \sum_{j,k,r} (A_j^t)^*\, A_k^t C_{jr} U_j(x)\, U_r(x)\, e^{i(\phi_k-\phi_j)t} \qquad (1.28)$$

after t transits. Using Eq. (1.24), the integrated intensity across the aperture after t transits is therefore

$$\int_{-1}^{1} |E_y(x, t)|^2\, dx = \sum_{j,k} (A_j^t)^*\, C_{jk} A_k^t e^{i(\phi_k-\phi_j)t}, \qquad (1.29)$$

where

$$C_{jk} = \int_{-1}^{1} U_j^*(x)\, U_k(x)\, dx = C_{kj}^*$$

and

$$C_{jj} = \int_{-1}^{1} |U_j(x)|^2\, dx.$$

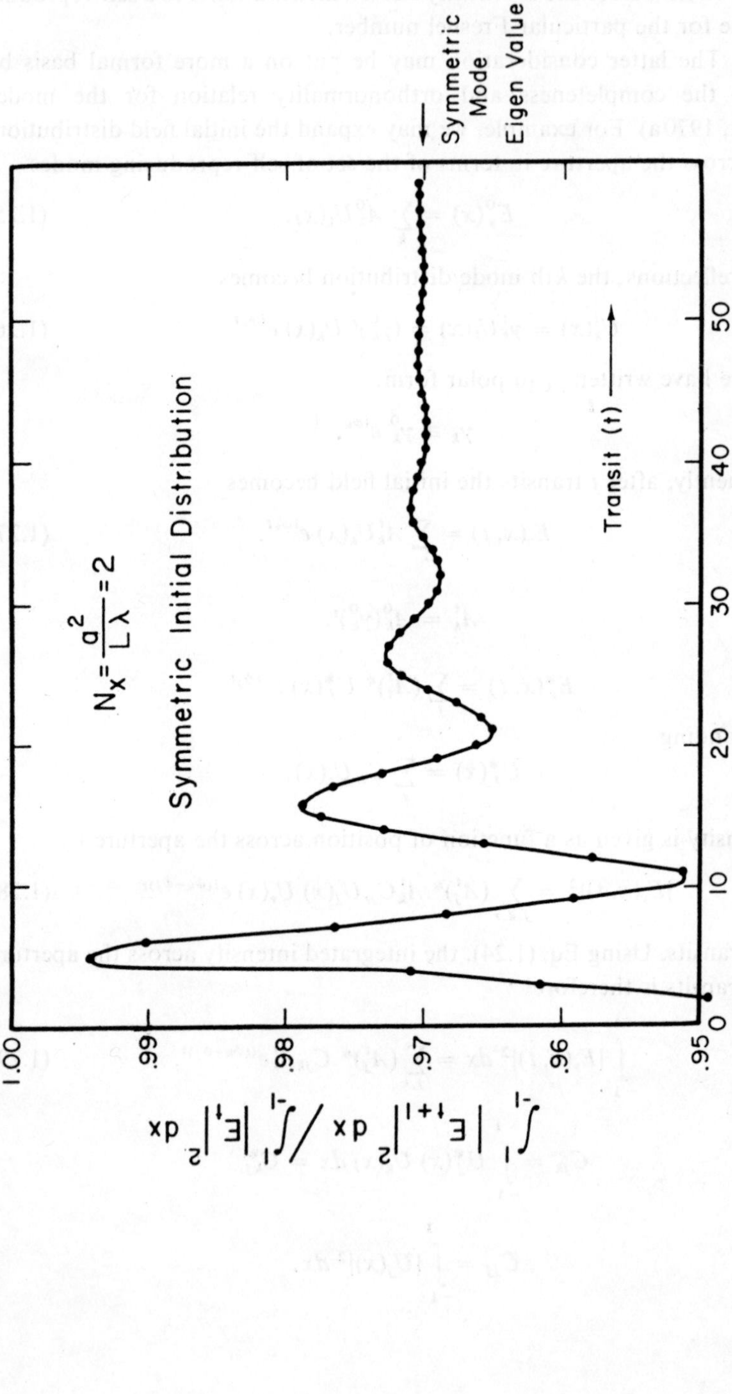

FIGURE 1-4. Beating effects in the average intensity loss per transit due to mixing of higher-order modes based on a calculation performed by the author (Bennett, 1970a).

(The coefficients C_{jk} would have satisfied $C_{jk} = \delta_{jk}$ if the self-reproducing mode distributions satisfied a Hermitian orthogonality condition.) The exponential factor in Eq. (1.29) gives rise to very observable beating effects in the energy loss per transit (see Fig. 1-4). By starting with field distributions of a definite symmetry, it is possible to make at least a rough estimate of the eigenvalues for the next higher-than-dominant modes of even and odd symmetry from these beating effects. This beating effect is illustrated in Fig. 1-4 for plane parallel infinite-strip modes for $N_x = 2$ and an arbitrary symmetric initial distribution of excitation.

Eventually, after a large number of transits, the field settles down to the dominant mode having the symmetry of the initially-assumed wave. A number of representative infinite-strip mode distributions are shown in Fig. 1-5 (Bennett, 1970a). As noted above, the mode is only strictly defined across the aperture. However, one can calculate the field outside the aperture to the extent that the field is negligible in the region where approximation (1.19) fails.

A source of practical difficulty arises in the region outside the aperture because of the finite size of the numerical integration interval. For the dominant modes, the field is a slowly varying function over the aperture and is modulated by

$$\cos[N(x_2 - x_1)^2] - i\sin[N(x_2 - x_1)^2].$$

At some point outside the aperture the frequency of this modulation term will become equal to the reciprocal of the numerical integration interval. In order to avoid the spurious, large oscillation which would occur in the value of the integral, the results in Fig. 1-5 were obtained through the use of a smoothly varying, truncating envelope which was chosen to satisfy several normalization requirements and the additional requirement that

$$\int_1^\infty |E|^2\,dx \bigg/ \int_0^1 |E|^2\,dx \tag{1.30}$$

equal the energy loss determined from the eigenvalue.

The magnitude of the eigenvalues for the dominant even- and odd-symmetry infinite-strip modes is shown in Fig. 1-6 as a function of Fresnel number. Because the dominant mode distribution is peaked toward the axis, the magnitude of the diffraction loss for this mode can be much smaller than might be expected from ordinary plane-wave diffraction theory. The latter property of these modes is, of course, of considerable practical value for the construction of lasers.

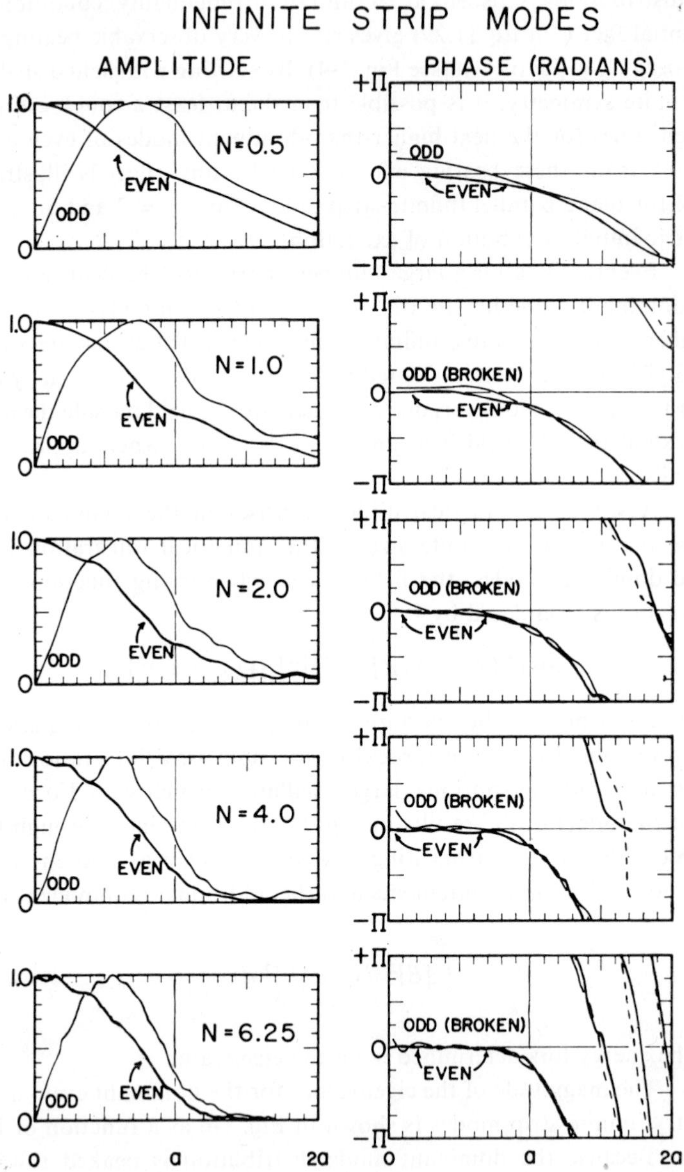

FIGURE 1-5. Infinite-strip modes computed by the author (Bennett, 1970 a) using the method of Fox and Li. The field outside the aperture was calculated using the approximation described in the text.

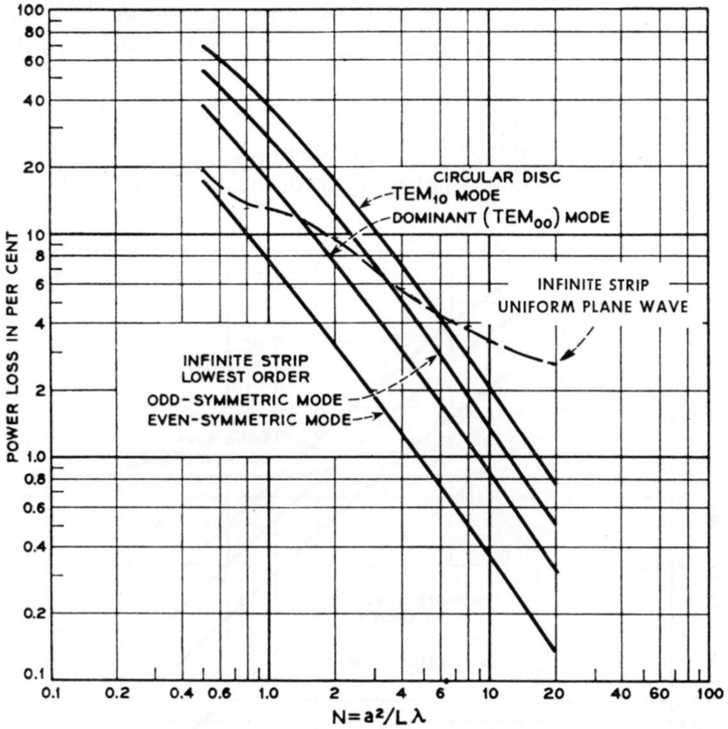

FIGURE 1-6. Power loss for the dominant even- and odd-symmetry infinite-strip modes determined by Fox and Li (1961, 1963). For comparison, values of the power loss for the dominant modes for plane circular mirrors are also shown. The dashed curve shows diffraction loss computed by the author for propagation of an infinite strip, uniform plane wave.

Also shown in Fig. 1-6 are similar results for plane circular mirrors having the same Fresnel number. Note that the dominant mode for a square-aperture cavity can be made up from the product of two infinite-strip modes about orthogonal axes. Thus the square-aperture TEM_{00} mode is given by

$$V_{00}(x, y) = U_0(x)\, U_0(y) \tag{1.31}$$

for which

$$\gamma_{00} = \gamma_0^2 .$$

Consequently, the dominant symmetric mode of the square-aperture cavity has approximately twice the diffraction loss of the infinite-strip mode at large values of N and therefore, from Fig. 1-6, has approximately the same loss as the circular plane mirror cavity with $N = N_x = N_y$.

More complex rectangular aperture modes can, of course, be made up by taking various linear combinations of products of different infinite-strip modes.

Values of the phase shift for the dominant infinite-strip modes are shown in Fig. 1-7, along with similar quantities for circular plane parallel apertures. Again note that the phase shift for the dominant TEM_{00} circular-

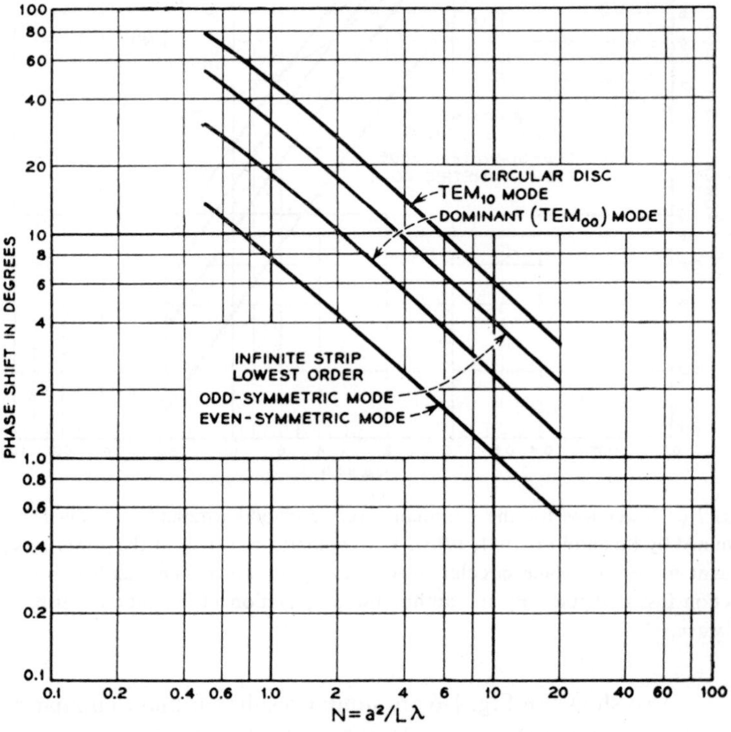

FIGURE 1-7. Phase shift per transit for the dominant infinite-strip and circular-aperture plane-parallel modes determined by Fox and Li (1961, 1963).

aperture mode is roughly twice that for the TEM_0 infinite-strip mode. The difference between the phase shifts for these modes removes the frequency degeneracy in the resonance condition. Namely, the resonance is shifted by

$$\delta\nu = \left(\frac{\delta\phi}{\pi}\right)\left(\frac{c}{2L}\right)$$

from the resonance determined by the geometric phase shift. Thus the mode spectrum for plane parallel mirror cavities exhibits additional structure.

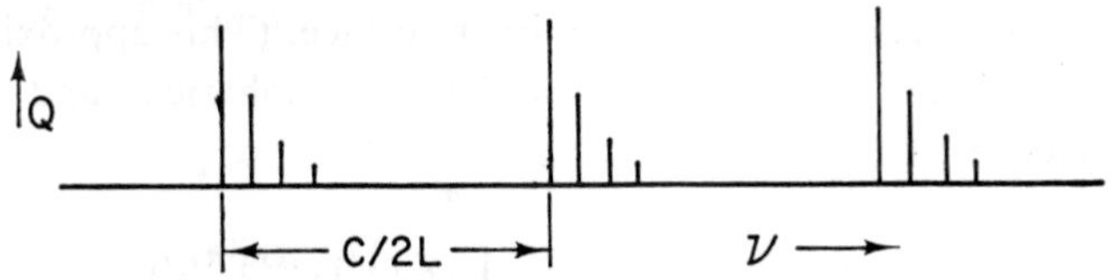

The existence of resonance frequencies separated by $\ll c/2L$ was demonstrated by internal beat detection with the first gas laser (Javan, Bennett, and Herriott, 1961).

1.7. Confocal Cavities

Analyses of confocal cavities have been given by Fox and Li (1961) for circular apertures, by Boyd and Gordon (1961) for square apertures, and in more generalized form in the square-aperture limit by Boyd and Kogelnick (1962).

The diffraction problem is again separable in orthogonal transverse coordinates (in both the rectangular and circular case) in much the same approximation as considered before for plane parallel mirrors. The normal modes are much more tightly confined to the axis in the confocal case and condition (1.19) is again more stringent than necessary for rectangular geometry.

The basic geometry for the confocal case is shown in Fig. 1-8. $\cos\theta \approx 1$ and the resulting integral equation is

$$\gamma_m \gamma_n f_m(x)\, f_n(y) \approx \int\int_{-a}^{+a} \frac{ik}{2\pi\varrho} e^{-ik\varrho} f_m(x')\, f_n(y')\, dx'\, dy'. \tag{1.32}$$

Second-order variations in ϱ may be neglected, and from the geometry

$$\frac{\varrho}{R} = 1 - \frac{xx' + yy'}{R^2} + O\left(\frac{x^2y^2}{R^4}\right), \tag{1.33}$$

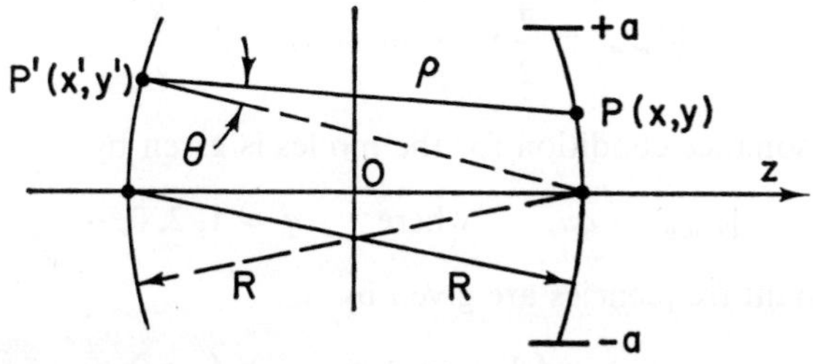

FIGURE 1-8. Relationship of variables used in the confocal cavity analysis by Boyd and Gordon (1961).

where x, y are measured along one mirror surface. (This approximation is equivalent to ignoring the difference between spherical and parabolic surfaces.) Hence

$$\gamma_m f_m(x) \approx \sqrt{\frac{ik}{2\pi R}}\, e^{-ikR/2} \int_{-a}^{+a} f_m(x')\, e^{ikxx'/R}\, dx'. \tag{1.34}$$

With the substitution

$$\xi = \sqrt{2\pi N}\,(x/a) \qquad \text{where} \qquad N = a^2/R\lambda,$$

this equation may be expressed in the form of the integral equation (Boyd and Gordon, 1961)

$$F_m(\xi) = \frac{1}{\sqrt{2\pi}\,\chi_m} \int_{-\sqrt{2\pi N}}^{+\sqrt{2\pi N}} F_m(\xi')\, e^{i\xi\xi'}\, d\xi' \tag{1.35}$$

studied by Slepian and Pollak (1961). They have shown solutions to be

$$F_m\left(2\pi N, \xi/\sqrt{2\pi N}\right) \propto S_{0m}\left(2\pi N, \xi/\sqrt{2\pi N}\right) \tag{1.36}$$

and

$$\chi_m = 2\sqrt{N}\, i^m R_{0m}^{(1)}(2\pi N, 1), \tag{1.37}$$

where S_{0m} and $R_{0m}^{(1)}$ are, respectively, the angular and radial wave functions in prolate spheroidal coordinates as defined by Flammer (1957). Similar solutions may be defined in the y direction, and the total field will be given as products of such solutions.

The eigenvalues for the product solution, $f_m(x)\, f_n(y)$, are of the form

$$\gamma_m \gamma_n = i\chi_m \chi_n\, e^{-ikR} \propto i^{(m+n+1)}\, e^{-ikR}. \tag{1.38}$$

Hence the phase shift for such a (TEM_{mn}) mode in one transit between the confocal surfaces is

$$|\phi_{m,n}| = \frac{\pi}{2}(1 + m + n) - 2\pi \frac{R}{\lambda}. \tag{1.39}$$

Hence the resonance condition for the modes is given by

$$|\phi_{m,n}| = q\pi, \qquad \text{where} \qquad q = 1, 2, 3, \cdots \tag{1.40}$$

and the resonant frequencies are given by

$$\nu_{m,n,q} = \left(\frac{1 + m + n}{2} + q\right)\left(\frac{c}{2R}\right), \tag{1.41}$$

where $R = L$, the mirror separation. Note that although successive modes ($\Delta q = \pm 1$) with the same m, n are spaced by $c/2L$, modes with the same q and n and $\Delta m = \pm 1$ are separated by half that amount. Hence adjacent modes of opposite symmetry are spaced by $c/4L$. There is also a high degree of degeneracy in these modes, as indicated by Eq. (1.41).

The diffraction loss per transit for the dominant confocal modes can be extremely small compared to either plane-wave diffraction loss or the loss for plane parallel mirrors. The latter is illustrated in Fig. 1-9 for circular aperture confocal and plane parallel mirrors.

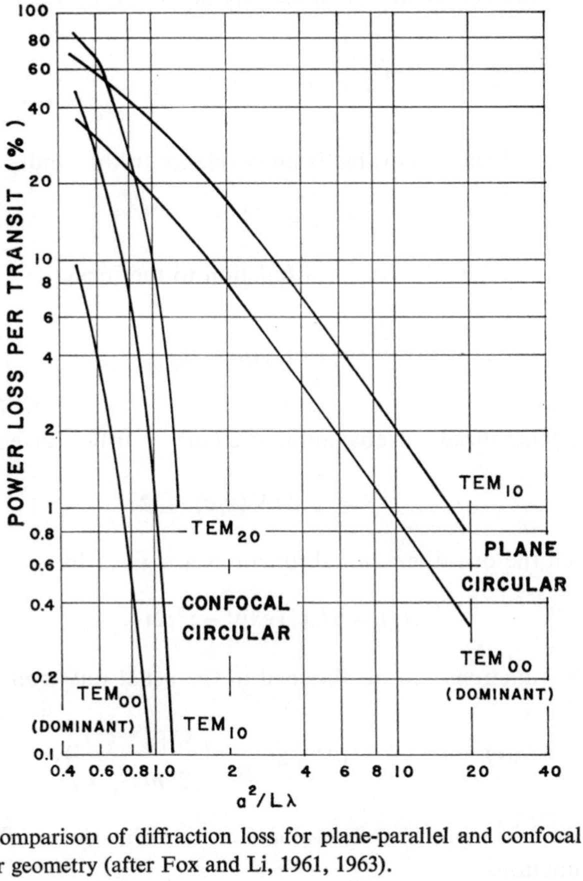

FIGURE 1-9. Comparison of diffraction loss for plane-parallel and confocal modes with circular mirror geometry (after Fox and Li, 1961, 1963).

For Fresnel numbers, $N \gg 1$, the diffraction loss is totally negligible in respect to reflectance loss from the best available mirrors. In this limit, as noted by Boyd and Gordon (1961) Flammer has shown that the confocal mode amplitude distributions are given closely by the products of Hermite polynomials and a Gaussian,

$$f_m(x) = H_m\left(\sqrt{2\pi N}\, x/a\right) \exp[-\pi N(x/a)^2], \tag{1.42}$$

where $N = a^2/\lambda R$ and $2a$ is the full aperture width in the x-direction. (These functions are, of course, the solutions to the one-dimensional harmonic oscillator in Schrödinger theory—a coincidence which doesn't appear to have any particular physical significance.) The large aperture limit for which Eq. (1.42) holds is equivalent to the neglect of diffraction loss. Hence in the limit for which the modes are described by Eq. (1.42),

$$\int_a^{\infty} |f_m(x)|^2\, dx \ll 1 \tag{1.43}$$

or the modulus of the eigenvalue is unity. Hence in this limit

$$\gamma_{m,n} \approx e^{i\phi_{m,n}}$$

and we may normalize the confocal function to the form

$$\int_{-\infty}^{\infty} f_m(\xi)\, f_n(\xi)\, d\xi = \delta_{m,n}, \tag{1.44}$$

where ξ is a generalized variable along the mirror surface given by

$$\xi = \sqrt{2\pi N}\,(x/a) \tag{1.45}$$

and for which the one-dimensional functions are given by

$$f_m(\xi) = H_m(\xi) \exp(-\xi^2/2). \tag{1.46}$$

Although the functions can be obtained in the usual way from the relation

$$f_m(\xi) = (-1)^m \exp(\xi^2/2) \frac{d^m}{d\xi^m}\left(\frac{\exp(-\xi^2)}{2^m m!\,\sqrt{\pi}}\right), \tag{1.47}$$

it is more useful in practice to evaluate them from the confluent hypergeometric functions

$$F(a, c, \xi^2) = 1 + \left(\frac{a}{c}\right)\xi^2 + \frac{a(a+1)}{2!\, c(c+1)}\xi^4 + \frac{a(a+1)(a+2)}{3!\, c(c+1)(c+2)}\xi^6 + \cdots. \tag{1.48}$$

Noting that (see, for example, Morse and Feshbach, 1953)‡

$$H_m(\xi) = (-1)^{m/2}\left(\frac{m!}{(m/2)!}\right)F\left(-\frac{m}{2}, \frac{1}{2}, \xi^2\right), \qquad m = 0, 2, 4, \ldots,$$

$$H_m(\xi) = 2(-1)^{(m-1)/2}\left(\frac{m!}{[(m-1)/2]!}\right)\xi F\left(-\frac{(m-1)}{2}, \frac{3}{2}, \xi^2\right), \tag{1.49}$$

$$m = 1, 3, 5, \ldots,$$

The series for $F(a, c, \xi^2)$ terminates when $a = -m/2$. It is therefore a very practical matter to evaluate these functions with great accuracy for the mth-order Hermite polynomials as a subroutine in a computer program. In this way the mode distributions for arbitrary order may be readily handled numerically as a function of ξ. The intensity distributions for the first ten confocal modes generated in this manner are illustrated in Fig. 1-10.

The above results hold for the fields on the confocal mirror surfaces, which are surfaces of constant phase. Boyd and Gordon (1961) showed that the transverse electric field in the self-reproducing modes vary throughout the rest of space as

$$E(\xi, \eta, \zeta) = \sqrt{\frac{2}{1+\zeta^2}}\, H_m\left(\xi\sqrt{\frac{2}{1+\zeta^2}}\right) H_n\left(\eta\sqrt{\frac{2}{1+\zeta^2}}\right)\exp\left(-\frac{\xi^2+\eta^2}{1+\zeta^2}\right)$$
$$\times \exp[-i\phi(\xi, \eta, \zeta)] \tag{1.50}$$

for the TEM_{mn} mode, where the generalized coordinates

$$\zeta = \frac{2z}{R_0}, \quad \xi = \sqrt{2\pi N}\left(\frac{x}{a}\right), \quad \eta = \sqrt{2\pi N}\left(\frac{y}{a}\right)$$

have been introduced and the phase term may be written

$$\phi(\xi, \eta, \zeta) = \frac{kR_0}{2}\zeta + \frac{\zeta(\xi^2+\eta^2)}{1+\zeta^2} - (1+m+n)\left(\frac{\pi}{4} - \psi\right)$$

with

$$\tan\psi = \frac{1-\zeta}{1+\zeta}.$$

In present work, the quantity z and the phase have been defined to be zero at the median plane, where $\psi = \pi/4$. The median plane ($z = 0$) occurs halfway between the two initial confocal surfaces having radii of curvature, R_0.

‡ Note that there is an error in Morse and Feshbach (1953) in the corresponding expression for the odd-order Hermite polynomials in the argument of the confluent hypergeometric function.

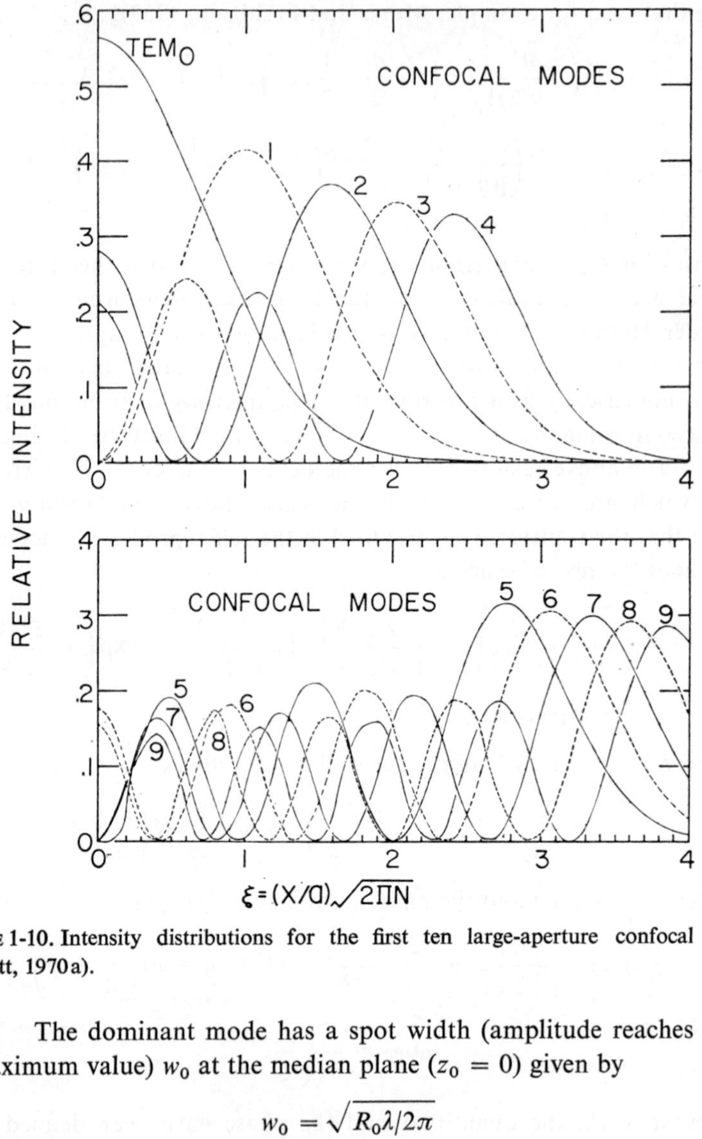

FIGURE 1-10. Intensity distributions for the first ten large-aperture confocal modes (Bennett, 1970a).

The dominant mode has a spot width (amplitude reaches $1/e$ of its maximum value) w_0 at the median plane ($z_0 = 0$) given by

$$w_0 = \sqrt{R_0\lambda/2\pi}$$

and increases with z_0 away from the median plane as

$$w_z \approx w_0\sqrt{1 + (2z_0/R_0)^2}.$$

Boyd and Gordon (1961) noted that both inside and outside the original confocal surfaces, the surfaces of constant phase were approximately spherical with radius of curvature, R_s, providing the slowly-varying term involving ψ and higher-than-quadratic terms in the transverse coordinates were neglected.‡

In the latter approximation, it is apparent from geometry that a spherical surface of radius, R_s, passing through the point x, z will intersect the z-axis at the point z_0 such that

$$z_0 - z \approx \frac{x^2}{2R_s}.$$

Hence the condition for a surface of constant phase in the Boyd and Gordon approximation becomes

$$z_0 - z \approx \frac{\zeta_0}{(1 + \zeta_0^2)} \cdot \frac{(x^2 + y^2)}{R_0} \equiv \frac{(x^2 + y^2)}{2R_s}$$

and the radius of curvature of a general surface of constant phase is given by

$$R_s \approx z_0 + \frac{R_0^2}{4z_0}$$

where z_0 is the intercept of this surface on the z-axis. Note that $R_s = R_0$ on the original confocal surface ($z_0 = R_0/2$) and that $R_s \to \infty$, both at the median plane ($z_0 = 0$) and infinitely far from the median plane ($z_0 \to \infty$). It thus becomes clear that, as illustrated schematically in Fig. 1-11, there is a continuous set of surfaces of constant phase which intersect the z-axis over the region $-\infty \leqq z_0 \leqq +\infty$ which could be used to generate the same set of confocal modes characteristic of the original pair of spherical mirrors separated by their common radii of curvature, R_0. Thus a spherical mirror with radius of curvature R placed a distance $R/2$ from a flat mirror is an easily-constructed and well-known form of the confocal equivalent cavity. There are, of course, an infinite number of other possible confocal-equivalent combinations. The relative phase shifts (relative in respect to the geometric phase shift and defined in respect to the median plane) are illustrated in a few representative cases in Fig. 1-11.

The question of "bending" confocal equivalent cavities seems not to have been specifically discussed in the literature. It is therefore worth noting that small angular displacements of the laser cavity axis can be achieved within the same approximation used by Boyd and Gordon to

‡ For consistency in the present work, different notation is used here than in Boyd and Gordon (1961) or Boyd and Kogelnick (1962).

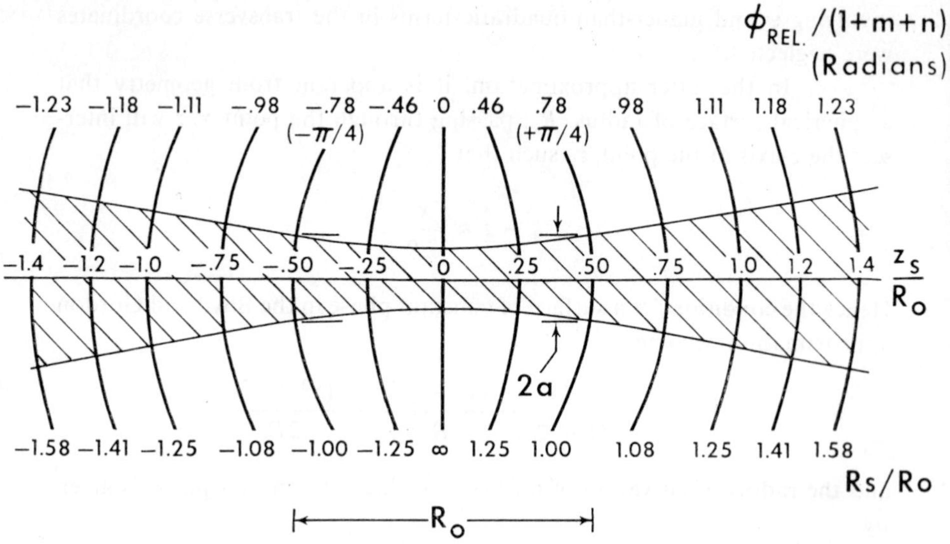

FIGURE 1-11. Representative surfaces of constant phase for confocal equivalent geometry.

describe spherical surfaces of constant phase if the mirror used has a radius of curvature which matches that of the surface of constant phase intersecting the laser axis at the mirror location. Thus, for example, a plane parallel mirror (or plane diffraction grating or prism) would be most appropriately used to bend the laser axis at the median plane; a mirror of radius, R_s, at z_0; etc.

Several specific confocal equivalent cavities are illustrated in Figs. 1-12 through 1-14 which have been investigated by Beiser (1968) and Pole and Myers (1966) for use in internally-scanned lasers. These are all very similar conjugate confocal cavities which are formally equivalent to two normal confocal cavities in series. Thus they have relative phase shifts (in respect to geometric phase) of

$$\Delta\phi = (1 + m + n)\pi$$

between conjugate surfaces. Hence the odd-symmetry modes have a relative phase shift of π-radians in respect to the even-symmetry modes. Since any intensity distribution on the surface of the first mirror can be expanded in terms of even and odd symmetric parts, it is clear that the intensity distribution on the second mirror is inverted in respect to that on the first mirror. Hence the relative phase shifts for the normal confocal modes make sense in terms of the known inverting properties of such an optical system.

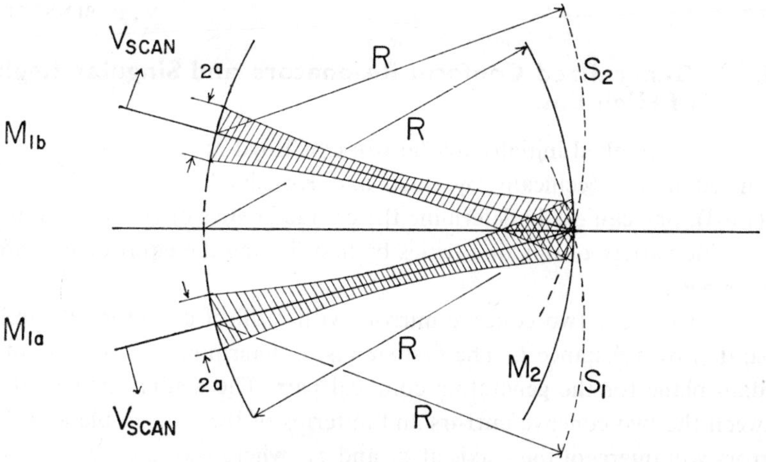

FIGURE 1-12. Split confocal-euqivalent resonator considered by Beiser for use in a scanning laser. In internal scanning applications two identical apertures of width "2a" would sweep in opposite directions as indicated by the arrows. (S_1 and S_2 are locations of mirror surfaces for confocal cavities with M_{2a} and M_{1b}, respectively.) (Bennett, 1970a).

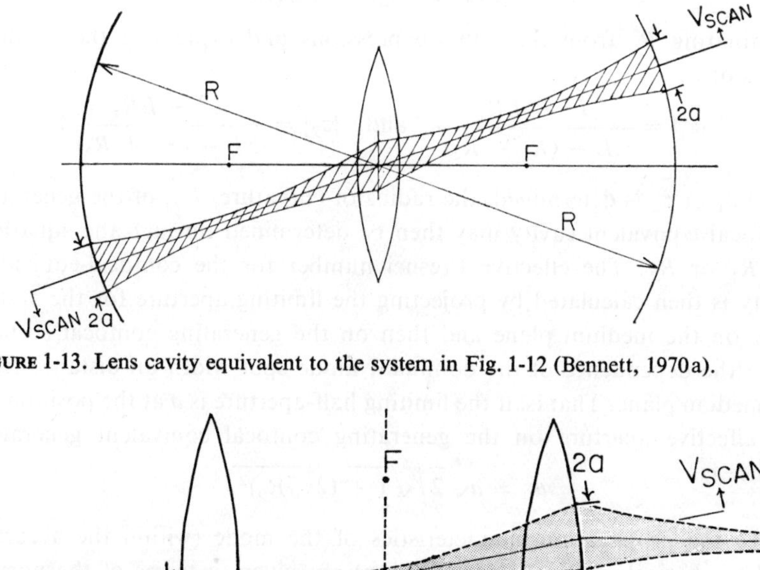

FIGURE 1-13. Lens cavity equivalent to the system in Fig. 1-12 (Bennett, 1970a).

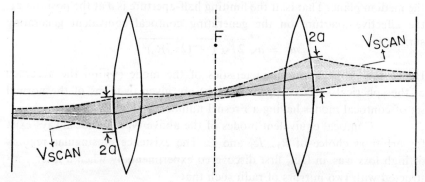

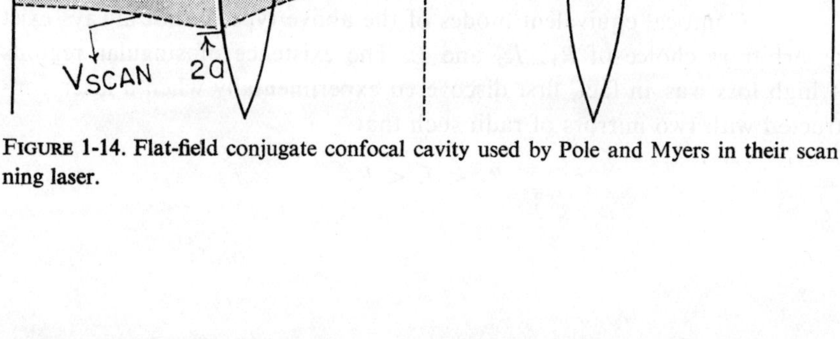

FIGURE 1-14. Flat-field conjugate confocal cavity used by Pole and Myers in their scanning laser.

1.8. Generalized Confocal Resonators and Singular Regions of High Loss

As implied initially by the work of Boyd and Gordon (1961) and discussed more specifically by Boyd and Kogelnick (1962) and Fox and Li (1963), one can easily determine the characteristics of the normal modes for a wide variety of mirror cavities by first finding the equivalent confocal mirror pair.

Consider two concave mirrors with radii of curvature R_1 and R_2 separated by a distance L. The first step is to determine the location of the median plane for the generating confocal pair. The median plane will fall between the two concave mirrors and in terms of the median plane the two mirrors will intercept the z-axis at z_1 and z_2, where

$$L = |z_1| + |z_2|$$

It follows from the discussion in the preceding section that

$$R_1 \approx |z_1| + R_0^2/4|z_1| \quad \text{and} \quad R_2 \approx |z_2| + R_0^2/4|z_2|.$$

Eliminating R_0 from these two expressions and expressing the result in terms of L,

$$|z_1| = \frac{L^2 - LR_2}{2L - (R_1 + R_2)} \quad \text{and} \quad |z_2| = \frac{L^2 - LR_1}{2L - (R_1 + R_2)}.$$

Once z_1 or z_2 is determined, the radius of curvature, R_0, of the generating confocal equivalent cavity may then be determined through the equations for R_1 or R_2. The effective Fresnel number for the confocal-equivalent cavity is then calculated by projecting the limiting aperture for the system back on the medium plane and then on the generating confocal surface, using the dependence of the dominant mode spot width on distance from the median plane. That is, if the limiting half-aperture is a at the position z_a, the effective aperture on the generating confocal equivalent generating surface is

$$a_0 = a\sqrt{2}/\sqrt{1 + (2z_a/R_0)^2}.$$

Hence the propagation characteristics of the mode (within the accuracy of the spherical surface approximation) are given in terms of the normal set of confocal modes having a Fresnel number $N = a_0^2/R_0\lambda$.

Confocal equivalent modes of the above type do not always exist for arbitrary choice of R_1, R_2 and L. The existence of singular regions of high loss was, in fact, first discovered experimentally when a laser constructed with two mirrors of radii such that

$$R_1 < L < R_2$$

did not oscillate. The source of the singularity in this case is apparent from the form of the equations given above for z_1 and z_2. As shown in considerable detail by Fox and Li (1963) and Boyd and Kogelnick (1962), there are a number of regions where confocal equivalent solutions simply do not exist and where the diffraction loss becomes very high. The problem scales with the parameters

$$g_1 \equiv 1 - L/R_1,$$

$$g_2 \equiv 1 - L/R_2.$$

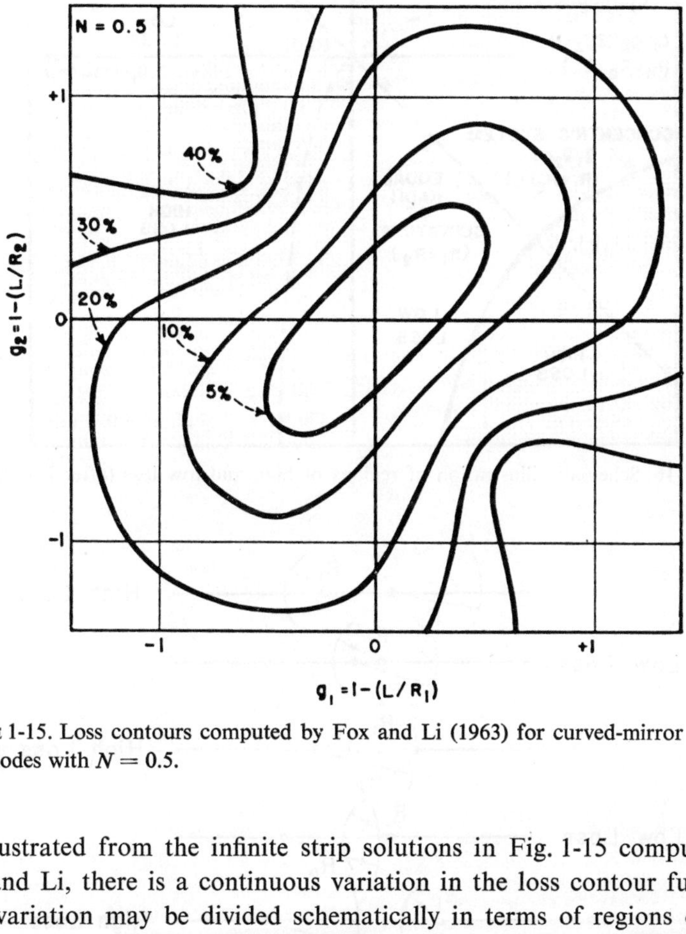

FIGURE 1-15. Loss contours computed by Fox and Li (1963) for curved-mirror infinite-strip modes with $N = 0.5$.

As illustrated from the infinite strip solutions in Fig. 1-15 computed by Fox and Li, there is a continuous variation in the loss contour function. This variation may be divided schematically in terms of regions of high and low loss as illustrated in Fig. 1-16. It is helpful in the analysis of confocal equivalent mode problems to consider the specific cases illustrated in Fig. 1-17.

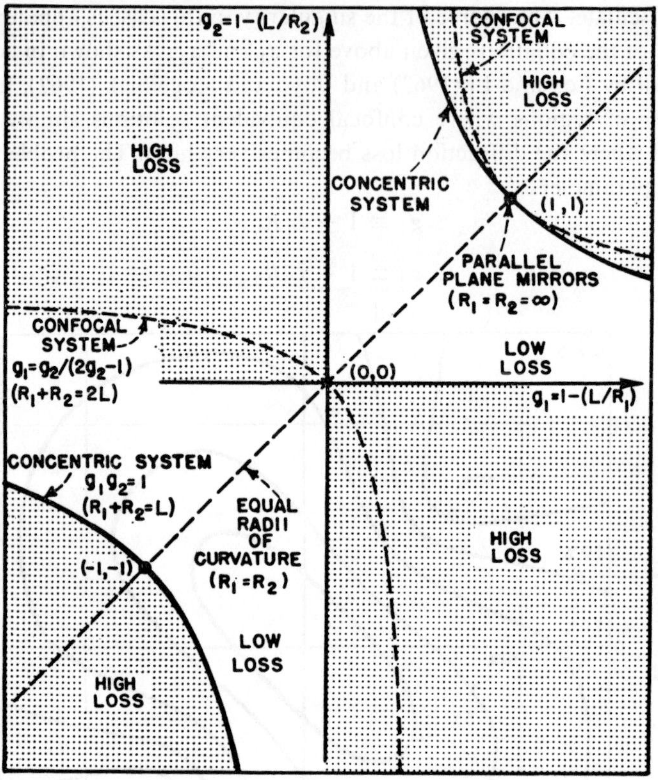

FIGURE 1-16. Schematic illustration of regions of high and low loss (after Fox and Li, 1963).

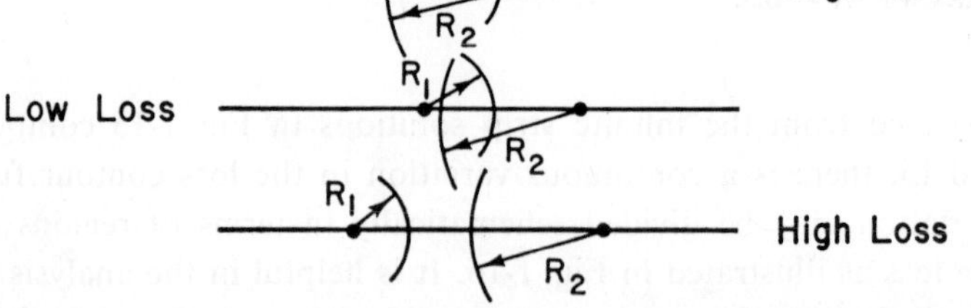

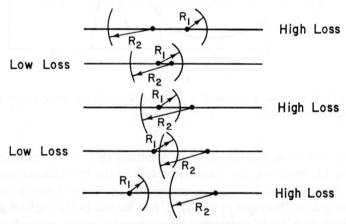

FIGURE 1-17. Some specific examples of high- and low-loss regions obtained by changing the spacing between mirrors of different, fixed radii of curvature (after Fox and Li, 1963).

1.9. Output Coupling

The existence of dissipative losses (typically $\lesssim$ 1 percent/per pass) in addition to output coupling loss from the mirror transmission coefficient implies that there will be an optimum coupling coefficient determined by the unsaturated single-pass gain in the laser medium. This optimum value can be estimated with sufficient accuracy for most applications from a relatively simple argument given originally by the author (Bennett, 1962a):

Let the total fractional energy loss per pass be

$$f = L_d + T$$

where L_d is the sum of the dissipative losses and T is the average coupling loss per pass (e.g., T = half the "dumping mirror" transmission coefficient if the beam is only taken out of one end and a high reflectance mirror is used at the other end of the cavity). Steady-state occurs in the laser when the actual gain per pass at the oscillation frequency, G, saturates at the total per pass, f. The reduction in gain, $(G - f)$, is accomplished through stimulated emission and the total laser power is therefore roughly proportional to $(G - f)$. Only a fraction, T/f, of this power actually gets out of the cavity in the laser beam. Hence

$$P_{\text{out}} \approx C(G - f)(T/f)$$

where C is an undetermined constant. The optimum coupling coefficient is determined by setting $\partial P_{\text{out}}/\partial T = 0$: hence,

$$T_{\text{opt}} \approx \sqrt{GL_d} - L_d$$

Consequently, the optimum output power is given by

$$P_{\text{opt}} \approx C(\sqrt{G} - \sqrt{L_d})^2$$

and

$$P_{\text{out}}/P_{\text{opt}} \approx \frac{(G - L_d - T)\,T}{(\sqrt{G} - \sqrt{L_d})^2\,(L_d + T)}$$

Rigrod (1963a, 1965) obtained essentially the same results through a more formal analysis of saturation effects in high-gain homogeneously-broadened lasers and he also performed a systematic experimental study of the effects of coupling in a helium-neon laser.

In reality, most gas lasers are inhomogeneously-broadened and a formal analysis of the optimum coupling problem which allowed for localized gain saturation would yield optimum coupling constants that varied with the frequency and line widths among other parameters. The present results should be reasonably accurate for single-mode oscillation near the peak of the line, if the maximum gain at the Doppler center is used for G in the above equations and power broadening of the hole widths can be ignored (see Chapter 2). The results also should be valid far above threshold in multi-mode oscillation if an average value of the gain $\left(\approx G_{max}/\sqrt{2}\right)$ over the Doppler line is used in place of G. The intermediate domain is very cumbersome. However, one seldom knows the actual dissipative loss within a real laser cavity to within a factor of $\sqrt{2}$ anyway.

1.10. Mode Selection

Although the use of Fabry–Perot cavities greatly reduces the mode density over that holding in a normal microwave-type cavity resonator, the density of hig-Q modes which is left is still bothersome for some applications of lasers. Obtaining a definite mode of linear polarization has become a routine practice through the incorporation of Brewster angle windows in the laser cavity (Rigrod *et al.*, 1962) and the remaining mode isolation problem generally takes two distinct forms:

a) isolation of a single transverse mode for applications which merely require a well defined transverse spatial phase distribution (irrespective of the number of different longitudinal modes spaced at $c/2L$ having the same transverse distribution).

b) isolation of a single longitudinal mode for applications where single-frequency operation is important. In this case, transverse mode suppression will also be required if oscillation on adjacent frequencies distributed over intervals $\ll c/2L$ can not be tolerated.

The most straightforward approach to the transverse mode isolation problem consists of choosing a cavity geometry such that the total loss in the TEM_{10} mode is about equal to the maximum single-pass gain in the particular laser transition. This choice assures that the dominant TEM_{00} mode is above threshold for oscillation and for the usual case in which the amplifier gain is peaked on the tube axis, spatial hole burning (or gain saturation) effects will tend to suppress the odd-symmetric TEM_{10} mode above threshold. The total single pass loss is, of course, comprised of the mirror transmission loss in addition to dissipative losses such as those represented by diffraction and scattering. Obviously one wants a cavity with the maximum difference between the TEM_{00} and TEM_{10} diffraction loss. From the data in Fig. 1-9, the most discrimination between these two modes is obtained from a confocal (or from a convocal-equivalent) cavity rather than from a plane-parallel cavity. In practice one usually has some estimate of the total scattering loss in the system and one can start the design process by choosing a Fresnel number for a confocal equivalent system such that the TEM_{00} diffraction loss is about equal to the expected dissipative loss. With a little care, it is usually possible to design the laser so that the bore diameter of the discharge tube becomes the limiting aperture of the system. For example, the median plane for the generating confocal resonator might be located at the middle of the discharge tube. The common bore diameter at each end then becomes limiting aperture for the system. The generating confocal resonator for the system would then correspond to a pair of spherical mirrors placed at each end of the tube with a common radius of curvature,

$$R_0 = a^2/N\lambda = \text{the bore length},$$

where a is the bore radius and N the desired Fresnel number. In practice one cannot place mirrors precisely at the ends of the discharge tube and the remaining problem consists of determining the appropriate positions and radii of curvatures of other mirror which would generate a confocal system equivalent to the hypothetical, desired one. One generally must do this computation subject to other practical constraints determined by the radii of curvature of available mirrors and permissible mounting locations for these mirrors. A method of computing mirror locations and radii of curvature for a specified confocal equivalent cavity is discussed in Sec. 1.8. One should, of course, check the components in the real system to make sure that the singular regions of high loss are avoided. There are, of course, an infinite number of other choices for the generating confocal resonator which would yield the same Fresnel number and utilize the bore diameter for the limiting aperture. For example, solutions may also be found which place the median plane of the generating confocal resonator outside of the tube, thereby permitting the use of a flat (or even convex) mirror on one end of the cavity and the obtainment of a greater mode volume (hence efficiency) throughout the discharge tube.

More esoteric methods for transverse mode discrimination of course exist in principle. For example, the mirror reflectance could be varied spatially across the end mirrors in such a manner as to optimize a particular mode and suppress the others. Equivalently, localized regions of loss could be introduced to discriminate in favor of one particular mode, etc. (see, for example, Kogelnik and Rigrod, 1962; Rigrod, 1963b). A related approach which has some practical merit was considered by LaTourette, Jacobs and Rabinowitz (1964) in which the laser output mirror consists of a highly-reflecting circular dot whose outer diameter is smaller than the limiting aperture of the system. The laser output is then diffraction coupled (around the edges of the "dot mirror") and a high degree of discrimination against the first odd-symmetry mode is obtained by virtue of the fact that its field distribution is peaked off the laser axis toward the edges of the dot (LaTourette *et al.*, 1964).

Various approaches to the longitudinal mode isolation problem have also been suggested including the incorporation of transmission prisms (White, 1964; Rigrod and Johnson, 1967), reflection gratings, additional mirror elements (Kleinman and Kisliuk, 1962; Smith, 1965, Rigrod, 1970), thin film absorbers (Smith, 1968) and transmission etalons (Kolomnikov, Troitskiy and Chebotayev, 1965) within the laser cavity.

Of these various approaches to the longitudinal mode isolation problem, the transmission etalon suggested initially by Kolomnikov *et al.* (1965) has turned out to be one of the easiest to incorporate in a wide variety of gas lasers. The basic arrangement of the apparatus is shown in Fig. 1-18 in which a relatively slight tilt of the etalon is introduced in respect to the laser axis and in which the etalon is located near the median plane (or plane mirror end) of the cavity to minimize alignment problems resulting from beam displacement. The multiply-reflected waves from the two

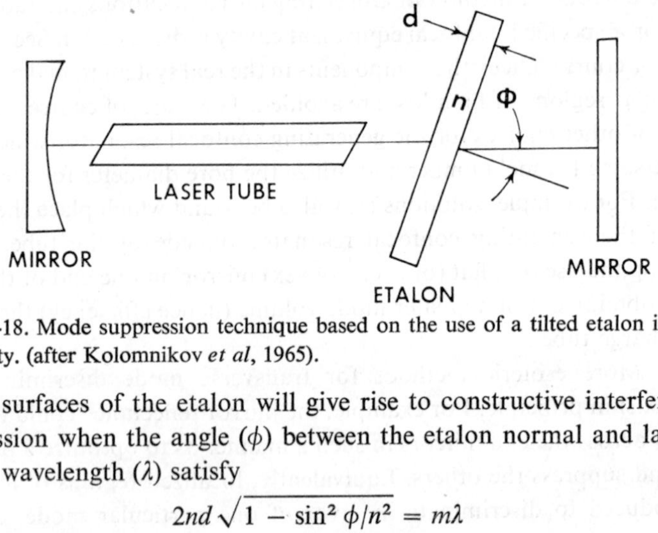

FIGURE 1-18. Mode suppression technique based on the use of a tilted etalon inside the laser cavity. (after Kolomnikov *et al*, 1965).

parallel surfaces of the etalon will give rise to constructive interference in transmission when the angle (ϕ) between the etalon normal and laser axis and the wavelength (λ) satisfy

$$2nd\sqrt{1 - \sin^2 \phi/n^2} = m\lambda$$

where

$$m = 0, 1, 2, 3, \ldots, 10^5, \ldots$$

Here, d is the etalon thickness and n its refractive index. The usefulness of the method is at first glance somewhat surprising because the finesse of an uncoated quartz flat is not great; hence the difference in etalon transmission loss between two laser cavity modes ≈ 150 Mhz apart is relatively slight when the etalon is chosen to have its transmission resonances spaced by an amount ($\approx c/2nd$) which is large compared to the Doppler width of the laser transition. The basic point is that although the difference in loss is slight, it is greatly exaggerated because it determines the point for oscillation threshold in the laser. As the laser pump rate is increased above threshold, hole burning effects (or gain saturation) help in the suppression of neighboring cavity resonances, with the result that most of the original multimode power can be obtained in an isolated single mode, far above threshold in lasers for which the longitudinal mode spacing is comparable to the Lorentz width for the transition.

The effect is illustrated in Fig. 1-19 with some unpublished data taken by B. Wexler, V. Sochor and the author on the strong 4880 Å argon ion laser transition. The normal incidence transmission resonance is generally very unstable and is best avoided in this method. The normal incidence resonance can be avoided altogether (as done in Fig. 1-19) by suitable adjustment of the angle orthogonal to ϕ in Fig. 1-18.

The curve in Fig. 1-19 shows the single mode output power of the laser as the etalon was tuned through six successive transmission resonances

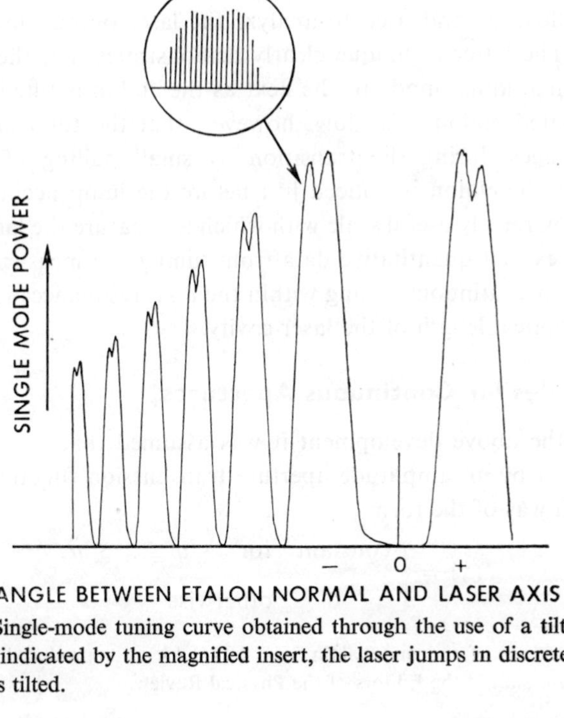

FIGURE 1-19. Single-mode tuning curve obtained through the use of a tilted etalon (see Fig. 1-18). As indicated by the magnified insert, the laser jumps in discrete steps of $c/2L$ as the etalon is tilted.

(typically in the order of a milliradian per Doppler width for a 1 cm flat near normal incidence). These resonances correspond to unknown and very large values of m in the resonance equation.

The asymmetry in the tuning data shown in Fig. 1-19 for individual resonances arises entirely from the dependence of transmission loss on etalon angle. The latter results from the relative displacement between the successive transmission paths in the etalon. The cavity transmission loss varies approximately as

$$f \approx f_0(1 + a\phi^2)$$

over the first transmission resonance where the term, $a\phi^2$, is typically in the order of a few percent. The quadratic nature of the leading term in the asymmetric correction can be easily understood in terms of the expansion coefficients for displaced modes derived below [see Fig. 1-25 for the plane parallel case and Eq. (1.70) for the confocal limit].

The curve in Fig. 1-19 was taken with a pen recorder having a fast enough time constant to illustrate the Lamb dip on the transition as each transmission order was passed. However, the time constant was not fast enough to show that the laser jumps successively in steps of $c/2L$ as the etalon is tuned through the Doppler line. The insert in Fig. 1-19 is based on the response of a swept Fabry–Perot interferometer observed with a storage oscilloscope and used to analyze the laser output under the same conditions. The latter technique clearly demonstrates that the laser jumps from one longitudinal mode to the next as the etalon is tilted. The finesse of the uncoated etalon is so low, however, that the total laser intensity scarcely changes during the transition. A small pulling of each cavity resonance by the etalon is noticed just before the jump occurs. The effect provides an extremely useful scale with which to measure the laser frequency and thereby extract quantitative data from tuning dip measurments. However, to obtain continuous tuning within the $c/2L$ resonances it is necessary to vary the optical length of the laser cavity.

1.11. Modes for Continuous Apertures‡

In the above development it was assumed that the optical wave was multiplied by an amplitude aperture transmission function after each transit which was of the form

$$T(x) = \begin{cases} \text{constant} & \text{for} \quad -a \leqq x \leqq a, \\ 0 & \text{otherwise.} \end{cases} \tag{1.51}$$

‡ Portions of the material in Secs. 1.11–1.13 are reproduced from Bennett (1970a) by permission of the Editors of the Physical Review.

As shown previously by the author (Bennett, 1970a), it is possible to set up an equivalent treatment of the self-reproducing mode problem in the limit where the aperture transmission function is not rectangular and in which the aperture function represents gain as well as loss.

In approaching this modification of the problem it is helpful to use the set of self-reproducing modes characteristic of the rectangular-aperture problem as base functions in a perturbation expansion. Confocal geometry is particularly amenable to solution in this case because the large-aperture normal modes can be written in closed form for arbitrary order.

Consider a confocal cavity having a rectangular reflection aperture defined as in Eq. (1.51) above. Next we shall assume that a continuous aperture transmission function exists in the region $-a < x < +a$. The amplitude response function will be defined by

$$T(x) = 1 + g(x), \tag{1.52}$$

where $g(x)$ represents an arbitrary variation in the transmission function within $-a \leqq x \leqq +a$. The normal set of self-reproducing modes characteristic of the response function in Eq. (1.52) must go continuously over into the normal set of confocal modes as $g(x) \to 0$. Because of the degeneracy inherent in the large-aperture confocal limit, any linear combination of the confocal modes with the same phase shift is also a self-reproducing mode for $g(x) = 0$. Consequently, any linear combination of these modes which is self-reproducing in the limit $g(x) \neq 0$ will also be a normal mode of the system as $g(x) \to 0$. In what follows it is convenient to use the generalized coordinate ξ, defined above, for which

$$\begin{aligned} T(\xi) &= 1 + g(\xi), \\ \xi &= \sqrt{2\pi N}\,(x/a). \end{aligned} \tag{1.53}$$

We next choose some initial field distribution, $E^0(\xi)$, localized within $-a < x < a$ and let

$$E^0(\xi) = \sum_n A_n^0 f_n(\xi). \tag{1.54}$$

After the first transit,

$$E^1(\xi) = [1 + g(\xi)]\, E^0(\xi) \equiv \sum_j A_j^1 f_j(\xi),$$

where

$$A_j^1 = A_j^0 + \sum_n A_n^0 g_{jn} \tag{1.55}$$

and

$$g_{jn} = \int_{-\infty}^{\infty} f_j(\xi)\, g(\xi)\, f_n(\xi)\, d\xi. \tag{1.56}$$

The normal stationary modes are found by solving Eq. (1.55) reiteratively and requiring that

$$A_j^{t+1} = \Gamma A_j^t$$

for all confocal modes j which contain a significant fraction of the total energy subject to some particular choice in the relative excitation coefficients, A_n. The relative expansion coefficients determined in this way and which satisfy the reiteration requirement will be defined to be α_j and result in the ith mode distribution of the static system

$$E^i(\xi) = \sum_j \alpha_j^i f_j(\xi), \tag{1.57}$$

characterized by a particular eigenvalue, Γ^i.

In the normal confocal case, there is a relative phase shift per transit of $\pi/2$ radians between the odd- and even-symmetry modes. In this case modes of opposite symmetry are not degenerate. If the aperture function is symmetric about $\xi = 0$, only modes of either even or odd symmetry will have nonzero expansion coefficients.

If the continuous aperture function does not have a definite symmetry about $\xi = 0$, there will not be self-reproducing modes at successive apertures in the sense of Eq. (1.56). In some special cases there may, however, be nonsymmetric modes which are self-reproducing (see below).

By choosing $E^0(\xi)$ above to correspond to either the TEM_0 or TEM_1 confocal mode initially, it seems clear that one may in general expect to be able to determine at least the set of coefficients and eigenvalues for the dominant even- and odd-symmetry modes of the static system with $g(\xi)$ present. The values of Γ^i may be either greater than or less than one, depending on whether $g(\xi)$ is chosen to represent gain or loss. The total energy across the aperture for a particular choice of α_j^i may be normalized such that

$$\int_{-\infty}^{\infty} |E^i(\xi)|^2 \, d\xi = \sum_j (\alpha_j^i)^2 = 1 \tag{1.58}$$

and the fractional energy loss (or gain) per transit for the ith self-reproducing mode of the static system is $1 - |\Gamma^i|^2$ (negative values of $1 - |\Gamma^i|^2$ corresponding to gain).

For the purpose of illustrating the method, consider the case of the confocal equivalent cavity in Fig. 1-12 which has the property that the relative phase shift between even- and odd-symmetry modes in one transit from left to right is π (i.e. it is equivalent to two confocal cavities in series for which the field is self-reproducing at the second aperture). Next assume that there is a Gaussian amplitude gain aperture localized within each of the rectangular apertures of width $2a$ shown at apposite ends of the cavity. For illustration, consider the case

$$T(\xi) = 1 + G_0 f_0(\xi), \tag{1.59}$$

where the applitude gain function is proportional to the normal dominant confocal mode of the problem which would occur in the limit $G_0 \to 0$.

The dominant odd- and even-symmetry self-reproducing modes for this gain aperture were determined by the author (Bennett, 1970a), using the expansion method outlined above. The amplitude expansion coefficients are summarized in Table 1-1 for the first ten expansion coefficients. Note that roughly 98 percent of the TEM_0 energy is contained in the dominant even-symmetry confocal mode (f_0) characteristic of the rectangular-aperture problem having the same Fresnel number used in the multiplicative amplitude gain function [$g(\xi) = G_0 f_0(\xi)$ in Eq. (1.53)]. Similarly, more than 99 percent of the TEM_1 energy is contained in the dominant odd-symmetry confocal mode (f_1) characteristic of the rectangular-aperture problem. The amplitude distributions for the dominant modes of opposite symmetry corresponding to the Gaussian gain aperture are shown in Fig. 1-20.

The variation of the eigenvalues for the normal modes of the Gaussian gain aperture are shown in Fig. 1-21 as a function of the peak fractional energy gain per pass. It is interesting to note that these results imply that the odd-symmetry mode has higher gain per transit than the even-symmetry mode for peak fractional intensity gains of $\lesssim 1/\sqrt{2}$ per transit. Hence it is only for very high gain coefficients that the even-symmetry mode is dominant in the active system. The situation is therefore significantly different from the situation characteristic of diffraction-limited loss in the rectangular-aperture problem.

Real gas lasers, of course, do not have rectangular transmission apertures. Typically, because of the boundary conditions at the walls of the tube containing the active medium, there is a tendency for the upper-state population distributions to follow a J_0- or J^2-type radial distribution. Hence results of the type derived above for the Gaussian gain aperture are apt to be more relevant to the situation in a real laser than those pertaining to the rectangular-aperture problem. However, it should be noted that even-symmetry amplitude gain distributions which correspond approximately to the dominant even-symmetry amplitude distribution of the rectangular-aperture problem give rise to dominant even- and odd-symmetry modes which closely resemble those of the rectangular aperture problem for the same Fresnel number. The main difference in respect to the rectangular-aperture problem regards the relative mode energy gain coefficients of the

TABLE 1-1. Amplitude expansion coefficients α_j^l of the dominant self-reproducing modes for the Gaussian gain aperture $T(\xi) = 1 + G_0 f_0(\xi)$. (From Bennett, 1970 a.)

Confocal mode:	f_0	f_1	f_2	f_3	f_4	f_5	f_6	f_7	f_8	f_9
TEM_0 (dominant even-symmetry)	0.991496	0	0.130161	0	0.006835	0	0.000141	0	0.000003	0
TEM_1 (dominant odd-symmetry)	0	0.997862	0	0.065356	0	0.000235	0	0.000006	0	0.000000_1

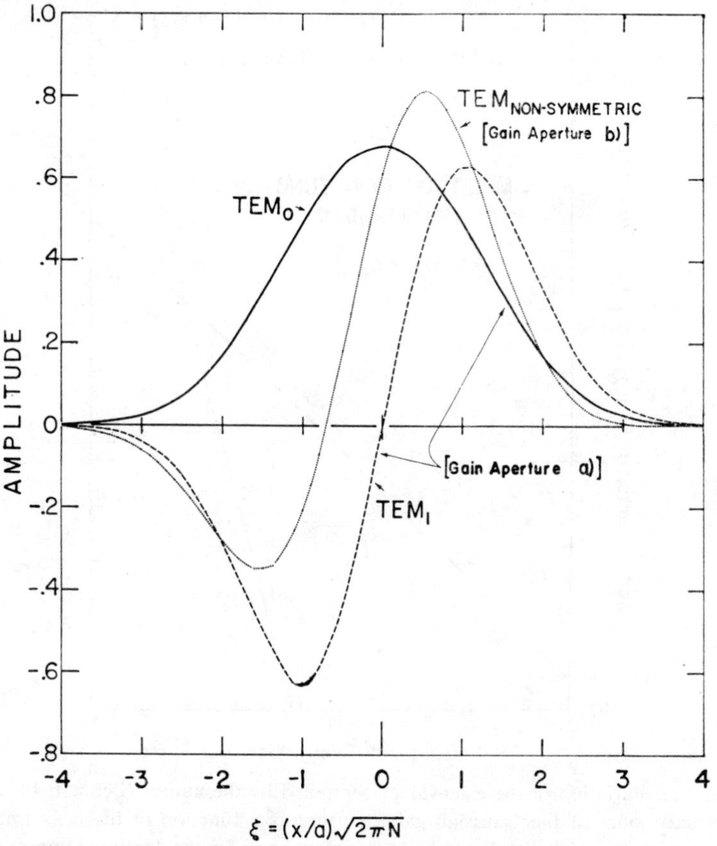

FIGURE 1-20. Dominant even- and odd-symmetry modes represented by the expansion coefficients in Table 1-1 for the Gaussian gain aperture $T(\xi) = 1 + G_0 f_1(\xi)$. The dominant non-symmetric mode of the aperture function $T(\xi) = 1 + G_0 f_1(\xi)$ is also illustrated (Bennett, 1970a).

TABLE 1-2. Amplitude expansion coefficients α_j^i of the dominant nonsymmetric self-reproducing mode corresponding to the aperture function in Eq. (1.60), $T(\xi) = 1 + G_0 f_0(\xi)$, as applied to the cavities in Figs. 1-20 and 1-21. (It is assumed that the direction of the asymmetry in the transmission function alternates, as does the direction of $+x$, between conjugate sufaces, from Bennett, 1970 a.)

Confocal mode	f_0	f_1	f_2	f_3	f_4
Dominant *TEM* mode	0.66744	0.706275	−0.232056	0.034168	−0.026056

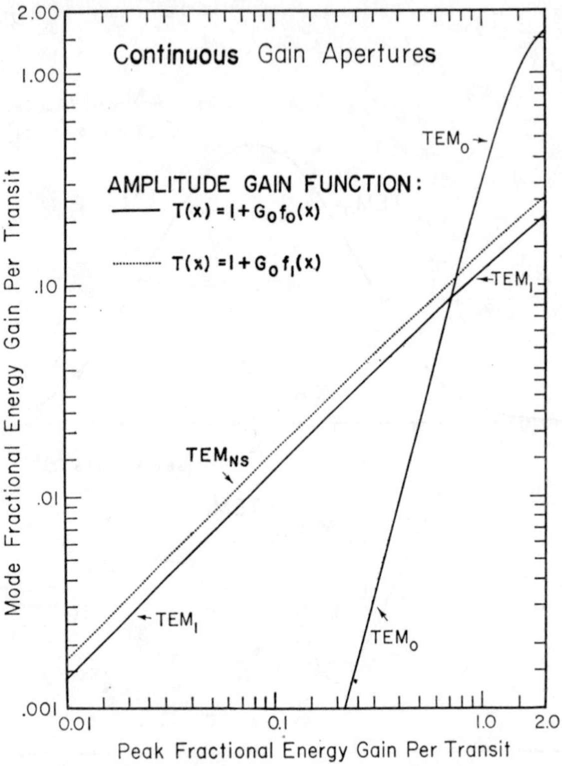

FIGURE 1-21. Variation of the eigenvalues computed by the author (Bennett, 1970a) for the normal modes of the Gaussian gain aperture as a function of the peak fractional energy gain per pass. The mode energy gain is also shown for the dominant nonsymmetric mode corresponding to the aperture function in Eq. (1.60) as applied to the cavities in Figs. 1-12 and 1-13.

dominant even- and odd-symmetry modes (as illustrated in Fig. 1-21). Similar results have been obtained by the author in the plane-parallel limit using direct numerical solutions to the diffraction integral problem, rather than the expansion technique outlined above for the confocal case.

In some number of real cases, nonsymmetric apertures may be anticipated. As mentioned above, such nonsymmetric aperture functions can generate self-reproducing modes in special cases. For example, consider the nonsymmetric amplitude gain transmission function

$$T(\xi) = 1 + G_0 f_1(\xi) \tag{1.60}$$

applied to the two apertures of width "$2a$" in Fig. 1-12, where $f_1(\xi)$ is the dominant odd-symmetry mode of the rectangular-aperture confocal problem and the directions of $+\xi$ are inverted on the two conjugate surfaces. Here the aperture amplifies for $\xi > 0$, and absorbs for $\xi < 0$ [see Fig. 1-30]. As illustrated in Table 1-2, the dominant mode in this case does not have a definite symmetry and roughly half of the energy is contained in the dominant even- and odd-symmetry confocal modes of the rectangular-aperture problem. For the same value of G_0, this mode ($TEM_{n.s}$) has approximately the same gain per pass as the TEM_1 mode of the corresponding symmetric Gaussian aperture problem discussed above. The mode amplitude distribution for the nonsymmetric case is shown in Fig. 1-20 and the variation of the eigenvalue is shown in Fig. 1-21.

1.12. Theory of Cavity Mode-Mixing Effects in Internally Scanned Lasers (Bennett, 1970 a)

It seems probable that an important eventual area of laser application will be the time-dependent display and storage of information. On the macroscopic scale, scanning lasers may eventually compete with the cathode-ray tube in the projection of rapidly varying waveforms. In the microscopic domain, diffraction-limited focusing of the output beam from a laser would, in principle, permit one to "read" and "store" vast amounts of information in a very small space. For these reasons there has been a considerable amount of recent interest shown in the development of a practical device which would permit nonmechanical, electronically controlled spatial scanning of a laser beam. The methods of approach to this problem may be divided into two broad areas: those dependent on external deflection of the laser beam and those relying on time-dependent changes of the spatial boundary conditions internal to the laser cavity. It is of interest to consider the cavity mode-mixing effects that might be anticipated in the latter approach to the scanning laser problem.

There are two basically different ways in which internal scanning might be accomplished which I shall briefly discuss.

1.13. Scanning through Transverse Mode-Locking (Bennett, 1970 a)

In the first case, a periodic scanning action might be obtained through simultaneous phase-locked oscillation in a large number of modes in a static cavity. If the oscillation frequencies are harmonically related and phase-locked, it is possible in principle to build up a periodic, spatially-sweeping beam in much the same way as one would mathematically accomplish the same objective with a periodic Fourier series. It is, for example, clear that at least some rudimentary degree of periodic scanning action can be obtained by the combination of only two different nondegenerate, simultaneously-oscillating modes in a laser with fixed transverse boundary conditions. Even the earliest helium–neon lasers exhibited a beat spectrum which indicated continuous and simultaneous oscillation of the dominant even- and odd-symmetry modes (Javan, Bennett, and Herriott, 1961).

The existence of simultaneous oscillation in two modes of opposite transverse symmetry implies that a spatial variation in intensity occurs across the beam profile which is periodic in the frequency difference between the two modes. Since these differences occur in the radio-frequency range, the effect might be of practical value in some particular areas of application. An illustration of some limitations to be expected with that approach is given in Fig. 1-22 for the infinite-strip, plane-parallel modes of the Fox and Li type. (A summary of the properties of these modes was given above.) The relative amplitude distributions for the two dominant modes of opposite symmetry for a Fresnel number of 6.25 are shown in the upper part of the figure. The time variation of the near field intensity distribution is illustrated in the lower portion of the figure corresponding to the case where both modes are assumed to be excited with equal intensity. The angle θ increases with time at the difference frequency between the two modes with the result that a binary type of spatial switching in the intensity distribution should occur at a fixed frequency determined by the cavity parameters. A similar time-varying spatial intensity distribution could presumably be obtained at the focus of a lens with f/L optics, thereby permitting a periodic binary scanning over distances in the order of a wavelength of light. Although such a scanning laser might have practical value for some specialized applications, it is clear that a very large number of modes would have to be phase-locked in a precisely defined manner to permit obtaining a continuous scanning action over a wide aperture with appreciable resloution. A number of very remarkable mode-locking effects have been reported over the last few years and the possibility of achieving some type of periodic transverse scanning action through similar nonlinear interactions should not be discounted. Preliminary theoretical (Soncini and Svelto, 1968) and experimental studies (Austen, 1968; Smith, 1968) of this type of periodic scanning action have, in fact, been given recently in the literature. Still more recently, Arakeljan, Karlov, and Prokhorov (1969) have reported transverse mode-locking and scanning action through use of saturable absorbers at 10.6 microns with the CO_2 laser. They obtained transverse mode-locking over sufficiently long periods of time (> 30 min) to give considerable encouragement for future possibilities of this scanning method. However, the method depends on complicated nonlinear effects which are not clearly understood at this time. In addition, although such a scanning approach would permit generation of a periodic pattern, it is also desirable to have a method which allows for arbitrary, nonperiodic deflection of a laser beam. Such internal scanning methods imply the need for externally-controlled time-dependent transverse apertures.

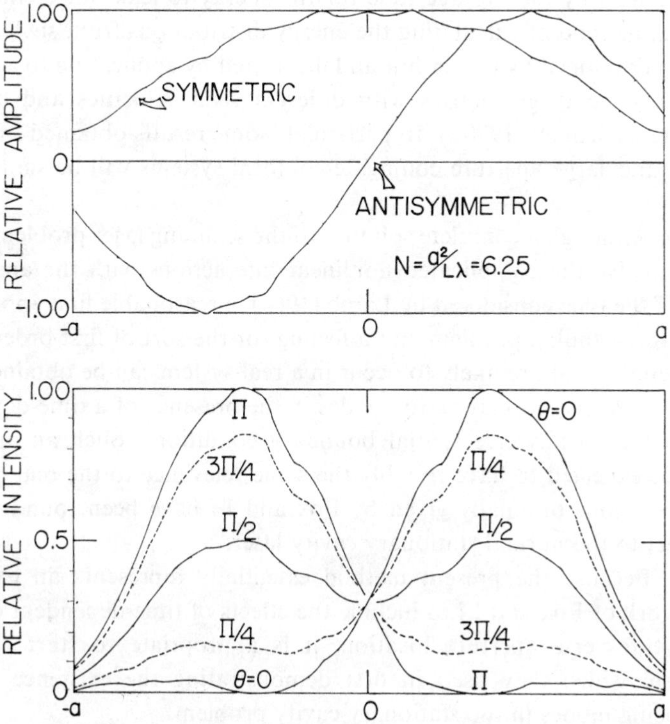

FIGURE 1.22. Illustration of the periodic, binary-type scanning action that might be obtained through simultaneous oscillation on the two dominant modes of opposite symmetry in an oscillating laser. (The example is for infinite-strip modes with $N = 6.25$.) (Bennett, 1970 a).

1.14. Swept Transverse Apertures (Bennett, 1970 a)

The second method of approach involves internal scanning of a transmission or gain aperture within the laser. Some experimental work involving swept reflectance apertures in pulsed lasers has been reported by Pole and Myers (1966, 1967). [See also Pole (1965).] More recently, the possibility of producing a similar scanning action through swept electron-beam excitation of the inversion density in gas lasers has been suggested by Leo Beiser (private communication). It is with some limiting aspects of the latter suggestion that the present method of analysis is concerned.

Consider a hypothetical laser cavity in which there is a continuous, constant transverse displacement of the aperture. On an intuitive basis it seems clear that such a system ought to provide spatial scanning of the output beam with scan velocities at least approaching the normal cavity mode width divided by the average time for the energy to leak out of the mode. A formal method of calculating the energy distributions from such a cavity has been developed by the author and illustrated by application to a number of limiting cavity geometries with different scan velocities and coupling parameters (Bennett, 1970a). In particular, some results obtained for plane-parallel and large-aperture conjugate confocal systems will be summarized here.

Although a complete solution to the scanning laser problem clearly would involve the complicated nonlinear interactions with the amplifying media of the type considered by Lamb (1964), a reasonable first approximation to the scanning problem and a feeling for the sort of first-order mode-mixing effects that are likely to occur in a real system can be obtained from an analysis of the normal cavity modes in the presence of a time-dependent change in the transverse spatial boundary conditions. Such an approach would be expected to have roughly the same relevance to the real problem as the solutions originally given by Fox and Li have been found to have in respect to the normal stationary cavity laser.

Because the present method essentially represents an extension of the work of Fox and Li to include the effects of time-dependent changes on the transverse aperture location, it is appropriate to start with the formalism which they used in first demonstrating the existence of self-reproducing modes in the stationary cavity problem.

It is a general conclusion of the present section that similar sets of self-reproducing modes will *not* exist in the scanning reference frame in the case where there is a continuous change of the transverse aperture location with time. However, the normal modes of the Fox and Li type form a very convenient set of base functions with which to describe the time-dependent problem when these functions are expressed in terms of a coordinate system which is scanning with the laser axis. The choice of these functions assures continuous behavior of the solutions as the scan velocity increases from zero. Also the use of these functions avoids the necessity of repeated solution of the diffraction integral in the scanning problem.

In the present method we start by assuming that there is an equal transverse displacement of the mirror aperture during each transit (Fig. 1-23).

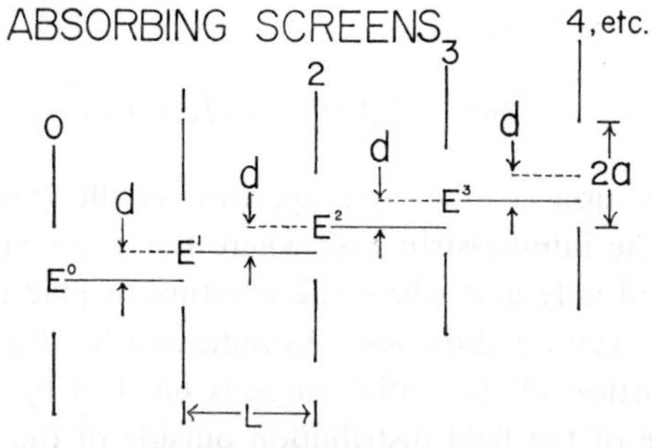

FIGURE 1-23. Model used by the author (Bennett, 1970a) to analyze the effects of a continuously-swept transverse aperture.

This situation corresponds physically to a continuous transverse scanning of the aperture at a constant scanning velocity of amount "d" per transit. We shall assume that the field is plane-polarized in the y-direction, perpendicular to the scan direction.‡ Initially we assume some particular field distribution, $E_y^0(x)$, is excited which we then expand in terms of the normal set of modes corresponding to the initial location of the optical axis,

$$E_y^0(x) = \sum_j A_j^0 f_j(x). \tag{1.61}$$

E_y^0 might, for example, have been entirely in the dominant even-symmetry mode. During the first transit, the jth mode, $f_j(x)$, becomes $\gamma_j f_j(x)$ due to diffraction and at the end of the first transit, the total field has become

$$E_y^1(x) = \sum_j A_j^0 \gamma_j f_j(x). \tag{1.62}$$

At the end of the first transit the field encounters the aperture, displaced in the transverse direction by amount d. At this point we expand the actual field across the new aperture in terms of the normal modes of the displaced system. Specifically,

$$f_j(x-d) = \sum_k B_{jk}(d)\, f_k(x), \tag{1.63}$$

where the expansion coefficients are a function of the aperture displacement and may be determined using the orthonormal proprties of the modes in

‡ More general polarization may be treated with the same approach by using appropriate linear combinations of different orthogonal polarizations. However, this procedure increases the number of coupled equations without adding any particularly different physical effects to the problem.

the new coordinate system,

$$B_{jk}(d) = \int_{-a}^{a} f_j(x - d)\, f_k(x)\, dx. \tag{1.64}$$

The situation at the first aperture is illustrated schematically in Fig. 1-24 for the infinite-strip case where the dominant even-symmetry mode is excited initially and where the aperture displacement is perpendicular to the polarization direction. As indicated by the shaded region in Fig. 1-24, one portion of the initial mode is blocked by the new aperture and a small piece of the field distribution outside of the original aperture location must be calculated from the diffraction integral before doing the expansion in Eq. (1.63). As can be seen qualitatively from Fig. 1-24 the main coupling in the expansion for small displacements will be between the dominant even- and odd-symmetry modes. This is illustrated quantitatively for a variety of different Fresnel numbers and plane-parallel geometry in Fig. 1-25.

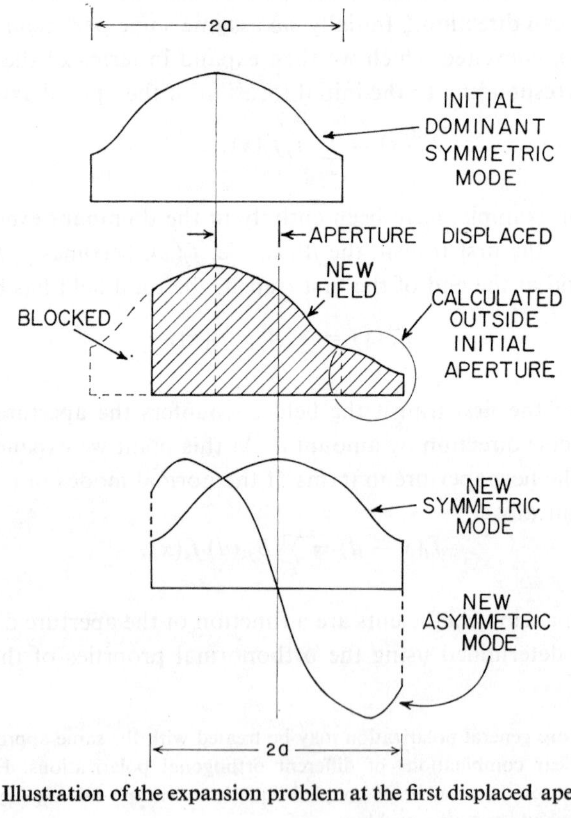

FIGURE 1-24. Illustration of the expansion problem at the first displaced aperture (Bennett, 1970a).

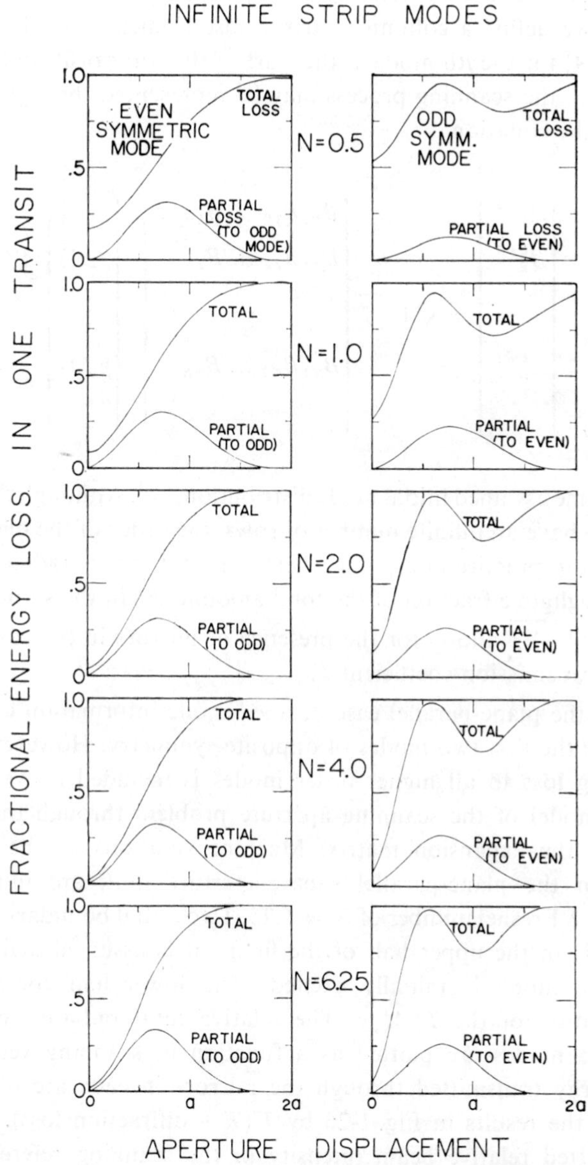

FIGURE 1-25. Fractional energy loss for the transit in which the step discontinuity occurs, for different even- and odd-symmetry plane-parallel infinite-strip modes (Bennett, 1970 a).

The continuous scanning problem is most readily treated in terms of a set of matrix equations relating the expansion coefficients on successive transits. If we define a column matrix whose elements are the expansion coefficients A_j^t for the jth mode at the start of the tth transit [in the manner of Eq. (1.61)], the scanning process may be represented through reiterative solution of the equation

$$\begin{pmatrix} A_1^{t+1} \\ A_2^{t+1} \\ \vdots \\ A_N^{t+1} \\ \vdots \end{pmatrix} = \sqrt{1-T} \begin{pmatrix} B_{11}B_{12} \ldots B_{1N} \ldots \\ B_{21}B_{22} \ldots B_{2N} \ldots \\ \vdots \quad \vdots \quad \vdots \quad \vdots \quad \vdots \\ B_{N1}B_{N2} \ldots B_{NN} \ldots \\ \vdots \quad \vdots \quad \vdots \quad \vdots \quad \vdots \end{pmatrix} \begin{pmatrix} \gamma_1 A_1^t \\ \gamma_2 A_2^t \\ \vdots \\ \gamma_N A_N^t \\ \vdots \end{pmatrix} \tag{1.65}$$

subject to some assumed initial mode distribution, A_i^0. Although the matrices in Eq. (1.65) have an infinite number of rows, the order of the matrices may be truncated in practice at a point where the total energy radiated into the mode is a negligible fraction of the total amount. As in the stationary case, the factor $\sqrt{1 - T}$ allows for the presence of mirrors in the cavity having a nonzero transmission coefficient T.

In the plane-parallel case detailed mode information can only be obtained for the first two modes of opposite symmetry. However, the total mode-mixing loss to all higher order modes is included irreversibly in a two-mode model of the scanning aperture problem through the diagonal elements of the expansion matrix. Machine solutions to the reiterative equations in the plane-parallel square-aperture limit are illustrated in Fig. 1-26 for a Fresnel number of $N = 6.25$. Two initial boundary conditions are assumed: in the upper half of the figure, it is assumed that the dominant TEM_{00} mode is initially excited. The lower half corresponds to initial excitation of the TEM_{10}. The relative total radiative energies in the first two modes are plotted as a function of scanning velocity. The relative energy transmitted through the mirrors in each are obtained by multiplying the results in Fig. 1-26 by $T/(T +$ diffraction loss). The average transmitted relative beam intensity in the scanning reference frame due to the first two modes is illustrated for $T = 0.01$ and $N = 6.25$ in Fig. 1-27. Fluctuations in intensity due to the difference in phase shift per transit between the two dominant modes of opposite symmetry are averaged out and in this approximation the mean relative profile in the scanning reference frame is given by

$$I_j = \sum_{j=1}^{2} TC_{jj}|A_j^t|^2, \tag{1.66}$$

where

$$C_{jj} = \int_{-1}^{+1} |f_j(x)|^2 \, dx$$

and is a factor which arises from the nonhermiticity of the normalization condition in the plane-parallel case.

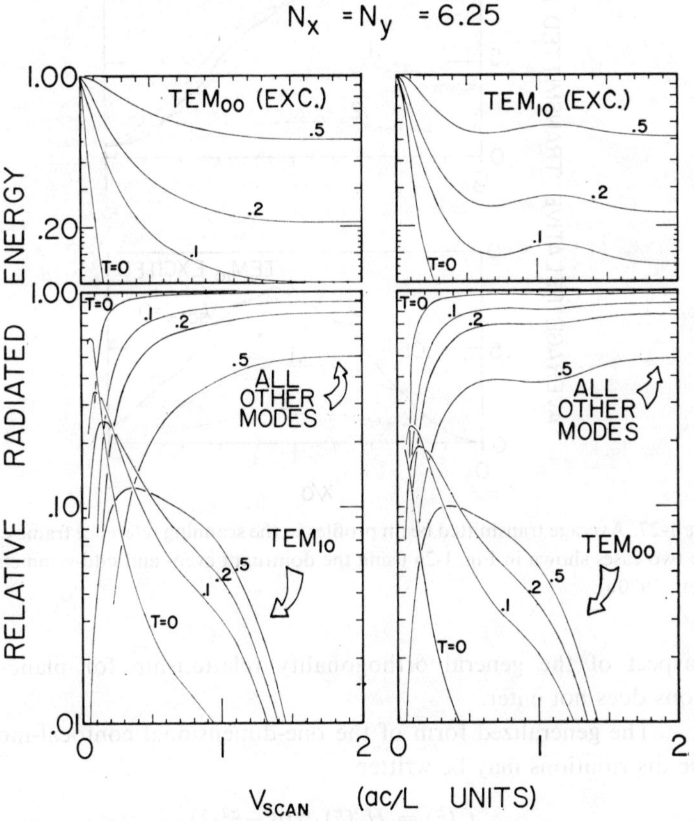

FIGURE 1-26. Solutions obtained by the author (Bennett, 1970a) to the mode coupling problem for square-aperture plane-parallel mirrors with $N_x = N_y = 6.25$. Results are given for the separate cases where the TEM_{00} and TEM_{10} modes were initially excited.

The problem is easiest to handle in the larger-aperture confocal limit because the matrix elements can be written in terms of the confluent hypergeometric functions for arbitrary indices and because the non-Hermi-

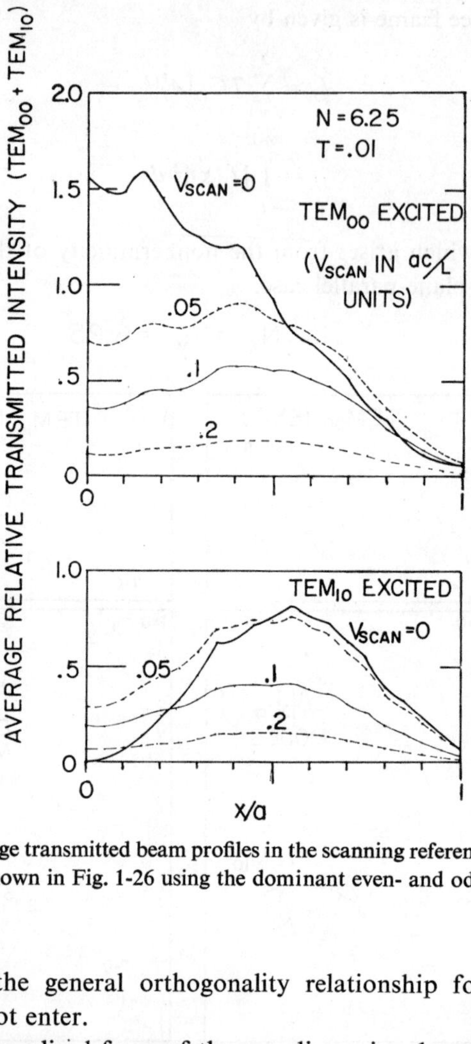

FIGURE 1-27. Average transmitted beam profiles in the scanning reference frame calculated for the two cases shown in Fig. 1-26 using the dominant even- and odd-symmetry modes (Bennett, 1970a).

tian aspect of the general orthogonality relationship for plane-parallel solutions does not enter.

The generalized form of the one-dimensional confocal-mode amplitude distributions may be written

$$f_m(\xi) = H_m(\xi)\exp(-\xi^2/2) \tag{1.67}$$

as before, where ξ is a generalized distance along the mirror surface given by

$$\xi = \sqrt{2\pi N}\,(x/a),$$

in terms of the actual physical displacement, x, along the constant phase mirror surface, and is valid for arbitrary Fresnel number, $N = a^2/R\lambda$, in the large-aperture limit. Similarly, it is convenient to formulate the confocal scanning problem in terms of a generalized displacement

$$\delta = \sqrt{2\pi N}\,(d/a) \tag{1.68}$$

of the center of the scanning aperture along the mirror surface in the time for the wave to propagate between the mirrors. The actual scan velocity is given by

$$V_{\text{scan}} = (\delta a/\sqrt{2\pi N})\,(c/L) \tag{1.69}$$

in terms of this generalized parameter. As previously noted, the Hermite polynomials are most conveniently given in terms of the confluent hypergeometric functions. As may be seen by starting with the generating function for the Hermite polynomials written in terms of a displaced origin, the elements of the expansion matrix may also be expressed in terms of the confluent hypergeometric functions. For the confocal formulation, the elements of the expansion matrix are given by (Bennett, 1970a)‡

$$B_{nm}(\delta) = \frac{e^{-\delta^2/4}}{(n-m)!}\sqrt{\frac{n!}{m!}}\left(\frac{-\delta}{\sqrt{2}}\right)^{n-m} F(-m, n-m+1, \delta^2/2) \tag{1.70}$$

for $n \geqq m$, and $B_{mn} = (-1)^{n+m} B_{nm}$. Conservation of energy is assured through the self-adjoint property of the expansion matrix,

$$\sum_m B_{mi} B_{mj} = \delta_{i,j}. \tag{1.71}$$

It is convenient to normalize the initial energy placed in the cavity so that

$$\sum_{j=0}^{\infty} (A_0^j)^2 = 1. \tag{1.72}$$

The total energy radiated may then be written

$$I_t = \sum_{j=0}^{\infty} I_j, \tag{1.73}$$

where the relative energy radiated in the jth mode coincident with the scanning reference frame is given by

$$I_j = T \sum_{t=0}^{\infty} (A_t^j)^2. \tag{1.74}$$

The exact expression for the transmitted relative intensity profile seen in the scanning reference frame is then

$$I(\xi) = T \sum_{t=0}^{\infty} \left[\sum_{j=0}^{\infty} A_j^t f_j(\xi) \right]^2. \tag{1.75}$$

‡ A similar expression is given in Morse and Feshbach (1953) which, however, is not written for normalized functions and which therefore does not satisfy the self-adjoint requirement of Eq. (1.71).

The conjugate confocal cavities in Figs. 1-12 and 1-13 are of particular interest in application to scanning lasers involving swept transmission apertures. Some computer solutions pertinent to these cavity geometries are shown in Fig. 1-28. In these cases, $\leqq 30$ modes were required in the reiterative matrix equations in order to obtain solutions for the average transmitted profile good to ≈ 0.5 percent. The problem scales roughly with the ratio of δ/T. Increasing the scan velocity decreases the peak intensity and produces an asymmetric broadening of transmitted beam profile. The upper curve represents the case where the dominant even-symmetry Gaussian mode is initially excited. The lower curve represents the initial excitation of the dominant odd-symmetry mode.

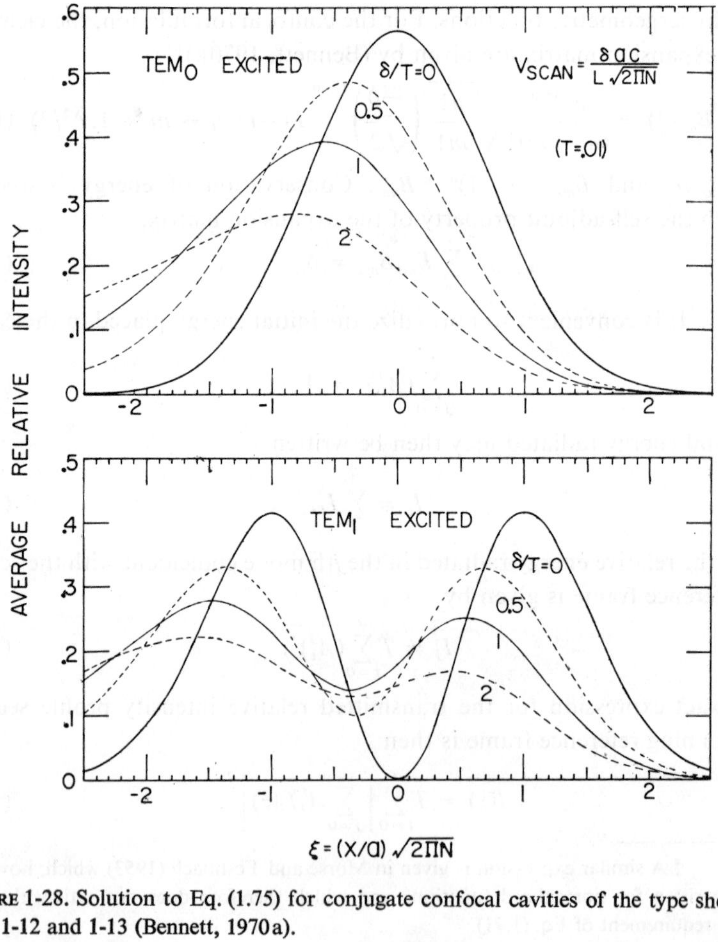

Figure 1-28. Solution to Eq. (1.75) for conjugate confocal cavities of the type shown in Figs. 1-12 and 1-13 (Bennett, 1970a).

1.15. The Scanning Problem for Continuous Apertures

Several suggestions have been made for realizing an internally-scanned laser in practice. One possibility consists of the use of a swept electron beam impinging on an electro-optic medium which would raise the reflectance at the point of impact. Another suggestion involves sweeping a gain aperture through electron beam excitation. In these cases a continuous (as opposed to rectangular) transmission or gain aperture will certainly be produced and the model discussed above, which is based on a scanned rectangular aperture, is clearly an approximation of the actual experimental situation.

It is of interest to formulate the continuous-aperture problem in such a way that the error in the previous approximation becomes apparent.

Consider the case where the amplitude aperture transmission function is given by

$$T(\xi) = 1 + g(\xi)$$

and the function $g(\xi)$ is moving along a confocal surface at the rate δ per transit. Such localized disturbances might be moving in opposite directions on the scanning axis illustrated in Fig. 1-12 or 1-13. If the localized variation in gain occurs well within the region $-a < x < a$, whether or not a rectangular aperture exists at $x = \pm a$ is immaterial. The localized function will steer the beam.

If we launch an initial wave fo the form

$$E^0(\xi) = \sum_n A_n^0 f_n(\xi), \tag{1.76}$$

where the A_n^0 might be chosen to correspond to one of the ith self-reproducing modes (α_j^i) derived above for stationary continuous apertures, the wave arrives at the second aperture modified by $T(\xi)$. Hence after the second aperture the nth-mode expansion term may be written

$$f_n(\xi - \delta)\, T(\xi) \equiv \sum_m \beta_{nm}(\delta)\, f_m(\xi). \tag{1.77}$$

Or from orthogonality

$$\begin{aligned}
\beta_{nm}(\delta) &= \int_{-\infty}^{\infty} f_n(\xi - \delta)\, T(\xi)\, f_m(\xi)\, d\xi \\
&= \int_{-\infty}^{\infty} f_n(\xi - \delta)\, f_m(\xi)\, d\xi + \int_{-\infty}^{\infty} f_n(\xi - \delta)\, g(\xi)\, f_m(\xi)\, d\xi \\
&= B_{nm}(\delta) + \sum_k B_{nk}(\delta) \int_{-\infty}^{\infty} f_k(\xi)\, g(\xi)\, f_m(\xi)\, d\xi\,.
\end{aligned} \tag{1.78}$$

Or, from Eq. (1.56) the matrix (β) is given by

$$(\beta) = (B) + (B)(g), \tag{1.79}$$

where $B_{nm}(\delta)$ is given by Eq. (1.70). Therefore by determining the matrix elements of g_{km} for a particular choice of $g(\xi)$, it is possible to determine both the normal modes of the stationary cavity (zero scanning velocity) and the expansion coefficients in the time-dependent mode-mixing problem for continuous apertures. Note that (g) is not a function of the scanning velocity, whereas (B) is. Note that

$$\beta_{nm}(\delta) \to B_{nm}(\delta) \qquad \text{as} \qquad g_{km} \to 0.$$

Therfore for $g(\xi)$ small, the previously described scanning calculations are a good approximation for the time-dependent problem. Clearly the results for arbitrary localized apertures must go continuously over to the previous solutions as the localized aperture function goes to zero.

The reiterative equations for the amplitude expansion coefficients in this case become

$$(A^{t+1}) = \sqrt{1 - T(\beta)}\,(A^t), \tag{1.80}$$

where

$$A_j^0 = \alpha_j^i.$$

The total intensity radiated in the jth confocal mode in the scanning reference frame in time τ is given by

$$I_j^\tau = T \sum_{t=0}^{\tau} (A_j^t)^2 \tag{1.81}$$

and the exact expression for the form of the total radiated distribution in the scanning reference frame is

$$I^\tau(\xi) = T \sum_{t=0}^{\tau} \left[\sum_{j=0}^{\infty} A_j^t f_j(\xi) \right]^2. \tag{1.82}$$

Because we would specifically like to consider the case where there can be a net gain in the system, the expressions for the intensity can approach infinity as $\tau \to \infty$. This merely means that if we leave the laser on for an infinite length of time, we will get an infinite amount of energy out of it.

As before, the total radiated intensity still settles down to a precisely-defined average distribution in the scanning reference frame at constant scan velocity. In the present case, that distribution can be determined by continuing the numerical iteration for a number of transits τ, such that the relative values of I_j are constant within an arbitrarily defined criterion (e.g. constant within one percent, etc.). From orthogonality of the confocal functions $f_j(\xi)$,

$$\int_{-\infty}^{\infty} I^{\tau}(\xi)\, d\xi = \sum_{j=0} I_j^{\tau}. \tag{1.83}$$

Hence the intensity distributions are obtained in a normalized form equivalent to that used in the treatment of the rectangular transmission function problem by evaluating the quantities

$$\bar{I}(\xi) \equiv I^{\tau}(\xi) \Big/ \sum_{j=0} I_j^{\tau} \tag{1.84}$$

from Eqs. (1.81) and (1.82).

Machine solutions were performed for this problem for several representative values of gain, transmission coefficient, and scanning velocity. Because the dominant self-reproducing modes of the stationary system in the Gaussian aperture problem are very close to the dominant confocal mode distribution of the appropriate rectangular-aperture problem, the results in the two cases are not very different. The latter is illustrated in Fig. 1-29 for several ratios of δ/T in the case where $T = 0.10$ and the maximum fractional gain on the axis was assumed to be 0.10. As seen by comparing Figs. 1-28 and 1-29, the two sets of results are qualitatively and quantitatively very similar. These results are approximately constant for constant ratios δ/T. However, a significant qualitative difference between the two sets of results is apt to occur due to the dominance of the odd-symmetry mode in the low-gain, Gaussian aperture limit.

Because of finite excited-state relaxation rates, there is apt to be an inherent asymmetry in the inversion density profile in scanning lasers utilizing electron-beam excitation. For example, overexcitation of the media can result in absorption in many gas-laser systems. Consequently, as the electron beam travels through the gas there will be a tendency for a region of absorption to travel along behind a region of gain. This is especially true of gas systems such as the high-gain metal vapor lasers which are normally only capable of transient inversion for time intervals in the order of the upper-state spontaneous radiative lifetime.

For these reasons, it is of interest to consider the response of the scanning system to a nonsymmetric gain profile of the type illustrated by Eq. (1.60) and Fig. 1-30. Numerical solutions to this problem have also been obtained and are shown in Fig. 1-31. As may be seen by comparing Figs. 1-29 and 1-31, the results in the nonsymmetric scanning case are

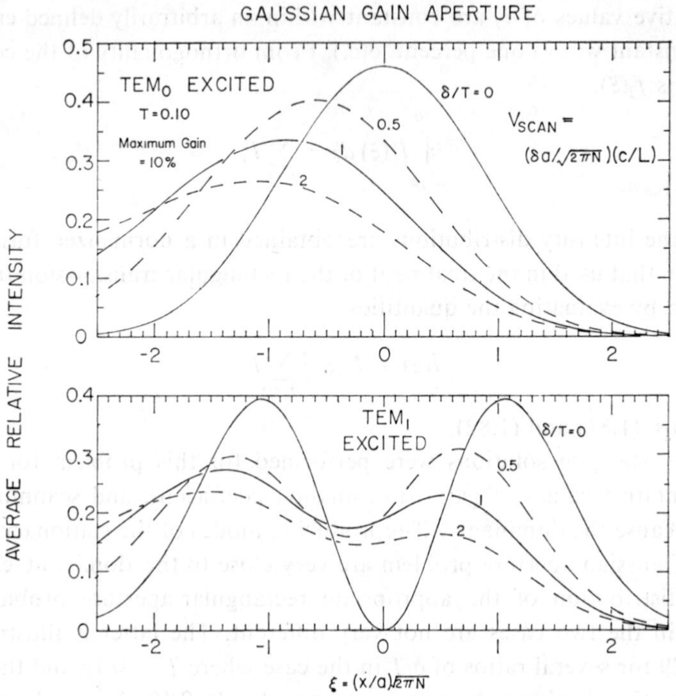

FIGURE 1-29. Solutions to Eq. (1.84) obtained by the author (Bennett, 1970a) for swept, Gaussian gain apertures in conjugate confocal cavities of the type illustrated in Figs. 1-12 and 1-13.

roughly intermediate between the two limiting cases evaluated for the symmetric gain aperture in Fig. 1-29.

The results of these calculations indicate that it should be possible to construct an electron-beam-scanned laser using the cw gain coefficients available from several gas lasers in the visible with scan velocities approaching $\approx 3 \times 10^7$ spot-width/sec. Such lasers could be expected at least to satisfy the resolution requirements of commercial television. However, the only experimental devices that have actually been reported are extremely rudimentary and have only been used in pulsed oscillation to generate a very crude scanning raster.

Much better inherent scanning speeds appear to be obtainable through swept electron beam excitation of thin film semi-conductor lasers of the type recently reported by Packard, Tait and Dierssen (1971). The above analysis of the mode-mixing problem indicates that the small "length" of the cavity permissible in this type of thin film laser should in principle permit

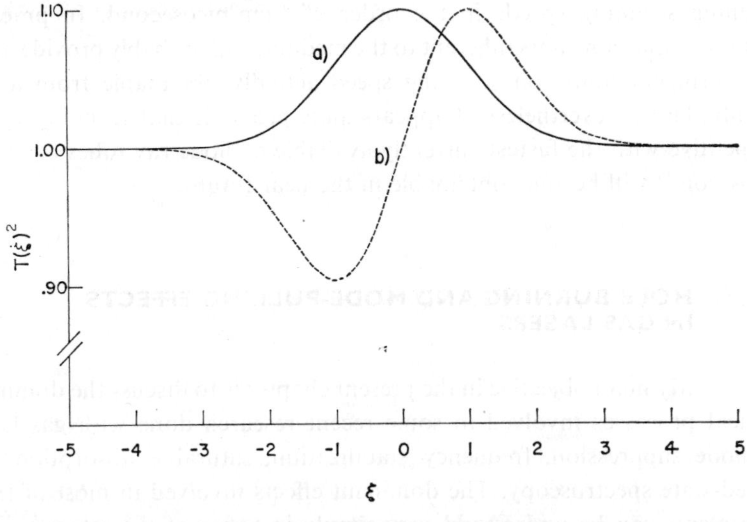

FIGURE 1-30. Two continuous-aperture functions considered in the internally-scanned conjugate confocal laser problem. The solid curve corresponds to the symmetric Gaussian gain aperture $T(\xi) = 1 + G_0 f_0(\xi)$ and the dashed curve corresponds to the nonsymmetric amplitude gain aperture $T(\xi) = 1 + G_0 f_1(\xi)$. The stationary-cavity, self-reproducing modes corresponding to these gain apertures (as applied to the conjugate confocal surfaces of the cavities in Figs. 1-12 and 1-13) are shown in Fig. 1-20 (Bennett, 1970a).

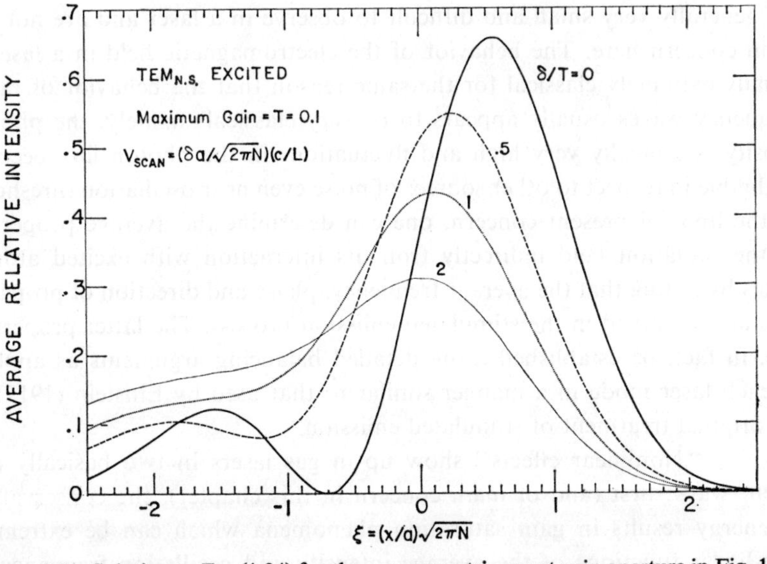

FIGURE 1-31. Solution to Eq. (1.84) for the nonsymmetric swept gain aperture in Fig. 1-30 as applied to the vacities in Figs. 1-12 and 1-13 (Bennett, 1970a).

obtaining scanning speeds in the order of 1 cm/picosecond. In practice, relaxation time constants inherent to the medium will probably provide much more stringent limits on scanning speed actually obtainable from a real thin film laser. Nevertheless, it appears quite probable that scanning speeds competitive with the fastest currently-available cathode ray tubes ($\gtrsim$ 1 cm/nanosecond) will become obtainable in the near future.

2. HOLE BURNING AND MODE-PULLING EFFECTS IN GAS LASERS

My main objective in the present chapter is to discuss the dominant physical processes involved in some recent research done with gas lasers on mode suppression, frequency stabilization, saturable absorption and excited-state spectroscopy. The dominant effects involved in most of these experiments can be understood very simply in terms of the interaction of excited states with single-frequency running waves and an analysis based on "hole-burning" developed initially by the author (Bennett, 1961 a, b; 1963).

I will treat optical radiation classically in this work, using Schrödinger theory to evaluate the behavior of excited atomic states. Although this approach will, by definition, ignore quantum effects which show up in the form of phase and amplitude fluctuations in the optical field, such effects are generally very small and difficult to observe in a laser and are not the main concern here. The behavior of the electromagnetic field in a laser is usually extremely classical for the same reason that the behavior of radio frequency waves usually appears to be very classical: namely, the photon density is typically very high and fluctuations in the photon flux become negligible in respect to other sources of noise even near oscillation threshold. In the limit of present concern, one can determine the average properties of the radiation field indirectly from its interaction with excited atomic states by noting that the average frequency, phase and direction of propagation are preserved in the stimulated emission process. The latter properties can, in fact, be established using detailed balancing arguments as applied to each laser mode in a manner similar to that used by Einstein (1917) in his original treatment of stimulated emission.

"Nonlinear effects" show up in gas lasers in two basically different ways: first (and of main concern in this chapter), the conservation of energy results in gain saturation phenomena which can be extremely non-linear functions of the average intensity and oscillation frequency of a laser. This phenomenon results in the well-known structure in the single-mode power tuning curve (or "Lamb dip") and all sorts of hysteresis effects in both single- and multi-mode laser oscillation. The dominant aspects of these problems do not depend on the generation of difference frequencies or phase-locking effects for a quantitatively-accurate description. Hence these effects can be calculated with a "rate equation" approach in a "linearized" theory at each field intensity in which the generation of sum and difference frequency terms in the macroscopic polarization of the medium is simply ignored. The reasons for adopting this procedure here are that the method is extremely simple, easily modified for widely different conditions, and readily identifies the main physical effects. For example, the basic physical cause of the Lamb dip and mode-competition effects is easily seen to be due to hole burning (Bennett, 1962a). Further the quantitative shape of the single-mode power tuning curve and single-mode frequency dependence in the Doppler broadened limit from the third-order Lamb theory (Lamb, 1964) can be obtained in just a few lines from the hole-burning model (Bennett, 1963). In addition it is possible to write a simple set of equations giving the dominant power and frequency characteristics for an arbitrary number of modes using the same method of analysis in the low power Doppler limit. These considerations become of considerable practical value in the analysis of new complex situations, such as those found in a gas laser containing a saturable absorber or mixture of isotopes with different linewidth parameters and resonant frequencies.

Of course, one has to know when *not* to use the hole-burning model. This circumstance arises primarily in problems involving non-linear effects of the "second" type in the macroscopic polarization. One must generally be wary of any situation in which two light waves with frequencies differing by less than the natural width interact with the same atom. The interference effects which would be present in this instance are ignored in the hole-burning model. In principle, one might expect to see some residual trace of such effects close to the center of the single-mode tuning curve of a gas laser due to the interference between the two oppositely-directed running waves, which when viewed in the atoms' Doppler-shifted reference frame appear to have different frequencies. Although such effects are clearly present in principle, they are largely washed out by the averaging effect of the velocity distribution in any real gas laser. Probably the best theoretical proof of the latter statement consists of noting that the single-mode results in the low power Doppler-broadened limit obtained from the hole-burning model (Bennett, 1963) are exactly the same as those obtained in the third-order Lamb theory (Lamb, 1964). However, the averaging effect of the velocity distribution may also be demonstrated directly (as done in Sec. 2.21) by comparison with exact numerical integrations of the Schrödinger equation. There are two instances which should be mentioned in which a clear inadequacy of the hole-burning model has been demonstrated: the locking effect seen in three-mode oscillation with the first helium-neon laser which occurs when two pairs of combination tones differ by much less than the natural width (see discussion in Lamb, 1964), and the correlation effects discussed at the end of this chapter which have recently been observed in three-level laser experiments in which two widely different optical frequencies interact through a common level in the same atom.

2.1. Background Considerations

It is useful to review some very simple resonance cases. Let us assume we have solved the Schrödinger equation

$$H_0\phi_n = i\hbar = \frac{\partial\phi_n}{\partial t} \tag{2.1}$$

for the "running electron" in a particular atom for which

$$\begin{aligned} \phi_n &= U_n(\mathbf{r})\, e^{-iE_nt/\hbar}, \\ H_0 U_n &= E_n U_n, \end{aligned} \tag{2.2}$$

and add a perturbing term $V(\mathbf{r}, t)$ to the Hamiltonian. Then

$$[H_0 + V(\mathbf{r}, t)]\Psi = i\hbar \quad \partial\Psi/\partial t \tag{2.3}$$

and we clearly may expand

$$\Psi(r, t) = \sum_m C_m U_m(r)\, e^{-iE_mt/\hbar}. \tag{2.4}$$

where, in general, the coefficients C_m are functions of the time even when $V(r)$ is not. For simplicity we shall only consider the case where the perturbing field is uniformly distributed over the space occupied by the atom. From the orthonormality of the solutions to Schrödinger's equation,

$$\int |\Psi(r, t)|^2\, dv = \sum_m |C_m(t)|^2 = 1. \tag{2.5}$$

Therefore, we may use a probability interpretation of the expansion coefficients and we note that the probability of finding the atom in state m, $P_m = |C_m(t)|^2$, is a function of the time in the presence of the perturbing potential. By that statement is meant that if we were to turn off the field quickly at time t and examine the atom, the relative probability of finding it in state m would be P_m. From the Schrödinger equation

$$\dot{C}_n = \frac{1}{i\hbar} \sum_m C_m V_{nm}(t)\, e^{i\omega_{nm}t}, \tag{2.6}$$

where

$$\begin{aligned} \omega_{nm} &= (E_n - E_m)/\hbar, \\ V_{nm}(t) &= \int U_m^* V(\mathbf{r}, t)\, U_n\, dv. \end{aligned} \tag{2.7}$$

2.2 Simple Two-Level Resonance Experiment

Consider an atom with two levels interacting with a uniform electric field oscillating at frequency, ω.

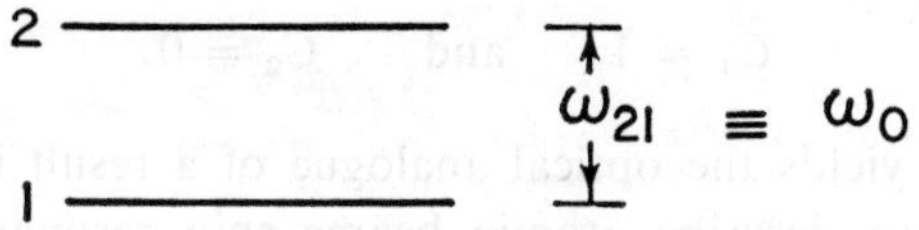

For electric dipole transitions, the perturbing potential is

$$V(r, t) = -\boldsymbol{\mu} \cdot \mathbf{E}_0 \cos \omega t$$

where $\boldsymbol{\mu}$ is the electric dipole moment operator ($\boldsymbol{\mu} = -e\mathbf{r}$) and

$$\begin{aligned} V_{11} &= V_{22} = 0, \\ V_{21}^* &= V_{12} = V_0 \cos \omega t = \frac{V_0}{2}(e^{i\omega t} + e^{-i\omega t}), \end{aligned} \tag{2.8}$$

in which

$$V_0 = -(\boldsymbol{\mu} \cdot \mathbf{E}_0)_{12}$$

and we will assume all other $V_{nm} = 0$. In practice, this assumption is justified if all other transitions are far off resonance. Then

$$i\hbar\dot{C}_1 = \frac{C_2 V_0}{2}\,[e^{i(\omega-\omega_0)t} + e^{-i(\omega+\omega_0)t}]. \tag{2.9}$$

When Eq. (2.9) is integrated, the terms $(\omega - \omega_0)$ and $(\omega + \omega_0)$ in the exponent effectively end up in the denominator. The second term $(\omega + \omega_0)$ is "anti-resonant" and may be neglected. Hence

$$\begin{aligned} i\hbar\dot{C}_1 &\approx \frac{C_2 V_0}{2}\, e^{i(\omega-\omega_0)t}, \\ i\hbar\dot{C}_2 &\approx \frac{C_1 V_0^*}{2}\, e^{-(i\omega-\omega_0)t}. \end{aligned} \tag{2.10}$$

Or

$$\ddot{C}_1 - i\Omega\dot{C}_1 + \left|\frac{V}{2\hbar}\right|^2 C_1 = 0, \tag{2.11}$$

where $\Omega = \omega - \omega_0$ and

$$C_1(t) = Ae^{\mu_+ t} + Be^{\mu_- t},$$
$$C_2(t) = \frac{2i\hbar}{V_0}(A\mu_+ e^{\mu_+ t} + B\mu_- e^{\mu_- t})\, e^{-i\Omega t}, \tag{2.12}$$

where

$$\mu_\pm = \frac{i\Omega}{2} \pm \frac{i}{2}\sqrt{\Omega^2 + \left|\frac{V}{\hbar}\right|^2} \equiv \frac{i\Omega}{2} \pm \frac{iR}{2}. \tag{2.13}$$

If we assume the atom is initially in state 1 (the lower state) at $t = 0$,

$$C_1 = 1 \quad \text{and} \quad C_2 = 0.$$

This assumption yields the optical analogue of a result initially obtained by Rabi (1937) to describe atomic beams spin resonance experiments; namely,

$$P_2(t) = |C_2(t)|^2 = \frac{|V_0/\hbar|^2}{\Omega^2 + |V_0/\hbar|^2} \sin^2(Rt/2) \tag{2.14}$$

where $P_2(t)$ is the probability of finding the atom in state 2 as a function of time.

The atom is actually in a mixed state

$$\Psi(\mathbf{r}, t) = C_1(t)\, U_1(\mathbf{r})\, e^{-iE_1t/\hbar} + C_2(t)\, U_2(\mathbf{r})\, e^{-iE_2t/\hbar} \tag{2.15}$$

in the presence of the field and neither $U_1e^{-iE_1t/\hbar}$ nor $U_2e^{iE_2t/\hbar}$ are eigenfunctions of the problem. The probabilities $|C_2(t)|^2$ and $|C_1(t)|^2$ $(= 1 - |C_2|^2)$ represent the average probability of finding the atom in states 2 or 1 if we were to turn the field off suddenly at time t and repeat the experiment many times.

The total energy of the whole system (atom plus radiation field) is constant. However, the time dependence of $|C_2(t)|^2$ implies that an exchange of energy between the atom and the field occurs. The mixed-state wave function is an eigenstate of the electric-dipole moment operator. The expectation value of the dipole operator is

$$\langle \boldsymbol{\mu} \rangle = \int \Psi^* \boldsymbol{\mu} \Psi \, dv = (C_1^* C_2 e^{-i\omega_0 t} + \text{c.c.})\, \boldsymbol{\mu}_{21}, \tag{2.16}$$

where

$$\boldsymbol{\mu}_{21} = \int U_2^* \boldsymbol{\mu} U_1 \, dv.$$

The interaction energy with the field is $-\langle \boldsymbol{\mu} \rangle \cdot \mathbf{E}$ and goes in and out of phase with the field at the rate R. Note that

$$-\langle \boldsymbol{\mu} \rangle \cdot \mathbf{E} = \begin{cases} + \text{ for } 0 \leqq Rt \leqq \pi & \text{(absorbing)} \\ - \text{ for } \pi \leqq Rt \leqq 2\pi & \text{(emitting)} \end{cases}, \text{ etc.} \qquad (2.17)$$

where from Eq. (2.13)

$$R = \sqrt{\Omega^2 + \left|\frac{V}{\hbar}\right|^2}.$$

If all atoms in the system had the same boundary conditions (or phase) at $t = 0$, one could see fluctuations in the electromagnetic field, even though there were no net average flow of energy. In practice, it is necessary to "prepare" the state to see such effects because of random processes such as collisions. [Such state preparation is, for example, the basis of the double field resonance technique in atomic beams (Ramsey, 1950), the spin-echo effect in nuclear magnetic resonance experiments (Hahn, 1950), the more recent photon echo experiments (Kurnit, Abella, and Hartmann, 1964; Abella *et al.*, 1966), and self-induced transparency effects at optical frequencies (McCall and Hahn, 1969)].

Note that the average probability of finding the atom in the upper state is

$$\overline{|C_2(t)|^2} = \frac{1}{2}\frac{|V/\hbar|^2}{(\omega - \omega_0)^2 + |V/\hbar|^2} \to \frac{1}{2}\begin{cases} \text{for } |V/\hbar| \to \infty \\ \text{or } \omega \to \omega_0 \end{cases} \qquad (2.18)$$

for very long time intervals compared to $1/R$. Applying the field tends to equalize the average probability of finding the atom in each state and this probability oscillates with a frequency that increases with the field. This decreases the apparent lifetime of each state and results in an increasing width of resonance lines with field, in experiments designed to measure ω_0. For example, consider the hypothetical resonance experiment‡ illustrated schematically in Fig. 2-1. Here an atom is put in the lower level (1) by the state selector and enters the perturbing field at time $t = 0$. After a very long random time, t, spent in the inducing field (during which the atom is not otherwise perturbed), the atom enters the analyzer where the time-average

‡ It is interesting to note that an atomic beams spin resonance experiment somewhat similar to the hypothetical one in Fig. 1-1 has actually been done (Goldenberg, Kleppner and Ramsey, 1961) in which ground state cesium atoms made ≈ 200 wall collisions before losing their polarization. The actual experiment was far more complex due to the depolarizing effect of wall collisions, the incorporation of the Ramsey double field technique, and beam velocity distribution effects.

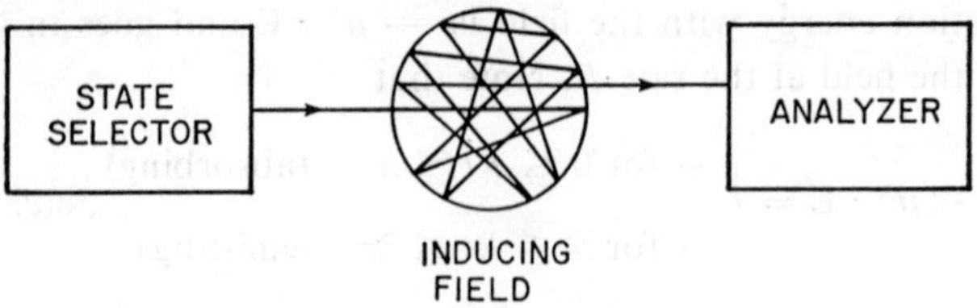

FIGURE 2-1. Hypothetical atomic beams resonance experiment used to illustrate power broadening. It is assumed that the atoms spend a long, random length of time in the "inducing field".

of the probability of finding the atom in the upper state (2) is examined. From Eq. (2.18) the average probability of finding the atom in state (2) is a Lorentzian function of the frequency with a width which increases with the field. The full width at half-maximum response is

$$\Delta\nu_L = |2V/\hbar| \tag{2.19}$$

and arises entirely from "power broadening", or the transition rate induced by the field itself. As $V \to 0$, $\Delta\nu_L \to 0$ in this case because the transit time in the field region was assumed very large and no allowance was made for spontaneous emission lifetimes or other sources of phase interruption in the two levels.

2.3 Inclusion of Phase Interruption

In real cases of interest there are always sources of phase interruption, or state destruction, present which occur through processes which are irreversible for all practical purposes within the time of the measurement. Specific examples in the case of a gas laser are spontaneous radiative decay, inelastic collisions which destroy the state, and elastic collisions which result in migration of the atom over the Doppler line. In the case of radiative decay, the phase interruption process itself provides a convenient means with which to detect the transition. Hence, one doesn't really need something as elaborate as an atomic beam apparatus (with its spatially-separated state analyzer) in order to do a resonance experiment. Because it is more like the situation in an actual gas laser, we will restrict the discussion from now on to situations such as those encountered in the early positronium ground state resonance experiments (see, for example, Deutsch and Brown, 1952) in which the states of the atom of concern are both created and destroyed in the same perturbing field. The following calculation (Bennett, 1962b) was in fact first worked out by the author in the analysis of an abortive experiment involving the $n = 2$ state of positronium attempted at Columbia University many years before the existence of the first gas laser.‡ An equivalent result for the transition probability was also obtained by Lamb and Sanders (1960) using the density matrix method as applied to an excited state resonance in atomic helium, and by Bloch (1946) using T_1, T_2 notation in the description of nuclear magnetic resonance experiments.

We can allow for the random relaxation processes by adding an average relaxation rate to Eqs. (2.6). That is, let

$$\dot{C}_n = \frac{1}{i\hbar} \sum_m C_m(t)\, V_{nm}(t)\, e^{i\omega_{nm}t} - C_n(R_n/2), \tag{2.20}$$

which has the desired property that for all $V_{nm} = 0$,

$$|C_n|^2 \propto e^{-R_n t}.$$

This modification of the time-dependent equations clearly results in the phenomenologically-sought exponential decay in the absence of the applied field. The justification of this treatment for *radiative* decay was discussed in connection with the hydrogen fine structure measurements by Lamb (1952). We will also extend the definition of the decay rates to include the effects of collisions through an expression of the type,

$$R_n = A_n + N\sigma V$$

in which A_n represents the total Einstein A coefficient for spontaneous radiative decay in all transitions from the n-th excited state and $N\sigma V$ is the appropriate collision rate per excited atom. The inclusion of elastic, as well as inelastic, collisions in such a term warrants separate discussion which is more appropriately brought in at a later point in the development. For the moment we shall merely emphasize that the introduction of decay rates, R_n, in Eq. (2.20) corresponds to the assumption of an irreversible average flow of probability out of the system of levels at rates

$$(\partial P/\partial t)_n = R_n |C_n(t)|^2.$$

Hence the total energy in the system of levels we are considering no longer is conserved. The inelastic decay rates represent a true flow of energy from the levels involved. The elastic phase interruption rates do not represent a net average flow of energy from the real system of atoms, even though they affect the transition probability in the same manner as do the inelastic terms

‡ A similar analysis for levels having different decay rates, but in which the approximation $R_2 \gg R_1$ is made, was given for the ground state of positronium by Hughes, Marder and Wu (1957).

in Eq. (2.20). Conservation of energy in respect to the elastic phase interruption rates is easily restored through appropriate modification of the average zero-field population densities as discussed in Sec. 2.4. [Eq. (2.40)].

Consider two levels connected by an electric-dipole transition:

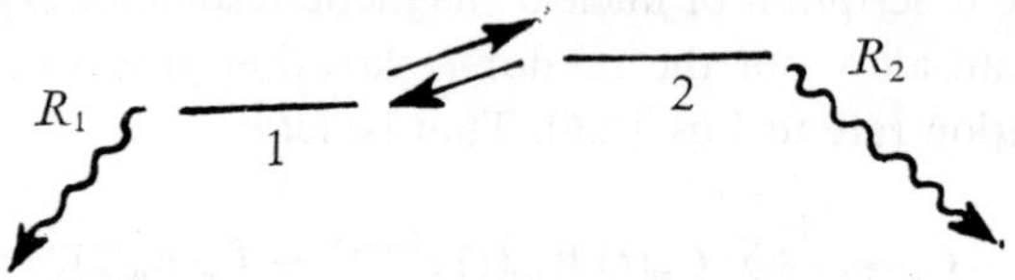

where

$$[V(\mathbf{r}, t)]_{12} = -(\boldsymbol{\mu} \cdot \mathbf{E}_0)_{12} \cos \omega t \equiv V(e^{i\omega t} + e^{-i\omega t}) \tag{2.21}$$

as before and

$$V \equiv -(\boldsymbol{\mu} \cdot \mathbf{E}_0/2)_{12} \equiv -\langle \mu_0 \rangle E_0/2 \tag{2.22}$$

where $\langle \mu_0 \rangle$ is the matrix element of the electric dipole moment operator in the (transverse) direction of quantization established by the linearly-polarized optical field; in other words, the optical polarizability of the excited atom on the transition.

Neglecting the antiresonant term,

$$\begin{aligned} \dot{C}_1 &= C_2\left(\frac{V}{i\hbar}\right)e^{i\Omega t} - C_1\left(\frac{R_1}{2}\right), \\ \dot{C}_2 &= C_1\left(\frac{V}{i\hbar}\right)e^{-i\Omega t} - C_2\left(\frac{R_2}{2}\right), \end{aligned} \tag{2.23}$$

where

$$\Omega = (\omega - \omega_0).$$

ω = the driving frequency and ω_0 = the resonance frequency of the atom.

We wish to solve these equations subject to two sets of boundary conditions:

(1) at $t = 0$, $C_1 = 1$ and $C_2 \equiv 0$,

where we want

$$P_{\text{abs}} \equiv R_2 \int_0^\infty |C_2(t)|^2 \, dt \tag{2.24}$$

or the total probability that an atom placed in that lower state initially has undergone a net absorption to the upper state (i.e., the atom removes one quantum, $h\nu$, from the field at the field frequency).

(2) at $t = 0$, $C_1 = 0$ and $C_2 = 1$,

where we want

$$P_{\text{stim}} \equiv R_1 \int_0^\infty |C_1(t)|^2 \, dt \tag{2.25}$$

or the total probability that an atom placed in the upper state initially undergoes a net stimulated emission to the lower state (i.e. the atom gives up one quantum, $h\nu$, of energy to the field at the field frequency). These two quantities, P_{stim} and P_{abs}, are very important both for understanding many ordinary resonance experiments and for understanding the behavior of lasers. We can obtain "exact" expressions for both of these quantities valid in both the low- and high-field limit.

Consider P_{stim} first. Here we need $C_1(t)$ for $C_1 = 1$ and $C_1 = 0$ at $t = 0$. From Eqs. (2.23)

$$\ddot{C}_2 + \left(\frac{R_1 + R_2}{2} + i\Omega\right)\dot{C}_2 + \left(\left|\frac{V}{\hbar}\right|^2 + i\Omega\frac{R_2}{2} + \frac{R_1 R_2}{4}\right)C_2 = 0, \tag{2.26}$$

where

$$C_1 = \left(\frac{i\hbar}{V}e^{i\Omega t}\right)\left(C_2\frac{R_2}{2} + \dot{C}_2\right). \tag{2.27}$$

The solutions for C_2 are of the form $e^{\mu_\pm t}$ where‡

$$\mu_\pm = -\frac{1}{2}\left[\frac{R_1 + R_2}{2} + i\Omega \pm \sqrt{\left(\frac{R_1 - R_2}{2} + i\Omega\right)^2 - \left|\frac{2V}{\hbar}\right|^2}\,\right]. \tag{2.28}$$

Hence

$$C_2 = Ae^{\mu_+ t} + Be^{\mu_- t},$$
$$C_1 = \left(\frac{i\hbar}{V}e^{i\Omega t}\right)\left[A\left(\mu_+ + \frac{R_2}{2}\right)e^{\mu_+ t} + B\left(\mu_- + \frac{R_2}{2}\right)e^{\mu_- t}\right] \tag{2.29}$$

and for $C_2 = 1$, $C_1 = 0$ at $t = 0$ (atom initially in the upper state)

$$A = \frac{\mu_- + R_2/2}{(\mu_- - \mu_+)} \quad \text{and} \quad B = \frac{\mu_+ + R_2/2}{(\mu_- - \mu_+)}. \tag{2.30}$$

Hence

$$C_2(t) = \frac{(\mu_- + R_2/2)\,e^{\mu_+ t} - (\mu_+ + R_2/2)\,e^{\mu_- t}}{(\mu_- - \mu_+)},$$
$$C_1(t) = \left(\frac{i\hbar}{V}\right)\frac{(\mu_+ + R_2/2)(\mu_- + R_2/2)}{(\mu_- - \mu_+)}\left[e^{\mu_+ t} - e^{\mu_- t}\right]e^{i\Omega t}. \tag{2.31}$$

‡ It should be noted that the roots $\mu_\pm$ in Eq. (2.28) are totally distinct from the electric-dipole moment operator, μ, in Eq. (2.22) upon which the perturbing term V depends.

Then

$$P_{\text{stim}} = R_1 \int_0^\infty |C_1(t)|^2\,dt$$

$$= R_1 \left|\frac{\hbar}{V}\right|^2 \left|\frac{(\mu_+ + R_2/2)(\mu_- + R_2/2)}{(\mu_- - \mu_+)}\right|^2 \int_0^\infty |e^{\mu_+ t} - e^{\mu_- t}|^2\,dt. \tag{2.32}$$

Then note that

$$\int_0^\infty |e^{\mu_+ t} - e^{\mu_- t}|^2\,dt$$

$$= 2(R_1 + R_2)\left\{\frac{1}{R_1R_2 + |2V/\hbar|^2 + [(R_1 + R_2)/2]^2 + \Omega^2 + |\xi|^2}\right.$$

$$\left. - \frac{1}{R_1R_2 + |2V/\hbar|^2 + [(R_1 + R_2)/2]^2 + \Omega^2 - |\xi|^2}\right\}, \tag{2.33}$$

where

$$|\xi|^2 = \sqrt{\left[\left(\frac{R_1 - R_2}{2}\right)^2 + \Omega^2\right]^2 - 2\left[\left(\frac{R_1 - R_2}{2}\right)^2 - \Omega^2\right]\left|\frac{2V}{\hbar}\right|^2 + \left|\frac{2V}{\hbar}\right|^4}$$

$$= |\mu_- - \mu_+|^2 \tag{2.34}$$

and

$$\left|\left(\mu_+ + \left(\frac{R_2}{2}\right)\right)\left(\mu_- + \frac{R_2}{2}\right)\right|^2 = \left|\frac{V}{\hbar}\right|^4. \tag{2.35}$$

Combining terms,

$$P_{\text{stim}} = \frac{R_1(R_1 + R_2)\,|2V/\hbar|^2}{4(\omega - \omega_0)^2 R_1R_2 + (R_1R_2 + |2V/\hbar|^2)(R_1 + R_2)^2}, \tag{2.36}$$

where $E = E_0 \cos \omega t$ and $2V = -\langle \mu_0 E_0 \rangle_{1,2}$. [Note that the result would hold for a dc field E_0 if $V \equiv -\langle \mu_0 E_0 \rangle_{1,2}$ in the above expression. The factor of 2 difference between the optical and dc result enters because of the neglect of the antiresonant term in Eq. (2.23)]. Also note that the result is valid for high fields and that because the final result for the transition probability depends on $|V|^2$ and because the anti-resonant term was neglected in Eq. (2.23), the final result does not depend on the phase of the perturbing term in Eq. (2.21). In particular, it should be emphasized that P_{stim} is a Lorentzian function of the frequency with full width at half-maximum given by

$$\Delta\nu_L = \left(\frac{R_1 + R_2}{2\pi}\right)\sqrt{1 + \frac{1}{R_1R_2}\left|\frac{\langle \mu_0 E_0 \rangle_{1,2}}{\hbar}\right|^2}. \tag{2.37}$$

Note that

$$\Delta\nu_L \to \frac{R_1 + R_2}{2\pi} \quad \text{as} \quad E_0 \to 0.$$

Hence the Lorentz width approaches the natural width estimated previously from the uncertainty principle in Eq. (1.1) as $E_0 \to 0$.

P_{abs} is obtained by evaluating

$$R_2 \int_0^\infty |C_2(t)|^2 \, dt$$

for the boundary conditions $C_1 = 1$, $C_2 = 0$ at $t = 0$. The result may be obtained from the expression for P_{stim} in Eq. (2.36) merely by interchanging the subscripts 1 and 2. That is

$$P_{\text{abs}} = \frac{R_2(R_1 + R_2)\,|2V/\hbar|^2}{4(\omega - \omega_0)^2 R_1 R_2 + (R_1 R_2 + |2V/\hbar|^2)(R_1 + R_2)^2}. \tag{2.38}$$

Hence P_{abs} is also a Lorentzian function of the frequency with the same full width at half-maximum response given by Eq. (2.37).

Hence if we do a resonance experiment in which we feed atoms into state 1 and look at the decay through the transition R_2 as a function of

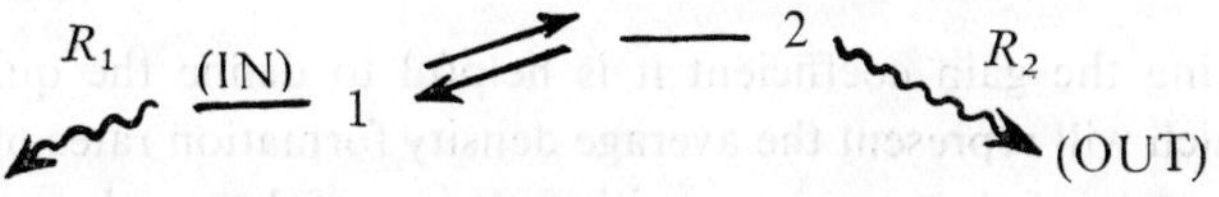

frequency, we see a Lorentzian line shape with full width at half-maximum given by Eq. (2.37).

The expression illustrates the importance of having relaxation mechanisms in the performance of resonance experiments. Aside from permitting the experimenter to detect an induced transition through the decay mechanism itself, the relaxation rates have to be above some critical value to permit seeing a resonance curve at all. For example, suppose $R_1 R_2$ were slow enough so that

$$1 \ll \frac{1}{R_1 R_2}\left|\frac{\langle\mu_0\rangle E_0}{\hbar}\right|^2,$$

then

$$\Delta\nu_L \to \frac{\langle\mu_0 E_0\rangle}{2\pi\hbar}\left(\frac{R_1 + R_2}{\sqrt{R_1 R_2}}\right).$$

If for example $R_1 \to 0$, the Lorentz width would approach infinity in the presence of the perturbing field. (Note that $R_2 \geq A_{21}$, the Einstein A

coefficient for the transition). In that event one would not be able to detect a resonance in an experiment of the present type by sweeping the frequency through a finite range. The example also illustrates the importance of making corrections for power broadening in experiments designed to measure $(R_1 + R_2)$ from the resonance curve width.

2.4 Gain (or Absorption) Coefficient for Running Waves (Bennett, 1965 b)

We can use the above results to determine the steady-state intensity-dependent gain or absorption coefficient for a running wave traversing a uniformly-excited medium consisting of atoms with the same resonant frequency and line breadth. These results as yet will apply only to a medium in which there is no random thermal motion of the atoms. A treatment of media in which significant Doppler shifts arise from the thermal velocities of the atoms will be given later.

If we consider a running wave at frequency ω and peak electric field amplitude E_0, the average light flux is

$$I = \frac{C}{8\pi} E_0^2 \quad \text{(ergs/cm}^2\text{-sec)}. \tag{2.39}$$

In evaluating the gain coefficient it is helpful to define the quantities F_1 and F_2 which will represent the average density formation rates of the lower and upper state of the atomic transition. It is useful to make a distinction here between the *total* lower- and upper-state phase interruption rates (R_1, R_2) and the *inelastic* portion of these decay rates, which we will define here as R_1^0 and R_2^0. The actual lower- and upper-state population densities, N_1^0 and N_2^0, are given at steady state in terms of the inelastic rates through equations of the type

$$F_1 + A_{21}N_2^0 = R_1^0 N_1^0 \text{ for the lower state}$$

$$F_2 = R_2^0 N_2^0 \text{ for the upper state,}$$

in which F_1, F_2 represent density excitation rates from sources of energy independent of the pair of levels, and A_{21} represents the Einstein A-coefficient for spontaneous emission on the laser transition.

By "elastic" is meant any source of phase interruption which does not affect the average level densities for given values of F_1 and F_2. Resonance-trapped radiative decay on a transition from one of the levels to the atom ground state (in which the excitation is not lost but reappears at a displaced point in the medium after the original state has been destroyed) represents one such example of an "elastic" phase interruption process within the present definition. Here the net decay rate (inelastic portion) may be substantially smaller than the total phase interruption rate from radiative decay (see Holstein, 1947; 1951). As discussed below, elastic collisions resulting from the atoms' random thermal motion may also be treated as a source of elastic phase interruption to first approximation within the present meaning.

In the following analysis, it is convenient to define two *effective* zero-field population densities, N_1 and N_2, by the relations

$$F_1 \equiv R_1 N_1 \text{ for the lower state}$$

$$F_2 \equiv R_2 N_2 \text{ for the upper state,}$$

where R_1 and R_2 are the *total* phase interruption rates (inelastic plus elastic) for the two levels. It is seen by inspection that

$$\begin{aligned} N_1 &= (R_1^0/R_1)\, N_1^0 - (A_{21}/R_1)\, N_2^0 \equiv F_1/R_1 \\ N_2 &= (R_2^0/R_2)\, N_2^0 \equiv F_2/R_2. \end{aligned} \tag{2.40}$$

We shall show that the small signal-gain coefficient depends on an effective zero-field inversion density,

$$N_2 - N_1 = (R_0^2/R_2 + A_{21}/R_1)\, N_2^0 - (R_1^0/R_1)\, N_1^0$$

rather than the actual zero-field inversion density, $N_2^0 - N_1^0$. This distinction, of course, vanishes in the limit that

$$R_1 \gg A_{21}, \quad R_1 \approx R_1^0 \quad \text{and} \quad R_2 \approx R_2^0.$$

It should also be emphasized that N_2, N_1 represent effective population densities in absence of the interaction with the radiation field. In the presence of the field we assume that the formation rates, F_1, F_2 remain the same, but the actual level densities are changed through stimulated emission and absorption.

In the presence of a running wave travelling in the z direction, the rate of flow of energy per unit volume from the atoms to the optical wave at some point z is

$$\frac{\partial \mathscr{E}}{\partial t} = h\nu\,(F_2 P_{\text{stim}} - F_1 P_{\text{abs}}).$$

Hence the spatial rate of growth of intensity flux of the running wave in moving from z to $z + dz$ in its direction of propagation is

$$\frac{\partial I}{\partial z} = h\nu(R_2 N_2 P_{\text{stim}} - R_1 N_1 P_{\text{abs}}). \tag{2.41}$$

where use has been made of Eq. (2.40).

In this expression we have assumed that the energy added to the wave through stimulated emission occurs in the same direction and with the same frequency and phase as the incident wave.

Dividing the previous expressions for P_{stim} and P_{abs} in Eqs. (2.36) and (2.38) by $R_1 R_2 (R_1 + R_2)^2$, substituting Eq. (2.22) for the electric-dipole perturbation and applying Eqs. (2.39)–(2.41) yields

$$\frac{\partial I}{\partial z} = \frac{I(N_2 - N_1)\, 8\pi\langle\mu_0\rangle^2/\hbar\lambda(R_1 + R_2)}{1 + 4(\omega - \omega_0)^2/(R_1 + R_2)^2 + 8\pi I\langle\mu_0\rangle^2/R_1 R_2 \hbar^2 c} \tag{2.42}$$

or letting‡

$$A_{21} \approx \frac{4}{\hbar\lambda^3}\langle\mu_0\rangle^2,$$

the Einstein A coefficient, we get

$$\frac{\partial I}{\partial z} \approx \frac{I(N_2 - N_1)\, 2\pi\lambda^2 A_{21}/(R_1 + R_2)}{1 + 4(\omega - \omega_0)^2/(R_1 + R_2)^2 + (2\pi\lambda^2 A_{21}/h\nu R_1 R_2)\, I}, \tag{2.43}$$

where N_2, N_1 are the excited-state densities in the absence of the running wave.

The expression for $\partial I/\partial z$ is of the form

$$\frac{\partial I}{\partial z} = \frac{aI}{1 + b + cI}, \tag{2.44}$$

where b is frequency-dependent. Or in going a distance L, the growth (or loss) in intensity of the running wave follows the transcendental equation

$$Ie^{[c/(1+b)]I} = I_0 e^{[c/(1+b)]I_0}\, e^{[a/(1+b)L]}. \tag{2.45}$$

The qualitative behavior of this relationship is illustrated in Fig. 2-2. That is, at low intensities, I/I_0 increases exponentially with distance with a gain

‡ In any real atom the Einstein A coefficient will be related to the electric dipole matrix elements through "line strength" factors which depend on level degeneracy and coupling approximations. Because the above discussion is inherently limited to a two-level situation, it seemed appropriate to omit such factors in relating the optical polarizability to the A coefficient. A theory of gain saturation in the presence of level degeneracy has recently been given by Dienes (1968).

(or loss) coefficient, $a/(1 + b)$. At high intensities the gain (or loss) coefficient saturates and there is a linear increase (or decrease) in intensity at the rate

$$\frac{\partial I}{\partial z} = \frac{a}{c}.$$

This saturation effect in the small signal gain (or attenuation) coefficient follows from the conservation of energy. The growth in the intensity of the wave clearly cannot increase beyond the rate at which energy is supplied to the medium through the excited-state formation rates. Saturation occurs by the changes of excited-state density which are produced by stimulated emission and absorption.

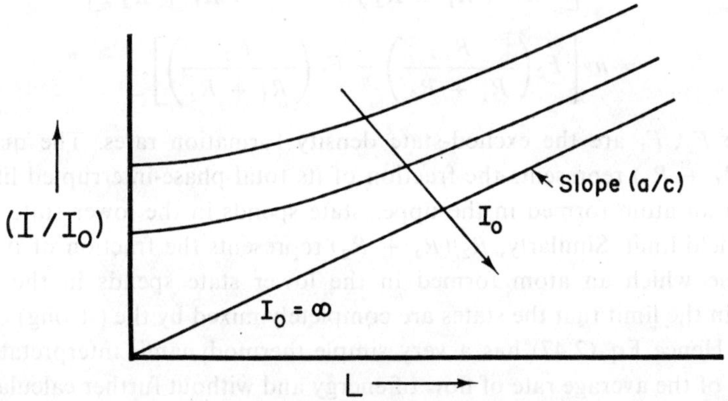

FIGURE 2-2. Qualitative illustration of the variation of intensity gain with I_0 from Eq. (2.44).

Although the intermediate intensity case is extremely nonlinear, one can easily see what goes on at the extremely low and high intensity limits. In fact, since the low- and high-intensity coefficients are related through Eq. (2.42), it is possible to predict the high-intensity behavior of the medium from its low-intensity properties.

At resonance $(\omega - \omega_0)$, the low-intensity gain coefficient is

$$(G)_{\max} = \frac{1}{I}\left(\frac{\partial I}{\partial z}\right)_{\max} = \frac{(N_2 - N_1)\, 8\pi\langle\mu_0\rangle^2}{\hbar\lambda(R_1 + R_2)} \approx (N_2 - N_1)\frac{2\pi\lambda^2 A_{21}}{(R_1 + R_2)}, \tag{2.46}$$

where if $R_2 = A_{21} \ll R_1$, the interaction cross section is $2\pi\lambda^2$, defined in respect to the actual zero-field population densities given in Eq. (2.40).

Note that the presence of elastic relaxation processes can reduce the small-signal gain coefficient through the determination of an effective zero-field inversion density, $N_2 - N_1$, which is smaller than the actual average zero-field inversion density, $N_2^0 - N_1^0$, which results from the net inelastic decay rates alone.

At high intensities, the gain coefficient saturates so that

$$P_\infty \equiv \left(\frac{\partial I}{\partial z}\right)_\infty = h\nu(N_2 - N_1)\left(\frac{R_1 R_2}{R_1 + R_2}\right). \tag{2.47}$$

This just says that the maximum power density gain at infinite field is

$$\begin{aligned} P_\infty &= h\nu\left[N_2 R_2\left(\frac{R_1}{R_1 + R_2}\right) - N_1 R_1\left(\frac{R_2}{R_1 + R_2}\right)\right] \\ &= h\nu\left[F_2\left(\frac{R_1}{R_1 + R_2}\right) - F_1\left(\frac{R_2}{R_1 + R_2}\right)\right], \end{aligned} \tag{2.48}$$

where F_2, F_1 are the excited-state density formation rates. The quantity $R_1/(R_1 + R_2)$ represents the fraction of its total phase-interrupted lifetime which an atom formed in the upper state spends in the lower state in the high-field limit. Similarly, $R_2/(R_1 + R_2)$ represents the fraction of its total lifetime which an atom formed in the lower state spends in the upper state in the limit that the states are completely mixed by the (strong) optical field. Hence Eq. (2.47) has a very simple thermodynamic interpretation in terms of the average rate of flow of energy and without further calculations, it is apparent that a knowledge of the parameters R_1, R_2, etc. will determine the maximum obtainable output power from the medium when it is used as a laser. Specifically,

$$P_\infty = 8\pi\langle\mu_0\rangle^2 R_1 R_2 c G_{\mathrm{max}} \approx h\lambda^3 A_{21} R_1 R_2 c G_{\mathrm{max}}. \tag{2.49}$$

At low intensities,

$$G(\nu) = \frac{(N_2 - N_1)\, 8\pi\langle\mu_0\rangle^2/[\hbar\lambda(R_1 + R_2)]}{1 + 4(\omega - \omega_0)^2/(R_1 + R_2)^2}, \tag{2.50}$$

where $\omega = 2\pi\nu$. Hence

$$\int_{-\infty}^{\infty} G(\nu)\, d\nu = (N_2 - N_1)\frac{\langle\mu_0\rangle^2}{2\hbar\lambda}\int_{-\infty}^{\infty} d\left(\frac{2\omega}{R_1 + R_2}\right)\Big/\left[1 + \frac{4(\omega - \omega_0)^2}{(R_1 + R_2)^2}\right],$$

where the integral may be evaluated through the substitution

$$\tan\theta = 2(\omega - \omega_0)/(R_1 + R_2).$$

Therefore

$$\int_{-\infty}^{\infty} G(\nu)\, d\nu = (N_2 - N_1)\, 2\pi\langle\mu_0\rangle^2/\hbar\lambda \approx (N_2 - N_1)\frac{\pi\lambda^2}{2} A_{21} \qquad (2.51)$$

Note that this result is independent of the resonant frequencies assumed for the atoms and depends only on the total density of atoms in the upper and lower state through Eq. (2.40). Therefore this result is of value in determining the normalization constants for more general spectral response functions which, for example, arise in the Doppler-broadened case and which hold in the limit that $\Delta\nu_L \ll \nu_0$.

Eq. (2.51) is equivalent to the well-known normalization integral obtained in the "old" quantum theory to describe resonance absorption experiments in the late 1920s (see, for example, the discussion in Mitchell and Zemansky, 1934).

2.5 Doppler-Broadened Gain and Absorption Coefficients

So far we have assumed that the excited atoms all have the same resonant frequency, and line breadth, a situation which is sometimes described by the term "homogeneous broadening". Radiative decay rates for optical transitions are typically in the range from about 10^7–10^8 cps. Consequently, in the absence of collisions $R_{1,2} \approx A_{1,2}$ and $\Delta\nu_{\text{nat}} \approx 10$ to 100 MHz.

In contrast, however, the Doppler shifts which arise from the atom's thermal velocities are much larger. For example, if we have an atom moving towards the source of the light wave at velocity v, the resonant frequency of the interaction is shifted toward higher frequencies by amount $\Delta\nu = v/\lambda$. The mean thermal velocity for atoms in a gas discharge is typically $\approx 10^5$ cm/sec. Consequently, for $\lambda \approx 1$ micron, $\Delta\nu_{\text{Doppler}} \approx 1000$ Mc/sec, and $\Delta\nu_L \ll \Delta\nu_D$. In the real case one, of course, has a continuous distribution of resonant frequencies in the excited medium whose individual Lorentz widths are very small compared to the full Doppler width of the line. Such a medium is sometimes referred to as an "inhomogeneously broadened medium" and is characterized by the fact that for a single-frequency wave incident on the medium, the interaction probability is different for atoms with different velocities in the direction of the running wave. Consequently, as we shall discuss later, either stimulated emission or absorption can burn a hole in the velocity distribution at high powers which is small compared to the Doppler width. Consequently, the exact nature of the line shape in an inhomogeneously broadened medium can become very complicated at high powers.

For a Maxwellian velocity distribution, the number of atoms with velocity in the z-direction between v_z and $v_z + dv_z$ is proportional to $e^{-Mv_z^2/2kT}$. Consequently, the number of atoms with center frequencies Doppler-shifted by $\Delta\nu = \nu_0 v_z/c$ as seen in the z-direction is proportional to

$$e^{-(4\ln 2)(\nu-\nu_0)^2/(\Delta\nu_D)^2},$$

where the full Doppler width at half-maximum is

$$\Delta\nu_D = \frac{2}{\lambda_0}\sqrt{\frac{2kT}{M}\ln 2}. \tag{2.52a}$$

in which k is Boltzmann's constant, T the absolute temperature and M the atomic mass. In the limit that $\Delta\nu_D \gg \Delta\nu_L$, the gain (or absorption) distribution near the center of the line is approximately Gaussian and near the center of the line we may say

$$G(\nu) = G_{\max}\, e^{-4\ln 2(\nu-\nu_0)^2/(\Delta\nu_D)^2} \approx G_{\max}\, e^{-(\nu-\nu_0)^2/(0.6\,\Delta\nu_D)^2}, \tag{2.52b}$$

where the value of $G_{\max}$ can be determined from the normalization integral in Eq. (2.51). That is,

$$\int_{-\infty}^{\infty} G(\nu)\,d\nu = \frac{\Delta\nu_D}{2\sqrt{\ln 2}} G_{\max} \int_{-\infty}^{\infty} e^{-x^2}\,dx \equiv (N_2 - N_1)\,2\pi\langle\mu_0\rangle^2/(\hbar\lambda)$$

$$\approx (N_2 - N_1)\frac{\lambda^2}{8\pi} A_{21}. \tag{2.53}$$

Noting that $\int_{-\infty}^{\infty} e^{-x^2}\,dx = \sqrt{\pi}$, the small signal gain coefficient is given by

$$G_{\max} = \sqrt{16\pi \ln 2}\,(N_2 - N_1)\,\langle\mu_0\rangle^2/(\hbar\lambda\,\Delta\nu_D)$$

$$\approx \sqrt{\frac{\ln 2}{16\pi^3}}\,(N_2 - N_1)\,\lambda^2 A_{21}/\Delta\nu_D \tag{2.54}$$

at the center of the Doppler-broadened line.

Note that the Gaussian functional form is only a good approximation, within the central portion of the line. The real distribution is a Gaussian distribution of Lorentzians. Far away from the center the Lorentzian varies as $1/(\nu_0 - \nu)^2$ and dominates over the Gaussian tail which goes to zero as $e^{-(\nu-\nu_0)^2}$ (see Fig. 2-3).

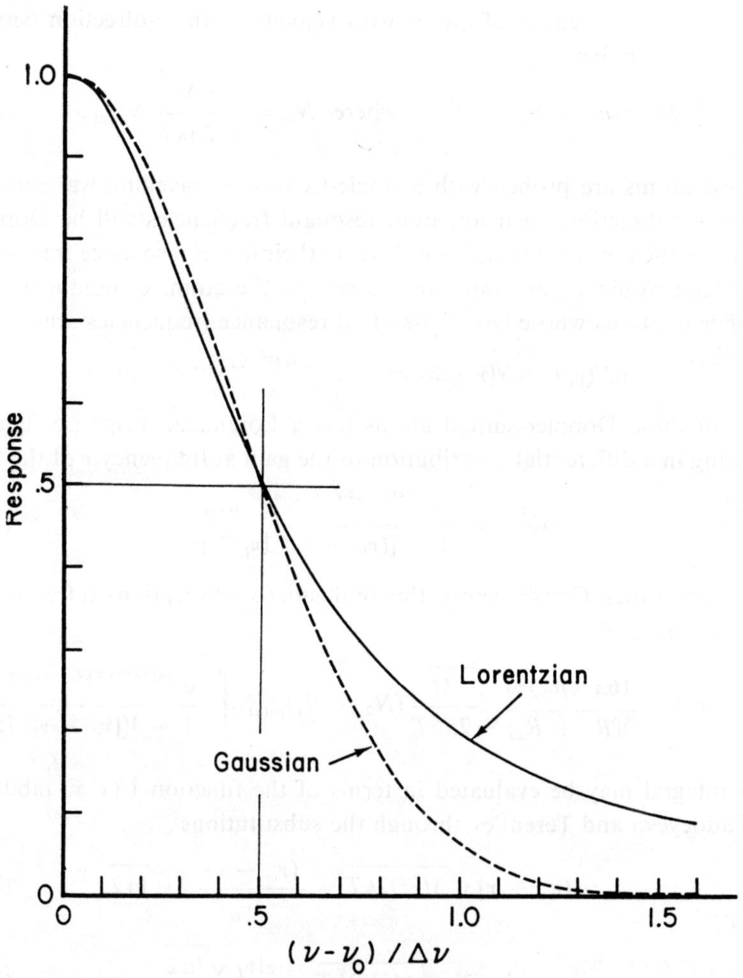

FIGURE 2-3. Comparison of Gaussian and Lorentzian profiles having the same full widths at half-maximum intensity.

2.6. Exact Gain (or Absorption) Expression for Doppler-Broadened Line

The line-shape function for a Doppler-broadened line was originally studied by Voigt (1912) and others (see discussion in Born, 1933 and in Mitchell and Zemansky, 1934) and can be expressed conveniently in terms of error functions of complex argument which have recently been tabulated extensively by Faddeyeva and Terent'ev (1961) with tabular steps that are extremely convenient for optical problems.

Let the density of atoms with velocity in the z-direction between v_z and $v_z + dv_z$ be

$$dN(v_z) = N(v_z)\,dv_z = N_m\,e^{-Mv_z^2/kT} \quad \text{where } N_m = \sqrt{\frac{M}{2\pi kT}}\,N_{\text{total}}. \tag{2.55}$$

If these atoms are probed with a single-frequency travelling wave moving in the $+z$ direction, their apparent resonant frequencies will be Doppler-shifted to the lower value $(\nu_m - v_z/\lambda)$ from their initial resonance frequencies ν_m, which would be measured in the rest of the atom. Consequently, the number of atoms whose Doppler-shifted resonance frequencies are $\nu_0 = \nu_m - v_z/\lambda$ is

$$dN(\nu_0) = N(\nu_0)\,d\nu_0 = N_m\lambda e^{-(M\lambda^2/2kT)\,(\nu_m-\nu_0)^2}\,d\nu_0. \tag{2.56}$$

Each of these Doppler-shifted atoms has a Lorentzian response function resulting in a differential contribution to the gain at frequency ν of the form

$$dG(\nu) \propto \frac{e^{-(M\lambda^2/2kT)\,(\nu_m-\nu_0)^2}\,d\nu_0}{1 + [(\nu_0 - \nu)/(\Delta\nu_L/2)]^2} \tag{2.57}$$

at low intensities. Consequently, the total gain (or absorption) at frequency ν is from Eq. (2.50)

$$G(\nu - \nu_m) = \frac{16\pi^2\langle\mu_0\rangle^2}{\hbar(R_1 + R_2)}\sqrt{\frac{M}{2\pi kT}}\,(N_2 - N_1)_{\text{total}}\int_{-\infty}^{+\infty}\frac{e^{-(M\lambda^2/2kT)\,(\nu_m-\nu_0)^2}\,d\nu_0}{1 + 4[(\nu_0 - \nu)/\Delta\nu_L]^2}.$$

This integral may be evaluated in terms of the function $U(x, y)$ tabulated by Faddeyeva and Terent'ev through the substitutions

$$x = (\nu_m - \nu)\sqrt{M\lambda^2/2kT} = \frac{(\nu_m - \nu)}{\Delta\nu_D}\,2\sqrt{\ln 2} \tag{2.58}$$

and

$$y = (\Delta\nu_L/2)\sqrt{M\lambda^2/2kT} = \frac{\Delta\nu_L\sqrt{\ln 2}}{\Delta\nu_D}. \tag{2.59}$$

Specifically,

$$G(\nu - \nu_m) = \frac{4\pi^2\langle\mu_0\rangle^2}{\hbar}\sqrt{\frac{M}{2\pi kT}}\,(N_2 - N_1)_{\text{total}}\,U(x, y) \tag{2.60}$$

where $(N_2 - N_1)_{\text{total}}$ is the total effective inversion density in the sense of Eq. (2.40). In practice one seldom knows the inversion density or the matrix element required in Eq. (2.60) with much accuracy. It is therefore much more useful in practice to express the small-signal gain coefficient in terms of the maximum value (G_m) at the center of the line:

$$G(\nu - \nu_m) = G_m U(x, y)/U(0, y) \tag{2.61}$$

A more complete discussion of the properties of Voigt profiles will be deferred to Sec. 2.16 where use is made of the Kramers-Kronig relations and the functions described in the next section.

2.7. Note on the Functions Tabulated by Faddeyeva and Terent'ev

Because many complicated line-shape functions can be evaluated specifically in terms of these quantities, it is worthwhile to summarize the properites of the functions tabulated by Faddeyeva and Terent'ev (1961).

Faddeyeva and Terent'ev consider the function

$$W(z) = e^{-z^2}\left(1 + \frac{2i}{\sqrt{\pi}}\int_0^z e^{t^2}\,dt\right) \tag{2.62}$$

for complex values of $z = x + iy$. They show that for $y > 0$ (where the pole at $t = x + iy$ is in the upper half-plane)

$$W(z) = U(x, y) + iV(x, y) = \frac{i}{\pi}\int_{-\infty}^{\infty} \frac{e^{-t^2}\,dt}{z - t} \tag{2.63}$$

and therefore that

$$U(x, y) = \frac{1}{\pi}\int_{-\infty}^{\infty} \frac{ye^{-t^2}\,dt}{y^2 + (x - t)^2}$$

and (2.64)

$$V(x, y) = \frac{1}{\pi}\int_{-\infty}^{\infty} \frac{(x - t)\,e^{-t^2}\,dt}{y^2 + (x - t)^2}.$$

The function $U(x, y)$ is an even function of x and $V(x, y)$ is an odd function of x. Note that in the limit $y \to 0$, it follows from Eq. (2.62) that

$$U(x, 0) = e^{-x^2}$$

and (2.65)

$$V(x, 0) = \frac{2}{\sqrt{\pi}}\,e^{-x^2}\int_0^x e^{t^2}\,dt.$$

As shown elsewhere in this paper, these functions may be used to describe the gain and phase-shift coefficients for a variety of different line shapes.

Faddeyeva and Terent'ev tabulate $U(x, y)$ and $V(x, y)$ to six-place accuracy for $0 < x < 3$ and $0 < y < 3$ in tabular steps of $\Delta x = \Delta y = 0.02$. The range $3 < x < 5$, $0 < y < 3$; $0 < x < 5$, $3 < y < 5$ is covered to six-place accuracy with tabular steps of $\Delta x = \Delta y = 0.1$. The range and tabular intervals are particularly useful for line-shape investigations in gas-laser transitions.

2.8. Qualitative Discussion of the Multi-Mode Laser Oscillator‡

As previously described, the self-reproducing modes of the Fox and Li type arise from diffraction in the successive propagation between the mirror apertures of the laser cavity. These modes are similar to the *TEM* modes encountered in the microwave spectrum, except that the transverse variation of the electric field is very small over a wavelength. These modes are characterized by different radial distributions of definite symmetry for the optical electric field in which the propagation is almost entirely along the axis of the laser. The different phase shifts from the diffraction integral result in frequency differences for the different order transverse distributions. However, different axial modes having the same radial distribution and diffraction loss are spaced at equal intervals of $c/2L$, where c is the phase velocity of light in the medium and L is the mirror separation.

The loss from diffraction in these modes is essentially independent of the frequency and energy leaks out of the system in a quasi-exponential rate. Consequently, if we excite one of these modes at $t = 0$, the energy will decay as

$$|E|^2 \propto e^{-\gamma t} \tag{2.66}$$

and consequently the electric field will vary roughly as

$$E(t) \propto e^{-\gamma t/2 + i\omega_0 t} \tag{2.67}$$

on successive transits. The Fourier transform of the field is thus

$$E_\omega = \frac{1}{\sqrt{2\pi}} \int_0^\infty E(t)\, e^{-i\omega t}\, dt = \frac{1}{\sqrt{2\pi}} \left[\frac{1}{\gamma/2 + i(\omega - \omega_0)} \right] \tag{2.68}$$

‡ Portions of the material in the sections on laser oscillation are reproduced from Bennett (1962a) by permission of the editors of *Applied Optics*.

and the spectral distribution of the energy is

$$I_\omega \propto |E_\omega|^2 \propto \left[1 + \left(\frac{\omega - \omega_0}{\gamma/2}\right)^2\right]^{-1}. \tag{2.69}$$

That is, the distribution is Lorentzian with a full width at half-maximum

$$\Delta\nu_{\text{cav}} = \gamma/2\pi,$$

where the rate of loss is

$$\gamma = f(c/L)$$

in terms of the fractional energy loss per pass, f. Hence the full cavity width is

$$\Delta\nu_{\text{cav}} = (f/\pi)\,(c/2L) \tag{2.70}$$

in the quasi-exponential loss region. The latter type of result, of course, applies to any cavity or resonant circuit for which the Q approximation holds. That is, in general, if the loss occurs exponentially with time, the frequency spectrum is Lorentzian with full width $\Delta\nu_c$ and

$$Q = \frac{\nu_0}{\Delta\nu_c} = \frac{2\pi(\text{energy stored})}{(\text{energy lost per cycle})} = \frac{2\pi\nu_0\mathscr{E}}{(fc/L)\,\mathscr{E}} \tag{2.71}$$

from which Eq. (2.70) follows.

The diffraction loss is typically < 0.1 percent for real lasers and the diffraction-limited Q's are typically greater than $\approx 10^9$ ($\Delta\nu_c \lesssim 1$ Mc/sec). Hence in practice the losses in the system are determined primarily by mirror reflectance losses which for the best mirrors are in the range $f > 0.001$ (i.e. greater than 0.1 percent per pass).

Below threshold for oscillation (i.e. $GL < f$), the atoms in the interferometer can radiate into the number of modes they normally see in free space. The radiation is isotropic and a very small fraction of the radiated light falls within the solid angle of a dominant diffraction-limited self-reproducing mode of the cavity. As the gain is increased to compensate for the loss, the field in the cavity builds up sufficiently to force atoms whose resonant frequencies are within a Lorentz width of the field frequency to emit their energy in the axial direction. This sudden switching of the direction of emission is one of the most striking characteristics of a laser. Although the output power varies continuously with inversion density as the system goes through threshold, the variation is so rapid as to give the appearance of a discontinuity in shape at threshold. That is, the power in the mode typically changes from about 10^5–10^{15} photons/sec in a very small range. It is this region in the vicinity of threshold in which quantum effects are the most important. Because the system is highly sensitive to all sorts of other fluctuations near threshold, quantum effects are very difficult to study in a laser.

In practice the sharpness of the slope discontinuity provides a very practical means of determining the small signal gain coefficient in single or multi-mode lasers in terms of the threshold loss in the system. That is, above threshold the saturated power out is also proportional to the small signal gain coefficient

$$P_{\text{out}} \propto (N_2 - N_1) \propto G \quad \text{or} \quad G(P) \approx f(1 + P/P_0)$$

and in many cases it is possible to find an experimental parameter (such as the discharge current) which is proportional to $(N_2 - N_1)$ near threshold. Consequently, in this case as illustrated in Fig. 2-4, the output power provides a measure of the small signal (unsaturated) gain in units of threshold loss. So long as the function is linear near threshold, the calibration provides a measure of G even in the region where the output power varies nonlinearly with the experimental parameter (e.g. discharge current, pumping power, etc.). However, the method only applies in the single-mode case where power broadening is insignificant, or in the multi-mode case where the entire Doppler line is being used up in oscillation. (That is, it does not apply where the width of the Doppler distribution involved in oscillation changes significantly with power level.)

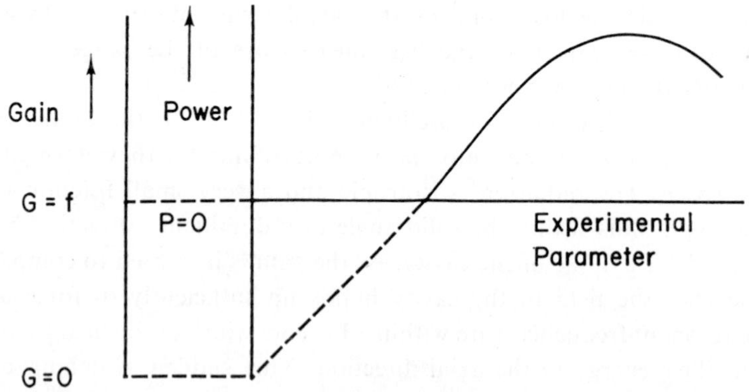

FIGURE 2-4. Approximate method of measuring the small signal gain coefficient in units of threshold loss from the output power of a multimode laser.

In practice, the frequency separations and radial field distributions of these modes still follow the diffraction calculations. Since widely different spatial regions are involved for the different radial mode distributions, different groups of atoms will supply energy to these modes. Consequently, it is not unusual to find at least the first two modes of opposite symmetry simultaneously oscillating at frequencies corresponding to the same longitudinal mode number $= L/(\lambda/2)$. In fact, some deliberate effort such as the introduction of apertures is frequently required to suppress higher order radial modes. In addition, because for the amplifying medium

$$\Delta\nu_{\text{Doppler}} \gg \Delta\nu_{\text{Lorentzian}},$$

simultaneous oscillation in a number of different axial modes is also generally obtained. The combined effects of simultaneous oscillation in different radial and axial modes showed up clearly in early studies of the first helium–neon laser.

2.9. Methods of Studying Multi-Mode Oscillations

The modes of oscillation were first studied using the photomixing techniques of the type employed earlier by Forrester *et al.* (1955) with incoherent light sources. That is, photodetectors act as square-law devices and the output current contains a component

$$i \propto \mathbf{E}_1 \cdot \mathbf{E}_2,$$

where $\mathbf{E}_1$ and $\mathbf{E}_2$ are the electric-field vectors for two optical waves of different frequency. Providing the two fields have parallel components, one may observe all possible difference frequencies up to the maximum frequency response characteristic of the photodetector. Although this difference frequency current disappears when the two fields are plane polarized at right angles the beat can always be restored in such a case by inserting a plane polarizer at 45° to the two fields (i.e. in this case parallel components of the two fields are transmitted through the polarizer). For example, a difference frequency spectrum of the type shown in Fig. 2-5 is obtained from a plane-parallel laser when operated at high power levels. That is, a peak at zero frequency corresponding to each line beating with itself, followed by a peak at ~ 1 Mc/sec corresponding to all possible differences between corresponding even- and odd-symmetric modes; a peak at $c/2L$ corresponding to the differences between the dominant modes of the same symmetry spaced by $c/2L$ surrounded by two satellites corresponding to the various odd-even and even-odd difference frequencies at $c/2L$, etc. As indicated by the arrows in Fig. 2-6, the 1 Mc/sec satellites are highly sensitive to the plate angular alignment and reach a minimum separation from the $c/2L$ peaks for plane-parallel alignment. Observation of these 1 Mc/sec satellites gave the first experimental proof of the existence of the Fox and Li modes. Similar results, showing a high degree of degeneracy and a spacing of $c/2L$ for the dominant modes, have since been obtained with confocal gas lasers and agree reasonably with the Boyd–Gordon theory.

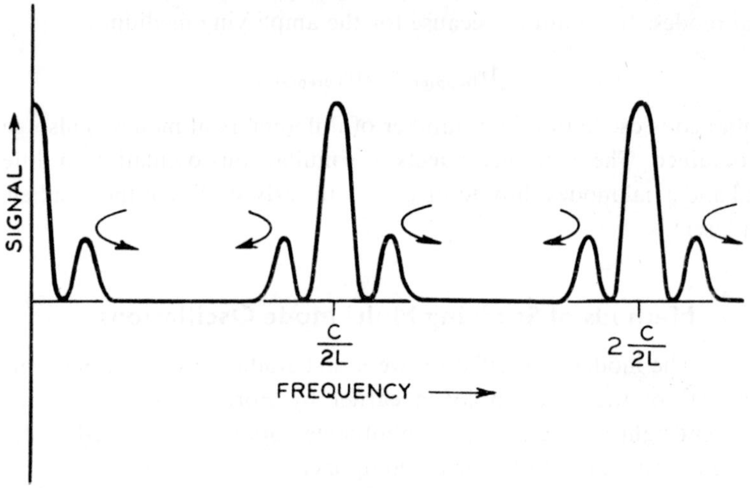

FIGURE 2-5. Beat frequency spectrum of the type observed with the first helium-neon laser (Bennett, 1962b).

Two additional experimental methods of studying the mode structure in an oscillating laser are worth mentioning briefly: *First*, the axial-mode separation in most lasers is relatively easy to study with a scanning Fabry–Perot interferometer, providing the laser is decoupled adequately from the additional cavity (Herriott, 1963; Fork, Herriott and Kogelnick, 1964). However, it is difficult to obtain sufficient resolution to permit distinguishing between different order radial modes of a near plane parallel laser in that manner without introducing considerable confusion due to overlapping orders of the scanning Fabry–Perot. *Second*, some laser transitions have high enough cw gain coefficients to permit constructing a short enough oscillator to guarantee oscillation on a single axial mode (i.e. $c/2L > \Delta\nu_D$). In these cases it is a very practical matter to use a short tunable laser to probe the beat spectrum of another oscillating laser directly with apparatus of the type shown schematically in Fig. 2-6 (Gordon and White, 1964). In principle, the tilted etalon mode-suppression technique (Fig. 1-18) could be synchronized with a piezo-electrically scanned mirror in long lasers to permit using this same heterodyning spectrum analyzer technique with very weak laser transitions as well.

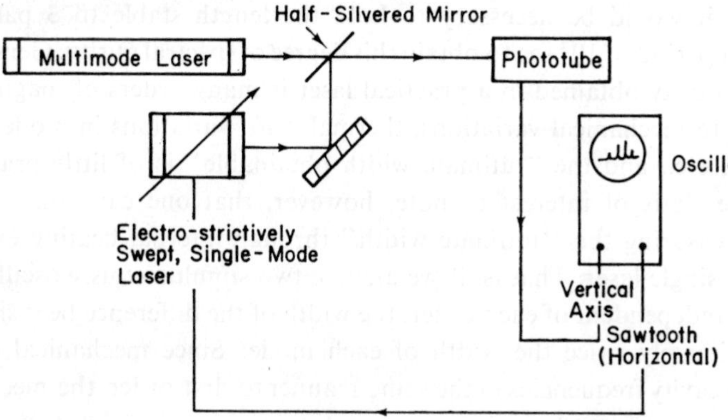

FIGURE 2-6. Method for probing the oscillating mode distribution on high-gain laser transitions (Gordon and White, 1964).

2.10. Line Narrowing and Spectral Purity

The presence of gain in the interferometer reduces the effective frequency width of the cavity from the value given in Eq. (2.70). Steady-state occurs in the system through saturation of the actual fractional energy gain coefficient (G_A) at the oscillation frequency such that for small gain

$$G_A(\nu_{osc}) = f/L, \tag{2.72}$$

where f is the fractional energy loss per pass. Because of spontaneous emission in the cavity mode, G_A saturates at a value just slightly less than f/L and the effective width of the cavity mode never quite reaches zero. Thus, in a sense, the laser is an extremely (but not infinitely) high-Q filter which selects out a narrow band of light "noise" emitted spontaneously in the cavity mode. The mechanism involved in this saturation process is discussed in more detail below. In the limit that $\Delta\nu_c \ll \Delta\nu_D$ the ultimate theoretical width of the oscillation, $\Delta\nu_{osc}$, due to spontaneous emission in a single mode is given by (Townes, 1961; Scully and Lamb, 1967)

$$\Delta\nu_{osc} \approx 8\pi h\nu(\Delta\nu_{cav})^2/P, \tag{2.73}$$

where P is the optical power in the mode. For $P \approx 1$ mW and $\Delta\nu_c \approx 1$ Mc/sec, Eq. (2.73) yields at 1 micron

$$\Delta\nu_{osc} \approx 10^{-3} \text{ cps},$$

corresponding to ≈ 3 parts in 10^{17} of the optical frequency. Since the oscillation frequency is determined to first order by the length of the interferometer, it would be necessary to hold the length stable to 3 parts in 10^{17} over a period of 10^3 sec to obtain this degree of spectral purity. Generally the actual purity obtained in a practical laser is many orders of magnitude worse due to mechanical variations, thermal drift, variations in mode pulling effects, etc., and the "ultimate width obtainable" is of little practical importance. It is of interest to note, however, that one can come fairly close to measuring this "ultimate width" through internal beating experiments in a single laser. That is, if we assume two simultaneously oscillating modes are independent of each other, the width of the difference beat should be approximately twice the width of each mode. Since mechanical drifts affect both cavity frequencies in the same manner to first order, the mechanical effects tend to cancel.

Using this approach with helium–neon lasers, several authors have found beat widths of less than a few cycles per second. The absolute stability in such experiments is always considerably worse, however, due to fluctuations in cavity dimensions and in the center of the spectral line.

2.11. Mode-Pulling Effects

The first study of mode-pulling effects in a laser was conducted by the author (Bennett, 1962b) using a 1 m plane-parallel magnetostrictively tuned helium–neon laser in which the plate separation was accurately determined and in which the mirror loss was known to be in the order of 1–2 percent per pass. This study was confined to the dominant beats at $c/2L$ and $2(c/2L)$ illustrated in Fig. 2-6 because of the strong dependence of 1 Mc/sec satellites on the plate angular alignment. The results of this study showed:

(a) the beats were not precisely spaced by $c/2L$, but were, less than $c/2L$ by about 1 part in 800 ($\approx$ 200 kc/sec) (see Table 2-1);

TABLE 2-1. Comparison of $c/2L$ beat at threshold with measured values of $c/2L$ in a plane-parallel laser (from Bennett, 1962b.)

$c/2L$ (in Mc/sec)	Measured beat (in Mc/sec)	Fractional difference between $c/2L$ and measured beat
161.316 ± 0.022	161.107 ± 0.010	$-(1.3 \pm 0.2) \times 10^{-3}$
158.713 ± 0.020	158.531 ± 0.010	$-(1.1 \pm 0.2) \times 10^{-3}$
156.189 ± 0.018	155.982 ± 0.010	$-(1.3 \pm 0.2) \times 10^{-3}$

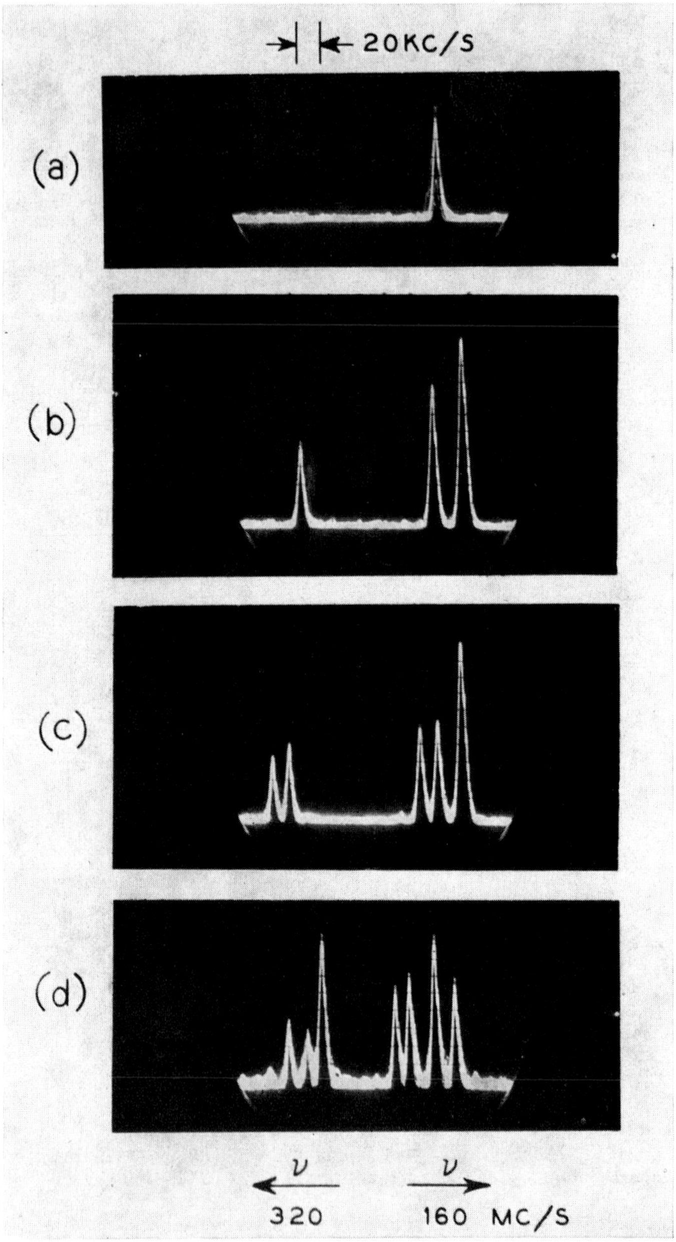

FIGURE 2-7. Power-dependent splitting of the central beats at $c/2L$ and $2(c/2L)$ observed by Bennett (1962b). The pumping rate increases from (a) through (d).

(b) careful examination of these beats showed a power-dependent splitting effect (= 20 kc/sec) from which the number of simultaneously oscillating modes could be deduced (see Fig. 2-7);

(c) the splitting of these beats varied with the plate separation and the frequencies of the beats exhibited an "anomalous" change ($\approx$ 20 kc/sec) with pumping power for most settings of the interferometer length.

These effects of course arise from the presence of the amplifying medium in the laser cavity.

Before describing any mode-pulling calculations in detail, it is helpful to consider the case of the plane-parallel Fabry–Perot qualitatively. The oscillating frequency is determined by a condition on the phase of the electric field. That is, resonance is described by the occurrence of a standing wave in the laser cavity. For a standing wave to occur at the optical frequency ν in the plane-parallel interferometer, the phase shift ϕ on a single pass through the interferometer must be an integral multiple of π. Neglecting the resonant nature of the mirrors, this phase shift is entirely produced by the time delay (nL/c) which results from a wave with phase velocity c/n travelling a distance L through an amplifying medium with index of refraction n. Hence

$$\phi = 2\pi\nu(nL/c) = m\pi,$$

where $m = 1, 2, \ldots 10^6, \ldots$ and

$$\nu = m(c/2L)(1/n).$$

Hence, in the evacuated case ($n = 1$) the resonances are uniformly spaced by $c/2L$. In the presence of the amplifying transition peaked at ν_D with full width $\Delta\nu_D$, n varies as shown qualitatively in Fig. 2-8. Hence, a cavity resonance located below ν_D has its frequency increased by the amplifying medium and a cavity resonance located above the line center has its frequency decreased during oscillation. The net result is a "pulling" of the cavity resonance frequency toward the line center during oscillation for homogeneously broadened lines. In general, the refractive index in the presence of the oscillation is a complicated, power-dependent nonlinear function of the frequency. Consequently, the oscillator modes are not uniformly spaced and splitting of the type shown in Fig. 2-7 results as the power of the laser is increased above threshold for two modes. With inhomogeneously broadened lines mode "pushing" effects can occur near the center of the line, which arise from hole-burning effects (see below).

Figure 2-7 illustrates data taken simultaneously on the dominant $c/2L$ (= 160 Mc/sec) and $2(c/2L)$ (= 320 Mc/sec) beats in a plane-parallel

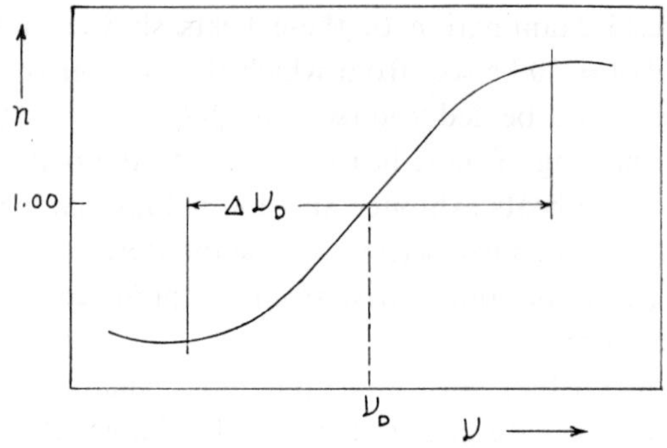

FIGURE 2-8. Qualitative variation of refractive index of an amplifying transition at low powers (Bennett, 1962a).

helium–neon laser as a function of increasing pumping power [Fig. 2-7 (a) and (d)]. The width of the beats in these data is determined entirely by the resolution (a few kc/sec) of the spectrum analyzer used and does not represent the "inherent" width of the oscillator modes in the laser. The frequency scale is the same for each set of beats as indicated by the 20 kc/sec interval shown, with the purely instrumental exception that increasing frequency occurs in opposite directions for the two sets of data. Since the amplifying line is peaked, there is an ordered appearance of oscillating cavity modes as the pumping supply is increased. In Fig. 2-7 (a) the gain is sufficient for two modes to be above threshold; consequently, there is one beat at $c/2L$ and none at $2(c/2L)$. In Fig. 2-7 (b) the gain has been increased to the point where three modes oscillated; hence two beats occur at $c/2L$—due to the nonuniform spacing of oscillating modes—and one beat is found at $2(c/2L)$. A similar analysis holds for Fig. 2-7 (c) and (d) in which four and five modes are oscillating, respectively. The splitting [e.g. Fig. 2-7 (b) for the $c/2L$ beat] is a function of the cavity length—hence, of the mode locations in respect to the center of the Doppler line. For three modes randomly located in respect to the line center, the splitting occurs. However, for three modes placed symmetrically about a symmetric line, it is obvious that the pulling on the outer modes must be equal and opposite (if both of the outer modes oscillate) and that the splitting in Fig. 2-7 (b) must vanish. This type of effect therefore permits generating an error signal which is a function of the cavity mode displacement in respect to the line center and which might, in principle, be used to stabilize the laser against slow frequency drifts. However, the effect is not very useful in practice for achieving frequency stabilization because of interference and hysteresis effects involving the holes burned in the Doppler line.

2.12. General Laser Frequency Equation (Bennett 1962a, b; 1963)

In order not to restrict the choice of cavity modes to the plane-parallel type, it is desirable to formulate the problem in terms of the shift in frequency from an arbitrary cavity resonance, ν_c. That is, we assume that a standing wave can occur in a normal mode at frequency ν_c in the absence of gain. The introduction of gain to the system shifts the oscillation frequency from ν_c to another frequency ν such that the net phase shift per pass is unchanged. In particular, the oscillation will occur at a frequency ν_o such that

$$\left(\frac{\partial \phi_c}{\partial \nu}\right)(\nu_o - \nu_c) + \Delta\phi_G(\nu_o) = 0, \tag{2.74}$$

where $\Delta\phi_G(\nu_o)$ is the change in single-pass phase shift at the frequency of oscillation due to insertion of a total fractional energy gain per pass, G. Equation (2.74) has a very simple graphic interpretation:

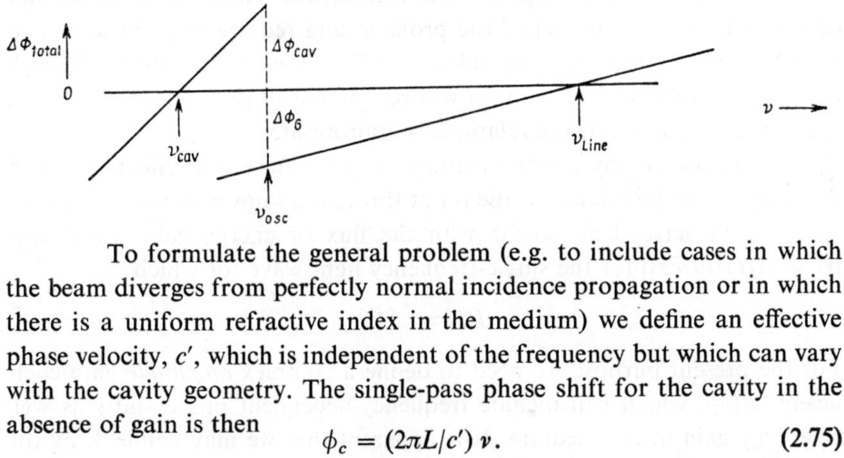

To formulate the general problem (e.g. to include cases in which the beam diverges from perfectly normal incidence propagation or in which there is a uniform refractive index in the medium) we define an effective phase velocity, c', which is independent of the frequency but which can vary with the cavity geometry. The single-pass phase shift for the cavity in the absence of gain is then

$$\phi_c = (2\pi L/c')\,\nu. \tag{2.75}$$

As shown above, the full width of the cavity resonance at half-maximum response may be expressed in terms of this effective velocity (c') and the fractional energy loss per pass (f) through a standard Q argument, yielding

$$\Delta\nu_c = (c'/2\pi L)\,f. \tag{2.76}$$

From (2.75) and (2.76) it follows that

$$\frac{\partial \phi_c}{\partial \nu} = \frac{2\pi L}{c'} = \frac{f}{\Delta\nu_c} = \text{constant}. \tag{2.77}$$

Substitution of (2.77) in (2.74) gives

$$\nu_{osc} = \nu_c - \frac{\Delta\nu_c}{f}\,\Delta\phi_G(\nu_{osc}), \tag{2.78}$$

where $\Delta\phi_G(\nu_{osc})$ is the actual change in phase shift at the oscillation frequency which results from introducing the amplifying medium. That is, $\Delta\phi_G(\nu_{osc})$ must be evaluated subject to saturation condition (2.72) and in the presence of the oscillation. With these restrictions and in the case of a single mode, Eq. (2.78) is exact.

$\Delta\phi_G(\nu_{osc})$, of course, depends on the shape of the amplifying transition in the presence of the oscillation. In the treatment given by the author (Bennett, 1962a, b; 1963), the shape of the gain curve is calculated quantum mechanically and the phase shifts resulting from this gain curve are obtained through use of the Kramers–Kronig relations.

2.13. The Kramers–Kronig Relations‡

The Kramers–Kronig relations permit relating the phase shifts to a knowledge of the amplification coefficients. These relations do not depend on a specific model of the problem and require only linearity and causality. Since frequency-dependent nonlinearities (which enter through the polarizability of the medium) will be ignored in the present treatment, use of the Kramers–Kronig relations is appropriate.

Consider any linear medium obeying cause and effect in which the output may be related to the input through a known transfer function.

We have dealt so far with the flux or energy gain coefficients $[G = (1/I)(\partial I/\partial z)]$ for the single-frequency light wave for which

$$I/I_0 = e^{GL}.$$

For the present purpose we need to define a complex *amplitude* gain coefficient, $K(\omega)$, which will include frequency-dependent phase shifts as well as energy gain in the medium. It is apparent that we may define K by the relation

$$\sqrt{\frac{I}{I_0}} = \frac{E}{E_0} = e^{GL/2 - i\phi L} \equiv e^{KL}, \tag{2.79}$$

‡ These relations, which result from a straightforward application of Cauchy's theorem, were introduced in the analysis of x-ray dispersion by Kronig (1926) and independently by Kramers (1927). Electrical engineers frequently refer to similar pairs of relations used in circuit analysis as "the Bode relations" (see Bode, 1945). Equivalent relations were also exploited in the early days of radar to understand systemmatic errors in time-delay measurement resulting from resonances in the water vapor molecule (see Van Vleck, 1948). They were applied by Gordon, Zeiger and Townes (1954) to determine cavity-pulling effects in the first ammonia maser and by the author (Bennett, 1962b) to evaluate hole-repulsion and mode-pulling effects in the first gas laser.

where

$$K(\omega) \equiv K_0(\omega) - i\phi(\omega)$$

and the real part of $K(\omega)$ is given by

$$K_0(\omega) = G(\omega)/2.$$

It is important to note the following implications of "cause and effect". Consider a pulse starting at $t = 0$ of the form

$$f(t) = \begin{cases} 0 \quad \text{for} \quad t \leqq 0, \\ e^{-\alpha t + i\omega t} \quad \text{for} \quad t \geqq 0. \end{cases} \tag{2.80}$$

Using the symmetric form of the Fourier transforms

$$f(t) = \frac{1}{\sqrt{2\pi}} \int_{-\infty}^{\infty} f(\omega)\, e^{i\omega t}\, d\omega,$$
$$f(\omega) = \frac{1}{\sqrt{2\pi}} \int_{-\infty}^{\infty} f(t)\, e^{-i\omega t}\, dt, \tag{2.81}$$

the Fourier transform of the pulse is

$$f(\omega) = \frac{-i}{\sqrt{2\pi}[\omega - (\omega_0 + i\alpha)]}. \tag{2.82}$$

Next let the pulse go through medium with complex gain coefficient $K(\omega)$. The spectral distribution of the output pulse is

$$f_{\text{out}}(\omega) = \frac{i}{\sqrt{2\pi}} \left[\frac{K(\omega)}{\omega - (\omega_0 + i\alpha)} \right]. \tag{2.83}$$

Consequently, the time-dependent form of the output is

$$f_{\text{out}}(t) = -\frac{i}{2\pi} \int_{-\infty}^{\infty} \frac{K(\omega)\, e^{i\omega t}\, d\omega}{\omega - (\omega_0 + i\alpha)} \tag{2.84}$$

and has a pole in the upper half-plane at $\omega = \omega_0 + i\alpha$.

The integral can be evaluated from separate contours which depend on whether $t > 0$ or $t < 0$.

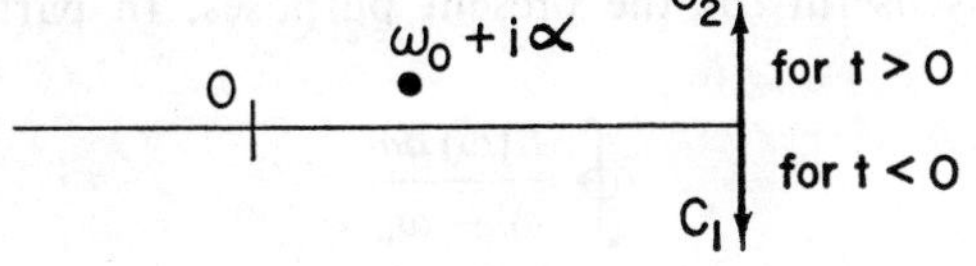

For $t > 0$, the contour integral in the upper half plane is

$$\oint_{C_2} = 0,$$

in the limit that $\text{Im}\omega \to +i\infty$. This permits obtaining the form of the output pulse for $t > 0$ by noting that

$$\oint K(\omega)\, f(\omega)\, d\omega = \int_{\substack{-\infty \\ \text{real axis}}}^{\infty} K(\omega)\, f(\omega)\, d\omega = 2\pi i \sum (\text{Residues}). \tag{2.85}$$

Note that for $t < 0$, the contour integral $\oint_{C_1} = 0$ as $\text{Im}\omega \to -i\infty$.

If there are no poles in the lower half-plane, for $t < 0$

$$\oint K(\omega)\, f(\omega)\, d\omega = -2\pi i \sum (\text{Residues}) = 0. \tag{2.86}$$

Therefore for $t < 0$

$$\int_{-\infty}^{\infty} K(\omega)\, f(\omega)\, d\omega = \oint_{C_2} K(\omega)\, f(\omega)\, d\omega = 0 \tag{2.87}$$

and we obtain the physically desirable result that there is no output before the input pulse has started. Hence if we take the limit as $\alpha \to 0$ in this system it is with the understanding that the pole at $\omega_0 + i\alpha$ must be in the upper half-plane—or that if a pole lies on the real axis, the contour must go under it ().

Causality also clearly *requires* that the complex form for $K(\omega)$ have no poles in the lower half-plane and that

$$K(\pm\infty, \pm i\infty) \to 0$$

faster than $1/\omega$.

From the properties of $K(\omega)$, we may obtain a form of the Kramers–Kronig relations useful for the present purposes. In particular, consider the integral

$$\oint \frac{K(\omega)\, d\omega}{\omega - \omega_0} \tag{2.88}$$

over the contour:

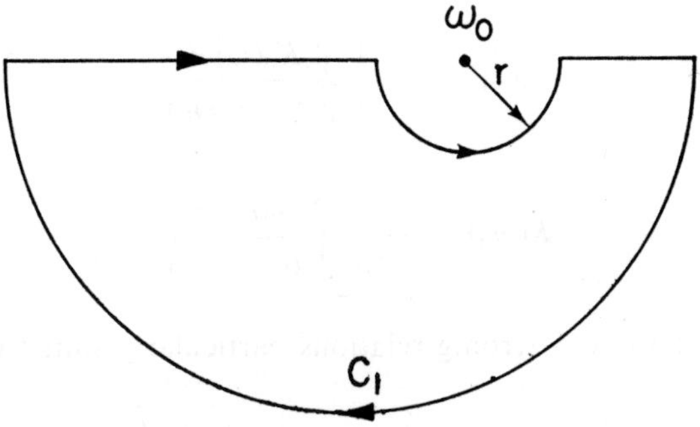

Because there are no poles in the lower half plane, and because $K(\omega)$, hence the integral, vanishes over C_1 as $|\omega| \to \infty$,

$$0 = \int_{-\infty}^{\infty} \frac{K(\omega)\, d\omega}{\omega - \omega_0} = \underset{\text{(principal part)}}{\fint_{-\infty}^{\infty}} \frac{K(\omega)}{\omega - \omega_0}\, d\omega + \lim_{r\to 0} \oint_{\text{at } \omega_0} \frac{K(\omega)\, d\omega}{\omega - \omega_0}. \tag{2.89}$$

For the circle of radius r about ω_0

$$K(\omega) = K(\omega_0) + \left(\frac{\partial K}{\partial \omega}\right)_{\omega_0} (\omega - \omega_0) + \frac{1}{2}\left(\frac{\partial^2 K}{\partial \omega^2}\right)(\omega - \omega_0)^2 + \cdots, \tag{2.90}$$

where

$$\omega = \omega_0 + re^{i\theta},$$

$$d\omega = ire^{i\theta}\, d\theta.$$

Therefore the right side of Eq. (2.89) is

$$\oint_{-\pi}^{0} \frac{K(\omega_0)}{re^{i\theta}}\, ire^{i\theta}\, d\theta + \int_{-\pi}^{0} \left(\frac{\partial K}{\partial \omega}\right)_{\omega_0} ire^{i\theta}\, d\theta + O(r^2) \tag{2.91}$$

and

$$\lim_{r\to 0} \int_{-\pi \text{ at } \omega_0}^{0} \frac{K(\omega)}{\omega - \omega_0}\, d\omega = i\pi K(\omega_0). \tag{2.92}$$

Therefore

$$i\pi K(\omega_0) = -\fint_{-\infty}^{\infty} \frac{K(\omega)\, d\omega}{\omega - \omega_0} \tag{2.93}$$

or letting

$$K(\omega) = K_0(\omega) - i\phi(\omega)$$

and equating real and imaginary parts we get

$$\phi(\omega_0) = -\frac{1}{\pi}\int_{-\infty}^{\infty}\frac{K_0(\omega)\,d\omega}{(\omega-\omega_0)},$$
$$K(\omega_0) = -\frac{1}{\pi}\int_{-\infty}^{\infty}\frac{\phi(\omega)\,d\omega}{(\omega-\omega_0)},\tag{2.94}$$

a form of the Kramers–Kronig relations particularly suited to the present problem.‡

2.14. Phase Shift for a Lorentzian Line

Let

$$K_0(\omega) = \frac{K_m}{1 + [(\omega-\omega_m)/(\Delta\omega/2)]^2}\tag{2.95}$$

as the form of the Lorentzian. Now consider the integral

$$\oint\frac{(K_0\omega)\,d\omega}{(\omega-\omega_0)}\tag{2.96}$$

over the contour:

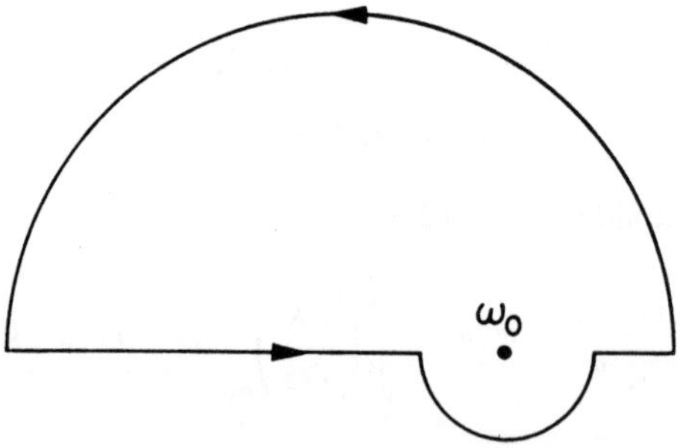

‡ These relations are frequently stated in the frequency-symmetric form

$$\phi(\omega_0) = -\frac{2\omega_0}{\pi}\int_0^{\infty}\frac{K_0(\omega)-K_0(\infty)}{\omega^2-\omega_2^0}\,d\omega,$$

$$K(\omega_0) = K_0(\infty) + \frac{2}{\pi}\int_0^{\infty}\frac{\omega\phi(\omega)}{\omega^2-\omega_2^0}\,d\omega.$$

However, we only consider line shapes for which $K_0(\infty) = 0$ and for which $\Delta\omega \ll \omega_0$, for which the two sets of expressions are equivalent.

First note that

$$\oint \frac{K_0(\omega)\,d\omega}{(\omega-\omega_0)} = \int_{-\infty}^{\infty} \frac{K_0(\omega)\,d\omega}{(\omega-\omega_0)} + i\pi K_0(\omega_0), \tag{2.97}$$

where the second term is the contribution from going under the pole on the real axis at ω_0. Substituting Eq. (2.95) in (2.96),

$$\oint \frac{K_0(\omega)\,d\omega}{(\omega-\omega_0)} = \oint \frac{K_m(\Delta\omega/2)^2\,d\omega}{(\omega-\omega_m+i\,\Delta\omega/2)\,(\omega-\omega_m-i\,\Delta\omega/2)\,(\omega-\omega_0)}, \tag{2.98}$$

which has two poles in the upper half-plane.‡ Hence from Cauchy's theorem, Eq. (2.98) becomes

$$\oint \frac{K_0(\omega)\,d\omega}{(\omega-\omega_0)} = 2\pi i \sum (\text{Residues}) = 2\pi K_0(\omega_0)\frac{\omega_m-\omega_0}{\Delta\omega_m} + i\pi K_0(\omega_0). \tag{2.99}$$

Equating the real parts in Eqs. (2.99) and (2.97) and utilizing the Kramers–Kronig relation Eq. (2.94),

$$\phi(\omega_0) = -\frac{1}{\pi}\int_{-\infty}^{\infty} \frac{K_0(\omega)\,d\omega}{(\omega-\omega_0)} = -2K_0(\omega_0)\frac{\omega_m-\omega_0}{\Delta\omega_m}$$

or expressed in terms of the energy gain coefficient,

$$\phi(\nu) = -G_0(\nu)\frac{\nu_m-\nu}{\Delta\nu_m}, \tag{2.100}$$

where

$$G_0(\nu) = \frac{G_m}{1+[(\nu-\nu_m)/(\Delta\nu_m/2)]^2}.$$

2.15. Homogeneously-Broadened Lasers

In the limit that

$$\frac{R_1+R_2}{2\pi} \gg \Delta\nu_{\text{Doppler}} \tag{2.101}$$

the line shape for the laser transition is Lorentzian and all atoms have the same probabilities for interacting with the optical field. This limit is characteristic of the original microwave ammonia maser (Gordon, Zeiger and Townes, 1954) and is probably satisfied by a number of far-infrared laser transitions, where both the Doppler widths are small and the radiative

‡ Note that, whereas $K_0(\omega)$ has a pole in the lower half-plane, $K(\omega)$ does not.

transition probabilities constitute the main source of phase interruption. For sufficiently high pressures, elastic collisions may increase the Lorentz width enough to satisfy Eq. (2.101). However, for most gas-laser transitions in the near visible part of the spectrum, Eq. (2.101) is not a very good approximation.

Oscillation threshold occurs when

$$G(\nu_{\text{osc}})\,L = f. \tag{2.102}$$

For homogeneously-broadened lasers with a Lorentzian line shape, the energy gain coefficient and phase shift are related by Eq. (2.100). Consequently, from Eq. (2.78) the oscillator frequency at threshold is given by solving

$$\nu_{\text{osc}} = \nu_{\text{cav}} + \Delta\nu_c \frac{G(\nu_{\text{osc}})\,L}{f}\left(\frac{\nu_m - \nu_{\text{osc}}}{\Delta\nu_m}\right)$$

for ν_{osc} subject to condition (2.102). Consequently, at threshold

$$\nu_{\text{osc}} = \frac{\nu_m\,\Delta\nu_c + \nu_c\,\Delta\nu_m}{\Delta\nu_m + \Delta\nu_c} \approx \nu_c + (\nu_m - \nu_c)\frac{\Delta\nu_c}{\Delta\nu_m} \quad \text{for} \quad \Delta\nu_c \ll \Delta\nu_m. \tag{2.103}$$

Above threshold steady state still requires that the gain at the oscillation frequency equal the loss. This equilibrium situation is satisfied through the stimulated emission process. Namely, as the pump rate is increased, the field in the cavity increases in such a way as to stimulate the excess upper-state atoms to emit their energy by just that amount required to keep the gain at ν_{osc} equal to the loss.

Because all atoms have the same Lorentzian probability of interacting with the field, the resultant gain curve above threshold is precisely equal to that obtained at threshold. Consequently, the oscillation frequency is also given by Eq. (2.103) above threshold, and there is no power-dependent pulling of the oscillator as the pump rate is increased. Similarly, because the gain remains clamped to the threshold value, additional modes with the same spatial field distribution will generally not go into oscillation with increasing pump rate. Consequently, such homogeneously-broadened transitions have the desirable property that all of the power can automatically be obtained in a single mode well above oscillation threshold.

The dependence of the output power on cavity tuning is a monotonic function of $|\nu_m - \nu_c|$ which goes through a maximum at $\nu_c = \nu_m$. The situation can be understood through the diagram in Fig. 2-9, where two Lorentzian gain curves are illustrated. The dashed curve $G(\nu - \nu_m)$ represents the gain coefficient that would have been obtained in the absence of the optical field with the same pump rate applied to the medium. The solid curve $G_A(\nu - \nu_m)$ represents the saturated small signal gain coefficient which is obtained in the presence of the field and which satisfies the steady-state requirement

$$G_A(\nu_{osc})\,L = f. \tag{2.104}$$

The power out of the laser is proportional to the difference in upper-state density required to reduce $G(\nu_{osc})$ to $G_A(\nu_{osc})$ at the oscillation frequency. So long as the power in the cavity is low enough to permit neglecting power broadening of the Lorentz width, the output power is simply given by

$$P \propto G(\nu_{osc})\,L - f, \tag{2.105}$$

where $G(\nu_{osc})$ is a Lorentzian function of the frequency proportional to the inversion density and with a Lorentz width equal to the natural width of the transition. However, as power broadening becomes important, the tuning curve exhibits a more complicated dependence, analogous to the nonlinear variation of the gain coefficient with field intensity described in Sec. 2.4. That is, the output power is proportional to the reduction in inversion density necessary to reduce $G(\nu_{osc})$ to the threshold value under conditions in which the Lorentz width itself is a function of the power. The tuning curve is therefore determined by a transcendental equation, which however still exhibits a monotonic dependence on $|\nu_m - \nu_c|$ with a maximum at the center of the line (i.e., there is no tuning dip at line center).

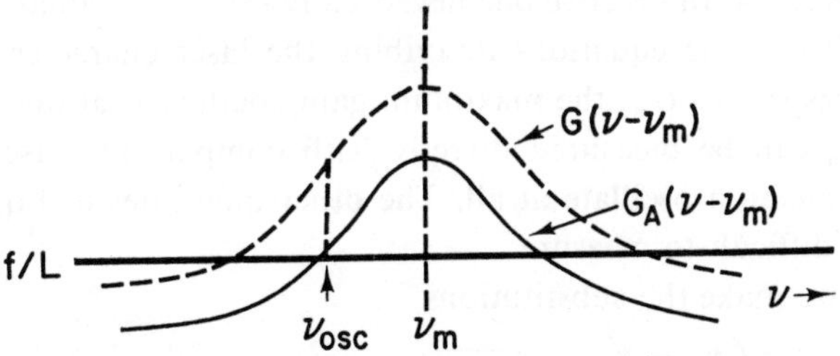

FIGURE 2-9. Lorentzian gain curves characteristic of a homogeneously-broadened laser. The dashed curve represents the gain in the absence of the cavity. The solid curve represents the steady gain in the presence of the cavity with fractional loss f/L per unit length.

2.16. Phase Shift for a Gaussian Distribution of Lorentzians

It was shown in Sec. 2.14 that for a Lorentzian gain coefficient

$$G(\nu) = \frac{G_0}{1 + [(\nu_0 - \nu)/(\Delta\nu_L/2)]^2}, \tag{2.106}$$

the phase-shift coefficient is given by

$$\phi(\nu) = -\frac{G_0[(\nu_0 - \nu)/\Delta\nu_L]}{1 + [(\nu_0 - \nu)/(\Delta\nu_L/2)]^2}. \tag{2.107}$$

The constant G_0 is proportional to the density of atoms with resonant frequencies centered at ν_0. Consequently, if we consider a Gaussian distribution of atoms with Doppler-shifted resonance frequencies, the gain coefficient is given by

$$G(\nu) = G_0' \int_{-\infty}^{\infty} \frac{\exp[-4\ln 2(\nu_m - \nu_0)^2/(\Delta\nu_D)^2]\, d\nu_0}{1 + [(\nu - \nu_0)/(\Delta\nu_L/2)]^2} \tag{2.108}$$

and the total phase shift coefficient at ν is

$$\phi(\nu) = -\frac{G_0'}{2} \int_{-\infty}^{\infty} \frac{[(\nu_0 - \nu)/(\Delta\nu_L/2)] \exp[-4\ln 2(\nu_m - \nu_0)^2/(\Delta\nu_D)^2]\, d\nu_0}{1 + [(\nu - \nu_0)/(\Delta\nu_L/2)]^2}, \tag{2.109}$$

where the constant G_0' can be related to the inversion density by the normalization condition

$$\int_{-\infty}^{\infty} G(\nu)\, d\nu = (N_2 - N_1)\, 2\pi\langle\mu_0\rangle^2/(\hbar\lambda) \approx (N_2 - N_1)\frac{\lambda^2 A_{21}}{8\pi} \tag{2.110}$$

derived in Sec. 2.4. In practice one never really needs to evaluate the integral in Eq. (2.110), if the equations describing the laser characteristics are all defined in respect to G_m, the maximum gain coefficient at line center. The quantity G_m can be measured directly with comparative ease if you can make the transition oscillate at all. The other quantities in Eq. (2.110) are much more difficult to measure.

If we make the substitutions

$$x = \left(\frac{\nu_m - \nu}{\Delta\nu_D}\right) 2\sqrt{\ln 2}, \qquad y = \left(\frac{\Delta\nu_L}{\Delta\nu_D}\right)\sqrt{\ln 2}, \qquad t = \frac{(\nu_m - \nu_0)\sqrt{2\ln 2}}{\Delta\nu_D}, \tag{2.111}$$

then

$$G(\nu_m - \nu) = G_0' \frac{\pi\,\Delta\nu_L}{2} U(x, y) = G_m U(x, y)/U(0, y), \tag{2.112}$$

$$\phi(\nu_m - \nu) = -G_0' \frac{\pi\,\Delta\nu_L}{4} V(x, y) = -G_m \frac{V(x, y)}{2U(0, y)} \tag{2.113}$$

and

$$\phi(\nu) = -\frac{G(\nu)\,V(x, y)}{2U(x, y)}, \tag{2.114}$$

where $U(x, y)$ and $V(x, y)$ are the real and imaginary parts of the function tabulated by Faddeyeva and Terent'ev (see Sec. 2.7).‡

The limit for the pure Gaussian can be obtained from the behavior of the real and imaginary parts of the function $W(z)$ in Eq. (2.62) of Sec. 2.7 in the limit that $y \to 0$ [Eq. (2.65)]. Therefore the gain coefficient for the pure Gaussian is

$$G_D(\nu) = G_m U(x, 0) = G_m e^{-x^2} \tag{2.115}$$

and the phase-shift coefficient for the pure Gaussian is

$$\begin{aligned}\phi_D(\nu) &= -\frac{G_m}{2} V(x, 0) \\ &= -\frac{G_m e^{-x^2}}{\sqrt{\pi}} \int_0^x e^{t^2}\,dt = -\frac{G(\nu)}{\sqrt{\pi}} \int_0^x e^{t^2}\,dt.\end{aligned} \tag{2.116}$$

The phase shift for the Doppler-broadened limit is closely approximated near line center by the empirical relation (Bennett, 1962b);

$$\phi_D(\nu) \approx -0.282 G_m \sin\left(\frac{\nu_m - \nu}{0.3\,\Delta\nu_D}\right) \tag{2.117}$$

determined through a "best fit" to a numerical integration of the Kramers–Kronig relations for a gaussian line. The "angle" in Eq. (2.117) is expressed in radians. As may be seen by comparison with the exact expression in Eq. (2.116), the error in approximation (2.117) is less than 1 percent for $|\nu_m - \nu| < 0.4\Delta\nu_D$; Eq. (2.117) is ≈ 6 percent low as the half-intensity points of the gaussian.

Oscillation threshold in the general case occurs at the frequency ν_{osc} or

$$x_{osc} \equiv \frac{\nu_m - \nu_{osc}}{\Delta\nu_D} 2\sqrt{\ln 2} \tag{2.118}$$

such that $G(\nu_{osc})\,L = f$ where

$$\Delta\phi(\nu_{osc}) = \phi(\nu_{osc})\,L = \frac{-G(\nu_{osc})\,L \cdot V(x_{osc}, y)}{2U(x_{osc}, y)}. \tag{2.119}$$

‡ Note by comparing Eqs. (2.114) and (2.116) that the leading term in the correction to the gaussian phase shift approximation represented by the exact expression for the Voigt profile phase shift is of order $(\Delta\nu_L/\Delta\nu_D)$.

Hence, from Eq. (2.78), at threshold,

$$\nu_{\text{osc}} = \nu_{\text{cav}} + \frac{\Delta\nu_c}{2} \frac{V(x_{\text{osc}}, y)}{U(x_{\text{osc}}, y)}. \tag{2.120}$$

In distinction to the pure Lorentzian case discussed in Eq. (2.103), the threshold oscillation frequency is determined by solution of a transcendental equation.‡

To the extent that we are primarily interested in the behavior near the line center, it becomes appropriate to expand both the gain and phase-shift expressions in a power series about the line center. In the case of a symmetric gain distribution, the gain is an even function of the frequency and the phase shift is an odd function of the frequency about the line center. Hence we may say

$$G(\nu) = G_m\left[1 - c\left(\frac{\nu_m - \nu}{\Delta\nu_m}\right)^2 + \cdots\right] \tag{2.121}$$

and

$$\begin{aligned}\phi(\nu) &= -aG_m\left[\left(\frac{\nu_m - \nu}{\Delta\nu_m}\right) - b\left(\frac{\nu_m - \nu}{\Delta\nu_m}\right)^3 + \cdots\right]\\ &= -aG(\nu)\left[\left(\frac{\nu_m - \nu}{\Delta\nu_m}\right) + (c - b)\left(\frac{\nu_m - \nu}{\Delta\nu_m}\right)^3 + \cdots\right].\end{aligned}$$

In solving the Eq. (2.120) for the threshold oscillation frequency $[G(\nu_{\text{osc}})L = f]$ we may use a reiterative expansion technique in either ν_c or ν_m depending on whether $\Delta\nu_c < \Delta\nu_m$ or $\Delta\nu_m < \Delta\nu_c$. The coefficients a and b may be related to c through the function $U(x, y)$ and $V(x, y)$ or, in more general cases, directly through the Kramers–Kronig relations. These quantities will all depend on the nature of the line shape. For example, *for the Gaussian limit* ($\Delta\nu_L \ll \Delta\nu_D$)

$$\begin{aligned} a &= 2\sqrt{\ln 2/\pi} \approx 0.94,\\ c - b &= 8(\ln 2)/3 \approx 1.84, \end{aligned} \tag{2.122}$$

whereas *for the Lorentzian limit* ($\Delta\nu_L \gg \Delta\nu_D$)

$$\begin{aligned} a &= 1,\\ c &= 4,\\ c - b &= 0. \end{aligned}$$

‡ Results similar to Eqs. (2.116) and (2.120) for threshold pulling effects were obtained in different formulations of the problem by Javan, Ballik and Bond (1962) and by Ballik (1964), respectively. The third-order Lamb (1964) theory also reduces to an expression equivalent to Eq. (2.120) at threshold.

2.17. Hole-Burning Effects

As pointed out by the author (Bennett, 1962b), the steady-state requirement in the typical inhomogeneously-broadened gas laser,

$$G_{\text{actual}}(\nu_{\text{osc}}) = f/L = \text{cavity loss coefficient},$$

is achieved above threshold by burning a hole in the line. That is, because $\Delta\nu_L \ll \Delta\nu_D$, this steady-state oscillation requirement is achieved through a localized reduction in the original gain coefficient to the threshold value, as shown in Fig. 2-10. The direct interaction, is, of course, with the atoms. However, there is a direct correspondence between distortion in the velocity distribution of the atoms and the resultant distortion in the gain profile.

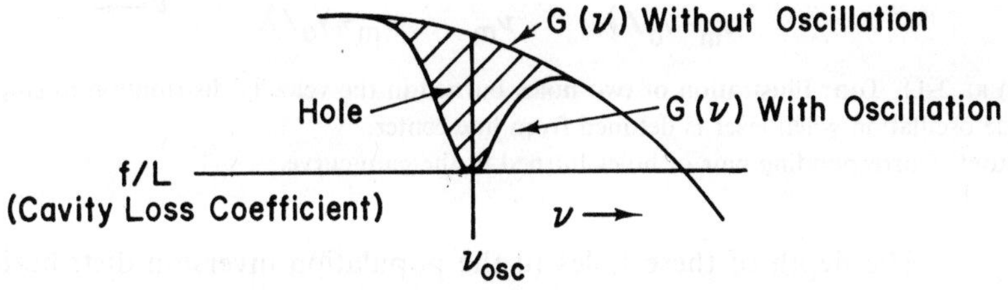

FIGURE 2-10. Hole burned in gain curve of Doppler-broadened laser transition to satisfy steady-state requirement above threshold (after Bennett, 1962b).

The actual situation in a single-mode gas laser is somewhat more complicated due to the existence of standing waves in the laser cavity. For each resonant frequency there are running waves travelling in opposite directions. If the frequency of the standing wave is detuned from the resonance frequency (ν_m) of the atom at rest, these two running waves interact with different portions of the velocity distribution due to the Doppler effect. Specifically, a running wave moving in the $+z$ direction with frequency $\nu > \nu_m$ will interact with atoms with v_z in the vicinity of

$$v_z = -\lambda(\nu - \nu_m)$$

and the running wave with this same frequency moving in the $-z$ direction will interact with atoms with v_z in the vicinity of

$$v_z = +\lambda(\nu - \nu_m).$$

Consequently there are two holes burned in the inversion velocity distribution when ν is detuned substantially from the center of the line (see top Fig. 2-11).

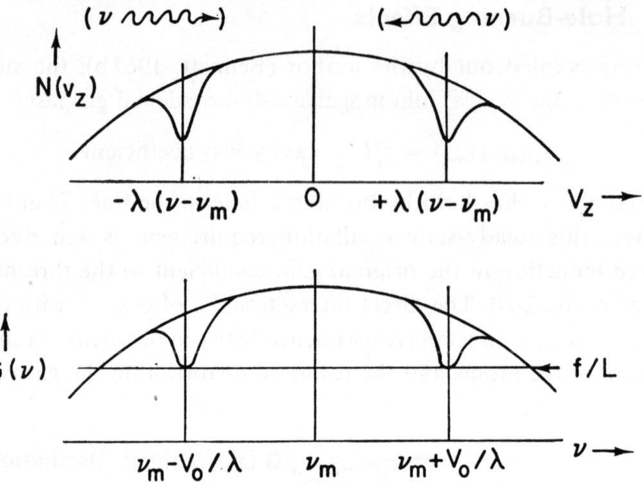

FIGURE 2-11. Top: Illustration of two holes burned in the velocity distribution in single-mode oscillation when laser is detuned from line center.
Bottom: Corresponding pair of holes burned in the gain curve.

The depth of these holes in the population inversion distribution is determined by the steady-state gain saturation requirement, Eq. (2.104). So long as the holes are widely separated compared to the Lorentz width, a hole in the velocity distribution of the population inversion density, ΔN, at $\nu_0 = v_z/\lambda$, gives rise to a decrease in the small signal gain distribution of the form

$$\Delta G(\nu) \propto \int_{-\infty}^{\infty} \frac{\Delta N(\nu_0)\, d\nu_0}{1 + [(\nu - \nu_0)/(\Delta\nu_L/2)]^2}. \tag{2.123}$$

Consequently, in this limit a saturated gain curve for the amplification coefficient in the $+z$ direction of the form shown in the bottom of Fig. 2-11 is obtained.

As the oscillator frequency is tuned towards the line center ($\nu \to \nu_m$), the holes overlap and both running waves compete for the same atoms in the central portion of the Doppler distribution. Consequently, the total power out of the laser (which is proportional to the net number of atoms stimulated to emit by the field) goes through a minimum at line center. This tuning dip was first predicted mathematically by Lamb (1964) in his analysis of the laser oscillator. However, the physical explanation of the effect was first given by the author in terms of hole-burning effects (Bennett, 1962a, 1963).

The hole-burning analysis is easiest to put on a quantitative basis in the complete Doppler-broadened limit where

$$\Delta\nu_D \gg \Delta\nu_L.$$

In this limit we may regard the Doppler distribution as being constant in the vicinity of the hole burned in the population inversion distribution. In the detuned case we may consider the two holes burnt in the Doppler distribution separately. Because the stimulated emission probability is Lorentzian in the frequency with a full width given by

$$\Delta\nu_L = \left(\frac{R_1 + R_2}{2\pi}\right)\sqrt{1 + \frac{1}{R_1 R_2}\left|\frac{\langle\mu_0\rangle E_0}{\hbar}\right|^2}, \tag{2.124}$$

a Lorentzian-shaped hole is burned in the ($\approx$ constant) inversion velocity distribution of full width at half-maximum given by

$$H \equiv \Delta v_z = \lambda\,\Delta\nu_L.$$

Hence the shape of the hole in the population inversion distribution in the vicinity of $v_0 = \lambda(\nu_m - \nu_0)$ is

$$\Delta N(v_z') \propto \frac{1}{1 + [(v_z' - v_0)/(H/2)]^2} = \frac{1}{1 + [(\nu' - \nu_0)/(\Delta\nu_L/2)]^2}. \tag{2.125}$$

One could at this point note that the form of the gain hole will be obtained by substituting Eq. (2.125) in Eq. (2.123), which at low fields results in a Lorentzian of full width $= 2\Delta\nu_L$ as shown explicitly on obtaining Eq. (2.139) below. One could also note at this point that the power gain through stimulated emission by the running wave is proportional to the change in the inversion density, and hence to the depth of the hole in the small signal gain curve at intensities low enough to permit neglecting the power broadening term in Eq. (2.124). Hence at this point one could immediately obtain the shape of the Lamb dip through application of Eqs. (2.141) and (2.142) without explicit evaluation of the hole depths (Bennett, 1963).

2.18. Population "Hole" Depths and "Ear" or "Bump" Heights

The discussion so far has been expressed in terms of the net change in inversion density by the optical field, rather than in the individual upper- and lower-state population densities. The changes in the individual population densities may be evaluated explicitly in terms of the net probability functions derived in Sec. 2.3 for stimulated emission and absorption. We will do this by introducing a frequency dependence of the excited state density formation rates, $F_1(\nu)$, $F_2(\nu)$, and of the original zero-field effective population densities, $N_1(\nu)$, $N_2(\nu)$, defined in analogy with Eq. (2.40). We shall also restrict the discussion here to the limit where the two regions of interaction with the oppositely-travelling running waves are widely separated.

It is helpful first to consider two separate cases:

Case I): First consider the situation in which only the upper state formation rate, $F_1(\nu)$, is present, in which case the effective upper state density in the absence of the field is given by

$$R_2 N_2(\nu) = F_2(\nu)$$

in analogy with Eqs. (2.40). However, note that densities are now expressed in units of $1/\text{cm}^3$ cps and that the density formation rates have dimensions $1/\text{cm}^3$ sec cps; i.e., these expressions now represent quantities per unit frequency interval in the central region of a very broad Doppler line. Application of the oscillating field changes both the upper and lower state effective densities in a manner that may be determined from the expression for P_{stim} in Eq. (2.36). The average rate of flow of excitation at the frequency, ν, in the vicinity of one interaction frequency (ν_0) with the oscillating laser may be represented schematically by the diagram:

$$F_2(\nu) \xrightarrow{\text{IN}} 2 \longrightarrow \xrightarrow{\text{OUT}} (1 - P_{\text{stim}})\, F_2(\nu) = R_2 N_2'(\nu)$$

mixing by the field ($\downarrow\ \uparrow$)

$$1 \longrightarrow \xrightarrow{\text{OUT}} P_{\text{stim}} F_2(\nu) = R_1 N'(\nu)$$

where the prime denotes the average population density in the presence of the field at frequency ν_0 and $P_{\text{stim}} = P_{\text{stim}}(\nu - \nu_0)$. Hence at steady-state in the presence of the field,

$$N_2'(\nu) = [1 - P_{\text{stim}}(\nu - \nu_0)]\, F_2(\nu)/R_2$$

$$N_1'(\nu) = P_{\text{stim}}(\nu - \nu_0)\, F_2(\nu)/R_1 .$$

The net changes in population densities near the interaction frequency in this case are

$$\begin{aligned} \delta N_2(\nu) &= N_2'(\nu) - N_2(\nu) = -P_{\text{stim}}(\nu - \nu_0)\, F_2(\nu)/R_2 \\ \delta N_1(\nu) &= N_1'(\nu) - N_1(\nu) = N_1'(\nu) = +P_{\text{stim}}(\nu - \nu_0)\, F_2(\nu)/R_1 . \end{aligned} \tag{2.126}$$

Hence, a "hole" appears in the upper state population distribution at the interaction frequency and a "bump" or "ear" appears on the lower state population distribution. Observation of such upper-state population "holes" (Bennett, Chebotayev and Knutson, 1967a; Bennett, 1968) and lower-state population "ears" (Schweitzer, Birky and White, 1967) or "bumps" (Cordover, Bonczyk, and Javan, 1967a; 1967b; Holt, 1967, 1968) has been reported through scanning Fabry–Perot analysis of spontaneous emission profiles on other transitions from the upper- or lower-state in the presence of laser oscillation. The actual spontaneous emission line shape observed in such experiments depends on the lorentz widths and wavelengths for the other transition, as well as upon correlation effects, and is discussed in a separate section below.

Case II): Next consider the situation in which only the lower state formation rate, $F_1(\nu)$, is present, in which case the effective lower state density in the absence of the field is given by

$$R_1 N_1(\nu) = F_1(\nu).$$

Application of the oscillating field again changes both the lower and upper state densities in a manner which can be determined from the expression for P_{abs} given in Eq. (2.38). In this case the average steady-state flow of excitation at the frequency (ν) near the laser frequency (ν_0) can be represented schematically by:

$$\begin{array}{lcl} & 2 \;\underline{\quad\uparrow\quad\downarrow\quad} & \xrightarrow{\text{OUT}} P_{\text{abs}} F_1(\nu) = R_2 N_2'(\nu) \\ & \text{mixing by field} & \\ F_1(\nu) \xrightarrow{\text{IN}} & 1 \;\underline{\quad\uparrow\quad\downarrow\quad} & \xrightarrow{\text{OUT}} (1 - P_{\text{abs}})\, F_1(\nu) = R_1 N_1'(\nu) \end{array}$$

where the prime again denotes the average population density in the presence of the field at frequency ν_0 and $P_{\text{abs}} = P_{\text{abs}}(\nu - \nu_0)$. Here at steady-state,

$$N_2'(\nu) = P_{\text{abs}}(\nu - \nu_0)\, F_1(\nu)/R_2$$
$$N_2'(\nu) = [1 - P_{\text{abs}}(\nu - \nu_0)]\, F_1(\nu)/R_1$$

and the net changes in the population densities near the interaction frequency are in this case,

$$\begin{aligned} \delta N_2(\nu) &= N_2'(\nu) - N_2(\nu) = N_2'(\nu) = +P_{\text{abs}}(\nu - \nu_0)\, F_1(\nu)/R_2 \\ \delta N_1(\nu) &= N_1'(\nu) - N_1(\nu) = -P_{\text{abs}}(\nu - \nu_0)\, F_1(\nu)/R_1 . \end{aligned} \tag{2.127}$$

Hence in the case of a stimulated absorption experiment, a "hole" would appear in the lower state population and a "bump" would appear in the upper state population distribution.

In the case of a real laser, we have the simultaneous involvement of both cases I) and II). The net average changes in level populations correspond to the sum of the appropriate expressions in Eq. (2.126) and (2.127). With both F_1 and F_2 present, the total net change in the average population densities at frequency ν due to the oscillating laser at ν_0 is given by:

$$\begin{aligned} \delta N_2(\nu) &= [-P_{\text{sim}}(\nu-\nu_0)\,F_2(\nu) + P_{\text{abs}}(\nu-\nu_0)\,F_1(\nu)]/R_2 \\ \delta N_1(\nu) &= [-P_{\text{abs}}(\nu-\nu_0)\,F_1(\nu) + P_{\text{stim}}(\nu-\nu_0)\,F_2(\nu)]/R_1 . \end{aligned} \tag{2.128}$$

Hence the net change in the effective inversion density at frequency ν produced by the laser field at the neighboring frequency ν_0 is given from Eqs. (2.36) and (2.38) without further approximation by

$$\begin{aligned} \delta N(\nu) &\equiv \delta N_2(\nu) - \delta N_1(\nu) \\ &= -\frac{D(\nu)}{1+[(\nu-\nu_0)/(\Delta\nu_L/2)]^2} \end{aligned} \tag{2.129}$$

where $D(\nu)$ is given in terms of the effective zero field inversion density, $[N_2(\nu) - N_1(\nu)]$, by the relation

$$\begin{aligned} D(\nu) &= \frac{\dfrac{1}{R_1R_2}\left|\dfrac{\langle\mu_0\rangle E_0}{\hbar}\right|^2 [N_2(\nu)-N_1(\nu)]}{1+\dfrac{1}{R_1R_2}\left|\dfrac{\langle\mu_0\rangle E_0}{\hbar}\right|^2} \\ &= \frac{2(R_1+R_2)^2}{\pi c(\Delta\nu_L)^2}\left(\frac{1}{R_1R_2}\right)\left|\frac{\langle\mu_0\rangle}{\hbar}\right|^2 I[N_2(\nu)-N_1(\nu)]. \end{aligned} \tag{2.130}$$

The intensity-dependent lorentz width, $\Delta\nu_L$, is given by Eq. (2.124) and the effective inversion density at zero field contained on the right side of Eq. (2.130) is given in terms of the formation rates and actual relaxation rates by Eq. (2.40).

The differential power gain coefficient at frequency ν_0 due to the interaction with the atoms at frequency ν is then,

$$dP(\nu_0) \equiv d(\partial I/\partial z)_{\nu_0} = h\nu_0[F_2(\nu)\,P_{\text{stim}}(\nu-\nu_0) - F_1(\nu)\,P_{\text{abs}}(\nu-\nu_0)]\,d\nu,$$

which in the manner of Sec. 2.4 [Eq. (2.42)] may be written

$$dP(\nu_0) = \frac{P_0(\nu)\,d\nu}{1+[(\nu-\nu_0)/(\Delta\nu_L/2)]^2}$$

where

$$P_0(\nu) = \frac{2\langle\mu_0\rangle^2 (R_1 + R_2)}{\pi\hbar\lambda(\Delta\nu_L)^2} [N_2(\nu) - N_1(\nu)] I = \frac{h\nu R_1 R_2 D(\nu)}{R_1 + R_2}.$$

If the effective zero-field inversion density varies appreciably over the lorentz width, the total power gain coefficient for the running wave at frequency ν_0 provided by the atoms in the inversion hole is

$$P_{\text{total}}(\nu_0) = \frac{h\nu R_1 R_2}{(R_1 + R_2)} \int_{-\infty}^{\infty} \frac{D(\nu)\, d\nu}{1 + [2(\nu - \nu_0)/\Delta\nu_L]^2}$$

where $D(\nu)$ is given by Eq. (2.130). If $D(\nu)$ is a gaussian function of the frequency of the form

$$D(\nu) = D_m \exp[-(M\lambda^2/2kT)(\nu_m - \nu)^2] = D_m \exp[-4 \ln 2(\nu_m - \nu)^2/\Delta\nu_D)^2] \tag{2.131}$$

the integral may be evaluated in terms of the function $U(x, y)$ tabulated by Faddeyeva and Terent'ev (1961) [see Sec. 2.7] through the substitutions

$$x = \sqrt{\frac{M}{2kT}}\lambda(\nu_m - \nu_0) = 2\sqrt{\ln 2}\,(\nu_m - \nu_0)/\Delta\nu_D$$

$$y = \sqrt{\frac{M}{2kT}}\lambda\,\Delta\nu_L/2 = \sqrt{\ln 2}\,\Delta\nu_L/\Delta\nu_D$$

yielding

$$P_{\text{total}}(\nu_0) = \frac{\pi h\nu R_1 R_2\, \Delta\nu_L}{2(R_1 + R_2)} D_m U(x, y) = \left(\frac{1}{4}\right) h\nu R_1 R_2 \left(\frac{\Delta\nu_L}{\Delta\nu_{LO}}\right) D_m U(x, y). \tag{2.132}$$

If we neglect power broadening of the Lorentz width,

$$P_{\text{total}}(\nu_0) = \left(\frac{1}{4}\right) h\nu R_1 R_2 D_m U(x, y). \tag{2.132b}$$

The dimensions for $P_{\text{total}}(\nu_0)$ in Eq. (2.132) are those of power flux gain per unit distance (i.e. energy per second per cm^2 per cm) when D_m is expressed in terms of atoms per cm^3 per cycle/sec at the center of the Doppler line. [D_m comes from Eq. (2.130) with $\nu = \nu_m$].

In the limit that the original (zero-field) population distribution is a slowly-varying function of the frequency over the lorentz width,

$$D(\nu) \approx D(\nu_0) \approx \text{constant},$$

$$P_0(\nu) \equiv P_0 = h\nu R_1 R_2 D(\nu_0)/(R_1 + R_2) = \text{constant},$$

and both the inversion hole and the differential power gain coefficient, $dP(\nu_0)$, are lorentzian functions of the frequency. In this limit, the integration over frequency for the power gain coefficient for one running wave is trivial and can be done directly in closed form:

$$P_{\text{total}}(\nu_0) \approx \int_{-\infty}^{\infty} \frac{P_0\, d\nu}{1 + [(\nu - \nu_0)\,(\Delta\nu_L/2)]^2} = P_0\left(\frac{\pi\,\Delta\nu_L}{2}\right)$$

$$= \frac{\pi(h\nu)\,\Delta\nu_L R_1 R_2}{2(R_1 + R_2)}\, D(\nu_0) \approx \left(\frac{1}{4}\right) h\nu R_1 R_2 D(\nu_0)$$

which agrees with the result in Eq. (2.132) for the gaussian zero-field inversion distribution in the limit that $y \to 0$.

2.19. Exact Expression for a Gain Hole Due to One Running Wave

Atoms whose resonance frequencies have been Doppler-shifted by amount v_z/λ to frequency ν' will have a small-signal gain coefficient with a Lorentzian response function whose full width is given by $\Delta\nu_L$ in Eq. (2.124) in the limit that $E_0 \to 0$. Let this width be,

$$\Delta\nu_{L0} = \frac{R_1 + R_2}{2\pi} \tag{2.133}$$

Consequently, the small-signal gain coefficient at frequency ν, that would have been produced by the atoms in the hole in the population inversion distribution, is from Eqs. (2.50) and (2.129),

$$\Delta G(\nu) = -\frac{8\pi\langle\mu_0\rangle^2}{\hbar\lambda(R_1 + R_2)} \int_{-\infty}^{\infty} \frac{[\delta N_2(\nu') - \delta N_1(\nu')]\, d\nu'}{1 + [(\nu' - \nu)/(\Delta\nu_{L0}/2)^2}$$

$$= \frac{8\pi\langle\mu_0\rangle^2}{\hbar\lambda(R_1 + R_2)} \int_{-\infty}^{\infty} \frac{D(\nu')\, d\nu'}{\{1 + [(\nu' - \nu)/(\Delta\nu_{L0}/2)]^2\}\,\{1 + [(\nu_0 - \nu')/(\Delta\nu_L/2)]^2\}} \tag{2.134}$$

In the case that $D(\nu)$ is of the gaussian frequency-dependent form given by Eq. (2.131), the integral in Eq. (2.134) for one gain hole may be done exactly and expressed in terms of the functions tabulated by Faddeyeva and Terent'ev (1961) [see Sec. 2.7]. The transformation of this integral can be accomplished through a straightforward, but rather tedious application of the method of partial fractions. The expansion of the integrand into its four terms having separate complex poles may be grouped in two pairs of expressions which are complex conjugates. Each pair contains an integral which may be written in the form for $W(z)$ given in Eq. (2.63). The presence of complex expansion coefficients introduces both the real $[U(x, y)]$ and imaginary $[V(x, y)]$ parts of $W(z)$ in the final answer. The exact expression for the gain hole is

$$\Delta G(\nu) = \frac{4\pi^2 \langle \mu_0 \rangle^2 \, \Delta\nu_D D_m}{\hbar\lambda(R_1 + R_2)\sqrt{\ln 2}}$$

$$\times \left[\frac{\left\{1 + \left(\frac{y_1}{y_2}\right)^2 \left[\left(\frac{S}{y_1}\right)^2 - 1\right]\right\} y_1 U(x_1, y_1) - 2\left(\frac{y_1}{y_2}\right)^2 S y_1^2 V(x_1, y_1)}{\left\{1 + \left(\frac{y_1}{y_2}\right)^2 \left[\left(\frac{S}{y_1}\right)^2 - 1\right]\right\}^2 + 4\left(\frac{y_1}{y_2}\right)^2 \left(\frac{S}{y_2}\right)^2} \right.$$

$$\left. + \frac{\left\{1 + \left(\frac{y_2}{y_1}\right)^2 \left[\left(\frac{S}{y_2}\right)^2 - 1\right]\right\} y_2 U(x_2, y_2) + 2\left(\frac{y_2}{y_1}\right)^2 S y_2^2 V(x_2, y_2)}{\left\{1 + \left(\frac{y_2}{y_1}\right)^2 \left[\left(\frac{S}{y_2}\right)^2 - 1\right]\right\}^2 + 4\left(\frac{y_2}{y_1}\right)^2 \left(\frac{S}{y_1}\right)^2} \right] \tag{2.135}$$

where the frequency displacement from the laser frequency (ν_0) is given by

$$S = \frac{2\sqrt{\ln 2}}{\Delta\nu_D}(\nu - \nu_0) = x_2 - x_1$$

and the other terms are defined by

$$x_1 = \frac{2\sqrt{\ln 2}}{\Delta\nu_D}(\nu_0 - \nu_m) \quad \text{and} \quad y_1 = \sqrt{\ln 2}\left(\frac{\Delta\nu_L}{\Delta\nu_D}\right);$$

$$x_2 = \frac{2\sqrt{\ln 2}}{\Delta\nu_D}(\nu - \nu_m) \quad \text{and} \quad y_2 = \sqrt{\ln 2}\left(\frac{\Delta\nu_{LO}}{\Delta\nu_D}\right).$$

The large bracketed in Eq. (2.135) may be rewritten,

$$\begin{aligned}
&[(y_2^2 - y_1^2 + S^2)\, y_1 y_2^2 U(x_1, y_1) - 2y_1^4 y_2^2 S V(x_1, y_1) \\
&\quad + (y_1^2 - y_2^2 + S^2)\, y_1^2 y_2 U(x_2, y_2) + 2y_1^2 y_2^4 S V(x_2, y_2)] \\
&\quad \times [(y_2^2 - y_1^2)^2 + 2(y_1^2 + y_2^2)\, S^2 + S^4]^{-1}.
\end{aligned}$$

However, no real simplification results without making approximations of the type discussed later on.

The depth, d_0, of the gain hole at resonance is obtained by evaluating Eq. (2.135) at $\nu = \nu_0$, hence $S = 0$ or $x_{\prime} = x_1$. Thus,

$$d_0 \equiv \Delta G(\nu_0) = \frac{4\pi^2 \langle \mu_0 \rangle^2 \Delta\nu_D D_m}{\hbar\lambda(R_1 + R_2)\sqrt{\ln 2}} \left[\frac{y_1 y_2 [y_2 U(x_1, y_1) - y_1 U(x_1, y_2)]}{(y_1 + y_2)(y_2 - y_1)} \right].$$

The latter may be expressed in terms of the power gain coefficient at ν_0 by use of Eq. (2.132):

$$d_0 = \frac{8\pi \langle \mu_0 \rangle^2 P_{\text{total}}(\nu_0)}{\hbar^2 c R_1 R_2} \left[\frac{y_2 - y_1 \left(\dfrac{U(x_1, y_2)}{U(x_1, y_1)} \right)}{(1 + y_1/y_2)(y_2 - y_1)} \right]. \tag{2.135b}$$

The exact expression for the gain hole simplifies considerably in various limiting cases. For example, if power broadening is negligible, the two lorentz widths are equal and Eq. (2.135) reduces to

$$\Delta G(\nu) = \frac{\pi \langle \mu_0 \rangle^2 D_m}{2\hbar\lambda} \times \left\{ \frac{U(x_1, y) + U(S + x_1, y) + \dfrac{2y^3}{S}[V(S + x_1, y) - V(x_1, y)]}{1 + \left(\dfrac{\nu - \nu_0}{\Delta\nu_{L0}} \right)^2} \right\} \tag{2.136}$$

where

$$y = y_1 = y_2 = \sqrt{\ln 2} \left(\frac{\Delta\nu_{L0}}{\Delta\nu_D} \right)$$

and use has been made of Eq. (2.133). The terms involving $U(x_1, y)$ in Eq. (2.136) are symmetric about the gain hole and of order $(\Delta\nu_{L0}/\Delta\nu_D)^2$. The terms involving $V(x_1, y)$ in Eq. (2.136) are asymmetric about the gain hole resonance and of order $(\Delta\nu_{L0}/\Delta\nu_D)^3$. At resonance ($S = 0$) the gain hole depth is given by

$$d_0 = \frac{\pi \langle \mu_0 \rangle^2 D_m U(x_1, y)}{\hbar\lambda} = \frac{4\pi \langle \mu_0 \rangle^2 P_{\text{total}}(\nu_0)}{\hbar^2 c R_1 R_2}. \tag{2.136b}$$

from Eq. (2.132b).

Note that the relationship between the gain hole depth at resonance and the power gain coefficient at resonance does *not* depend explicitly on the

Doppler width in the limit that power broadening can be neglected. For that reason, the equations developed below for the Doppler broadened limit ($\Delta\nu_D \gg \Delta\nu_L$) are much more accurate than might otherwise be suspected. The errors in replacing Eq. (2.136) by a Lorentzian of are order $(\Delta\nu_L/\Delta\nu_D)^2$.

2.20. Gain Hole Depth and Width in the Doppler-Broadened Limit

In the Doppler broadened limit, where $\Delta\nu_{LO} = \Delta\nu_L \ll \Delta\nu_D$, Eq. (2.136) for the gain hole becomes

$$\Delta G(\nu) = \frac{\pi\langle\mu_O\rangle^2 D(\nu_O)}{\hbar\lambda}\left[1 + \left(\frac{\nu - \nu_O}{\Delta\nu_{LO}}\right)^2\right]^{-1} \tag{2.137}$$

where $D(\nu_O)$ is given by Eq. (2.131).

Equation (2.137) could, of course, have been obtained directly in the Doppler broadened limit by taking $D(\nu') \approx D(\nu_O)$ out of the integral in Eq. (2.134) and noting directly that for $\Delta\nu_{LO} = \Delta\nu_L$, the integral

$$I \equiv \int_{-\infty}^{\infty} \frac{d\nu'}{\{1 + [(\nu' - \nu)/(\Delta\nu_{LO}/2)]^2\}\,\{1 + [(\nu_O - \nu')/(\Delta\nu_{LO}/2)]^2\}}$$

$$= \frac{\Delta\nu_{LO}}{2}\int_{-\infty}^{\infty}\left\{\frac{dt}{[1 + (i + \gamma)^2]\,(1 + t^2)} + \frac{dt}{[1 + (i - \gamma)^2]\,[1 + (t + \gamma)^2]}\right\},$$

$$= (\pi\,\Delta\nu_{LO}/4)\bigg/\left[1 + \left(\frac{\nu - \nu_O}{\Delta\nu_{LO}}\right)\right]^2, \tag{2.138}$$

where substitutions of the type,

$$t = \frac{\nu' - \nu_O}{\Delta\nu_{LO}/2} \quad \text{and} \quad \gamma = \frac{\nu_O - \nu}{\Delta\nu_{LO}/2}$$

and $\tan\theta = t$, $\tan\psi = (t + \gamma)$ were made.

Hence from either Eq. (2.137) or (2.138) it is apparent that the gain distribution that would have been produced by atoms in the population inversion hole is a Lorentzian function of the frequency, but with a full width at half-maximum gain which is *twice* the normal Lorentz width for the transition (i.e., $2\Delta\nu_{LO}$).

Collecting terms, the gain hole due to one running wave in the Doppler broadened limit is given by

$$\Delta G(\nu) = \frac{d}{1 + [(\nu - \nu_0)/\Delta\nu_{L0}]^2}, \tag{2.139}$$

where the depth, d, at ν_0 is related to the depth in the inversion hole $[D(\nu_0)]$ and the total power gain coefficient (P_{total}) for the running wave by

$$d = \frac{\pi\langle\mu_0\rangle^2}{\hbar\lambda} D(\nu_0) = \frac{4\pi\langle\mu_0\rangle^2 P_{\text{total}}}{\hbar^2 c R_1 R_2} \tag{2.140}$$

at low powers. As discussed in connection with Eq. (2.136) these expressions are good through terms of order $(\Delta\nu_L/\Delta\nu_D)$.

These results may be used to obtain an accurate quantitative expression for the tuning dip in the standing wave case in a remarkably simple manner. Consider the situation depicted in Fig. 2-12 where the oscillator frequency is detuned from line center. Two holes are burned in the velocity distribution and spaced symmetrically about the line center by the

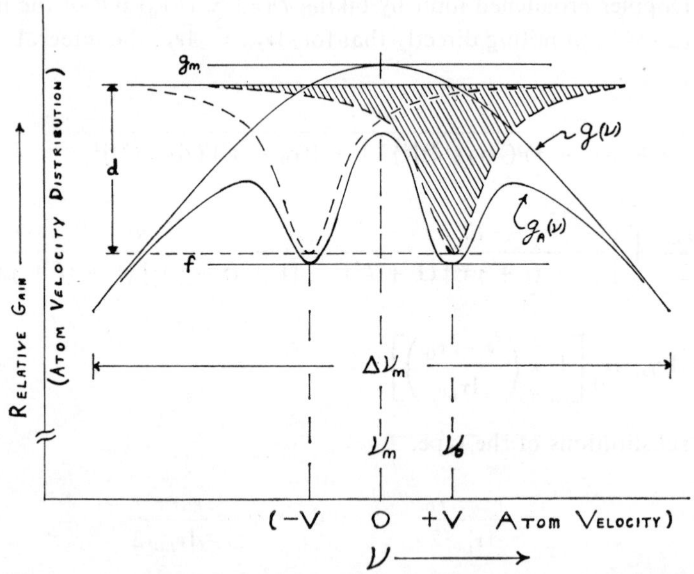

FIGURE 2-12. Single-pass gain curve for one oscillation frequency at ν_0 in the limit that the two Lorentzian holes of depth d begin to overlap. $g(\nu)$ represents the fractional energy gain per pass in the absence of oscillation and $g_A(\nu)$ represents the actual gain in the presence of the oscillation. f represents the fractional energy loss per pass. The power output in the $-x$ direction is proportional to the shaded area in the diagram for $\nu_0 > \nu_m$. The power can decrease by a factor approaching two as the laser is tuned through line center (from Bennett, 1963).

amount $|\nu_m - \nu_{osc}|$. Correspondingly, two holes occur in the gain curve seen by running waves going in either the $+z$ or $-z$ directions. The total reduction in gain at each gain hole is made up by contributions from the separate population holes. It is clear that so long as the two holes are widely-separated compared to the Lorentz width they will have an additive effect in reducing the gain at any point on the Doppler profile. The remarkable aspect of the situation is that, due to the averaging effect of the velocity distribution, the combined effect of the two holes is still *additive to a very good approximation even when the two gain holes are strongly overlapping.* (This point is discussed in more detail below). If we let "d" stand for the depth of the Lorentzian gain hole due to the interaction with *one* running wave alone, the condition for steady-state in the standing wave case (see Fig. 2-12) is given by

$$d\left[1 + \frac{1}{1 + [2(\nu_m - \nu_{osc})/\Delta\nu_{LO}]^2}\right] = G(\nu_{osc}) - f/L. \qquad (2.141)$$

The total power out of the oscillator in each running wave is proportional to "d" through Eq. (2.140). Hence

$$P \propto d = \frac{G(\nu_{osc}) - f/L}{1 + \{1 + [2(\nu_m - \nu_{osc})/(\Delta\nu_{LO})]^2\}^{-1}}, \qquad (2.142)$$

a result which was presented by the author at the Paris 1963 Conference on Quantum Electronics (Bennett, 1963). This result is precisely the same as that obtained by Lamb from his third-order theory (Lamb, 1964), also described at the Paris conference. That is, the power can go down by a factor approaching two as the laser is tuned through line center. (Strictly-speaking, $L = L_G$, the length of the gain section).

The tuning dip was reported in experimental observation by McFarlane, Bennett, and Lamb (1963), using a short magnetostrictively tuned helium–neon laser at 1.15 microns (see Fig. 2-13). The dip and its pressure dependence were subsequently studied in considerable detail on this transition by Szöke and Javan (1963; 1966) and it has now become a well-known phenomenon in large numbers of gas-laser transitions.

In this same approximation ($\Delta\nu_L \ll \Delta\nu_D$ and E_0 small)‡ one can calculate the small signal gain coefficient at a frequency $\nu' \neq \nu_{osc}$ in the presence of the oscillation. Specifically, from Eq. (2.139),

$$G_{actual}(\nu') \approx G(\nu')$$
$$- d\left\{\left[1 + \left(\frac{\nu' - \nu_{osc}}{\Delta\nu_L}\right)^2\right]^{-1} + \left[1 + \left(\frac{2\nu_m - (\nu' + \nu_{osc})}{\Delta\nu_L}\right)^2\right]^{-1}\right\}, \qquad (2.143)$$

‡ In what follows we will drop the distinction between $\Delta\nu_{LO}$ and $\Delta\nu_L$ unless power broadening is specifically emphasized.

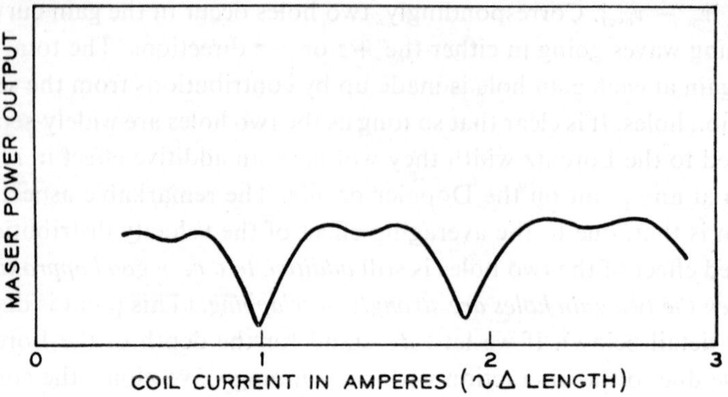

FIGURE 2-13. Single-mode tuning dip observed in a helium-neon laser (McFarlane, Bennett, and Lamb, 1963).

where d is given by Eq. (2.142) and $G(\nu')$ is given by Eq. (2.52). For example, if ν_{osc} is tuned to the center of the line ($\nu_{osc} = \nu_m$)

$$G_{\text{actual}}(\nu') \approx G_m e^{-\beta^2(\nu_m-\nu')^2} - \frac{G_m - f/L}{1 + [(\nu' - \nu_m)/\Delta\nu_L]^2}, \tag{2.144}$$

where

$$\beta = \frac{2\sqrt{\ln 2}}{\Delta\nu_D}$$

This function is qualitatively similar to the power tuning dip curve itself and has maxima for $\nu' \neq \nu_m$ if

$$1 - \frac{f/L}{G_m} > 4 \ln 2(\Delta\nu_L/\Delta\nu_D)^2, \tag{2.145}$$

which occur at

$$(\nu' - \nu_m) \approx \pm\sqrt{\frac{\Delta\nu_L\,\Delta\nu_D}{2\sqrt{\ln 2}}} \tag{2.146}$$

well above threshold ($G_m L \gg f$). Consequently, as the gain is increased above threshold, multimode oscillation will tend to occur on adjacent axial modes if

$$c/2L \lesssim \sqrt{\frac{\Delta\nu_L\,\Delta\nu_D}{2\sqrt{\ln 2}}}. \tag{2.147}$$

As previously described by the author (Bennett, 1963), one can also determine a quantitative expression for the oscillation frequency from the hole-burning analysis. For the inhomogeneously-broadened laser above threshold, one must allow for the changes in phase shift produced by the holes in the gain curve. In particular, for one mode above threshold Eq. (2.78) may be written

$$\nu_{\text{osc}} = \nu_{\text{cav}} - \frac{\Delta\nu_c}{f}(\Delta\phi_{\text{gain}} + \Delta\phi_{\text{holes}})\,L, \tag{2.148}$$

where the phase shift coefficient has been broken up into two parts. The first, $\Delta\phi_G$, represents the phase shift due to the original line profile in the absence of oscillation [Eq. (2.116), in the limit $\Delta\nu_D \gg \Delta\nu_L$]. The other represents the phase shift produced by the holes inthe gain curve. Again, in the limit $\Delta\nu_L \ll \Delta\nu_D$, we may assume that the holes in the gain profile are Lorentzian. Consequently, the phase shift produced by these holes is just the phase-shift expression for the Lorentzian gain distribution contained in Eq. (2.139). Hence the phase shift coefficient at frequency ν_0 due to a hole at frequency ν is [from Eq. (2.100) with $2\Delta\nu_L$ replacing $\Delta\nu_m$],

$$\Delta\phi_{\text{hole}}(\nu_0) = +\frac{d(\nu - \nu_0)/2\Delta\nu_L}{1 + (\nu - \nu_0)^2/(\Delta\nu_L)^2}, \tag{2.149a}$$

where the plus sign occurs because we are concerned with the phase shift arising from an absence of gain (i.e., a gain hole of depth, d). Note that because this expression is an odd function of the frequency difference from the hole center, there is no phase shift at the center of one gain hole due to itself. That is, the entire contribution to the phase shift from gain holes at one hole location is all due to the other ("mirror image") hole in the limit that $\Delta\nu_L \ll \Delta\nu_D$. [There would, of course, be an effect of one hole on itself if the hole were asymmetric as in Eq. (2.136).] Note that the two holes in single mode operation are spaced by $2(\nu_{\text{osc}} - \nu_m)$ [see Fig. (2-12)]. Hence the phase shift at the oscillator frequency (ν_{osc}) due to the "mirror image" hole at $\nu = 2\nu_m - \nu_{\text{osc}}$ is

$$\Delta\phi_{\text{hole}}(\nu_{\text{osc}}) = \frac{(\nu_m - \nu_{\text{osc}})\,d/\Delta\nu_L}{1 + [2(\nu_{\text{osc}} - \nu_m)/\Delta\nu_L]^2}. \tag{2.149b}$$

Combining Eqs. (2.142), (2.149), and (2.148) yields a transcendental equation for the oscillator frequency whose full non-linear properties will be discussed in more detail in a separate section below. For the moment we shall note that for $\Delta\nu_c \ll \Delta\nu_L \ll \Delta\nu_D$, one may let $\nu_{\text{osc}} = \nu_{\text{cav}}$ in the bracketed expressions appearing on the right hand side of Eq. (2.148). If we further expand the phase shift for the Gaussian line in the manner of

Eq. (2.121) valid near the line center, the oscillator frequency is given by (Bennett, 1963)

$$\nu_{\text{osc}} \approx \nu_{\text{thresh}} + \left(\frac{G(\nu_{\text{cav}})\,L - f}{f}\right)$$
$$\times\ \Delta\nu_c\left[a\left(\frac{\nu_m - \nu_{\text{cav}}}{\Delta\nu_D}\right) + a(c-b)\left(\frac{\nu_m - \nu_{\text{cav}}}{\Delta\nu_D}\right)^3 - \frac{(\nu_m - \nu_{\text{cav}})/2\Delta\nu_L}{1 + 2[(\nu_m - \nu_{\text{cav}})/\Delta\nu_L]^2}\right], \tag{2.150}$$

where the values of the coefficients a, b, c are given by Eq. (2.122) for a Gaussian line. Equation (2.150) has been factored in such a way as to illustrate that with increasing gain above threshold, the oscillator frequency moves away from line center for

$$|\nu_m - \nu_{\text{osc}}| \lesssim \sqrt{\frac{\Delta\nu_D \Delta\nu_L}{2a}} \tag{2.151 a}$$

and is attracted toward line center for

$$|\nu_m - \nu_{\text{osc}}| \gtrsim \sqrt{\frac{\Delta\nu_D\,\Delta\nu_L}{2a}}, \tag{2.151 b}$$

where $a \approx 0.94$.

Far away from line center [Eq. (2.151 b)] the phase shift from the entire line exhibits the normal pulling effect and the resonance moves toward the center with increasing power. Near the line center [region (2.151 a)], the "mirror image" hole in the gain curve results in a situation where the bulk of the line is in the opposite direction from the original center of the resonance and there is a net pushing of the resonance from line center with increasing power. This is in reality the same type of "hole repulsion" effect described originally by the author (Bennett, 1962b) in the multimode case discussed in the first paper on hole-burning effects. The phase shifts arising from holes in the gain curve are such that two holes tend to repel each other. Near the line center this repulsion effect dominates over the normal attraction by the bulk of the gain curve, whereas for the region described by condition (2.151 b), the opposite situation prevails. The results in Eq. (2.150) are equivalent to those obtained by Lamb (1964) in the same limit.

2.21. Accuracy of the Additive Approximation

The additive approximation used in Eq. (2.141) provides such a simple, direct method for determining both the power and frequency characteristics of a gas laser in the Doppler-broadened limit that the accuracy of the method warrants further discussion. It is clear that the method is at least as accurate as the third-order self-consistent theory of Lamb in predicting the average frequency and power of a single-mode laser because it gives the same answers for these quantities in the corresponding limit. As mentioned above, the accuracy of the method in the standing wave case arises from the averaging effect of the velocity distribution.

In the presence of the standing wave in the laser cavity any atom with a non-zero component of velocity in the z direction will see two different frequencies in its rest frame, even in single mode. Hence, in the presence of the standing wave, one should really do the calculation in Sec. 2.3 with the atom simultaneously subject to two separate perturbing terms at different frequencies. The basic problem with that approach is that it is not possible to obtain general solutions equivalent to Eqs. (2.36) and (2.38) in closed form when two or more frequencies are simultaneously present. That consideration is, of course, the reason that prompted Lamb to use a time-dependent perturbation expansion in his treatment of the problem. One can always do a numerical integration of the Schrödinger equation without recourse to a perturbation expansion method and, as noted by Rautian and Sobel'man (1963), the integrals involved simplify considerably in certain limiting cases.

The errors involved in adding the separate probabilities for induced emission by the two oppositely-travelling running waves are clearly least important when the oscillator is tuned far from the center of the line.

Similarly, it is also apparent that the worst case for the additive approximation occurs when the oscillator frequency is tuned precisely to the center of the Doppler line. Here every atom with velocity v_z in the z-direction sees two frequencies displaced from its resonance frequency (ν_0) by equal and opposite amounts due to the Doppler effect. Thus the perturbation contains frequencies

$$\nu = \nu_0 \pm v_z/\lambda$$

with equal field amplitudes. As noted by Rautian and Sobel'man (1963), the time-dependent solutions to the Schrödinger equation simplify considerably in this case in the limit that the lower- and upper-state phase interruption rates (R_1, R_2) are equal. The limit,

$$R_1 = R_2 \equiv R,$$

is fairly realistic for many laser problems (especially those including a large contribution from collision broadening).

The equations for the general two-frequency case analogous to those treated in Sec. 2.3, Eq. (2.23) are

$$\dot{C}_1 = C_2\left[\frac{V_a}{i\hbar}e^{i\Omega_a t} + \frac{V_b}{i\hbar}e^{-i\Omega_b t}\right] - C_1R_1/2$$

$$\dot{C}_2 = C_1\left[\frac{V_a}{i\hbar}e^{-i\Omega_a t} + \frac{V_b}{i\hbar}e^{i\Omega_b t}\right] - C_2R_2/2$$

where the antiresonant terms are neglected,

$$\Omega_a = \omega_a - \omega_0,\ \Omega_b = \omega_0 - \omega_b \text{ and } \omega_b < \omega_0 < \omega_a,$$

and V_a, V_b are the separate interaction terms for each running wave.

In the special case that

$$V_a = V_b \equiv V_0,\quad \Omega_a = \Omega_b \equiv \Omega \quad\text{and}\quad R_1 = R_2 \equiv R,$$

these equations become

$$\dot{C}_1 = C_2\left(\frac{2V_0}{i\hbar}\right)\cos(\Omega t + \phi_0) - C_1R/2$$

$$\dot{C}_2 = C_1\left(\frac{2V_0}{i\hbar}\right)\cos(\Omega t + \phi_0) - C_2R/2$$

where ϕ_0 is a phase angle introduced to allow for the randomness of excited state formation time in respect to the optical field, and have solutions,

$$C_1 = -i\sin\left[\frac{4V_0}{\hbar\Omega}\cos\left(\frac{\Omega t}{2} + \phi_0\right)\sin\frac{\Omega t}{2}\right]e^{-Rt/2}$$

$$C_2 = \cos\left[\frac{4V_0}{\hbar\Omega}\cos\left(\frac{\Omega t}{2} + \phi_0\right)\sin\frac{\Omega t}{2}\right]e^{-Rt/2}$$

satisfying the boundary conditions $C_1 = 0$ and $C_2 = 1$ at $t = 0$, as may be seen by direct substitution. In direct analogy with Eq. (2.25), the total probability that an atom placed in the upper state at $t = 0$ undergoes stimulated emission is

$$P_{\text{stim}} = R\int_0^\infty |C_1(t)|^2\,dt.$$

In the present problem the answer depends on the phase angle, ϕ_0, and we clearly want an average over phase angles of the type,

$$\overline{P_{\text{stim}}} = \left(\frac{1}{\pi}\right)\int_0^\pi P_{\text{stim}}\,d\phi_0.$$

Substituting the solution for $|C_1(t)|^2$, it is seen that

$$\overline{P_{\text{stim}}} = \frac{1}{2} - \frac{1}{2}\int_{t=0}^{\infty} e^{-Rt}\, d(Rt) \frac{1}{\pi}\int_{0}^{\pi} \cos\left[\frac{8V_0}{\hbar\Omega}\sin\frac{\Omega t}{2}\cos\left(\frac{\Omega t}{2}+\phi_0\right)\right] d\phi_0 .$$

Noting further that

$$J_0(x) = \frac{1}{\pi}\int_0^{\pi} \cos(x\cos\theta)\, d\theta ,$$

we get the result that

$$\overline{P_{\text{stim}}} = \overline{P_{\text{abs}}} = \frac{1}{2} - \frac{1}{2}\int_0^{\infty} J_0\left(\frac{8V_0}{\hbar\Omega}\sin\frac{\Omega t}{2}\right) e^{-Rt}\, d(Rt) \qquad (2.152)$$

which is analogous to, but in slightly more useful form for the present purposes than, the result presented by Rautian and Sobel'man (1963).

The additive approximation in the present instance then corresponds to replacing the exact expression of Eq. (2.152) by the sum of two separate expressions of the type obtained from Eq. (2.36) with $R_1 = R_2 \equiv R$. We shall define the result of that operation to be

$$P_{\text{approx}} = \frac{\left|\dfrac{2V_0}{R\hbar}\right|^2}{1 + (\Omega/R)^2 + \left|\dfrac{2V_0}{R\hbar}\right|^2} \qquad (2.153)$$

A comparison of the exact expression in Eq. (2.152) with the approximate expression in Eq. (2.153) is given in Fig. 2-14 for various values of the saturation parameter as a function of frequency.‡ In the present problem,

$$\frac{\Omega}{R} = \frac{\nu_0 - \nu_m}{\Delta\nu_{LO}/2} = \frac{2v_z}{\lambda\Delta\nu_{LO}} \text{ and } 2V_0 = -\langle\mu_0\rangle\, E_0 ,$$

where E_0 is the amplitude of the electric field and $\langle\mu_0\rangle$ is the matrix element of the electric dipole moment operator in the direction of the field, as previously defined.

‡ The author is indebted to Dr. C. J. Elliott for performing the numerical integrations of Eq. (2.152) used in Fig. 2-14.

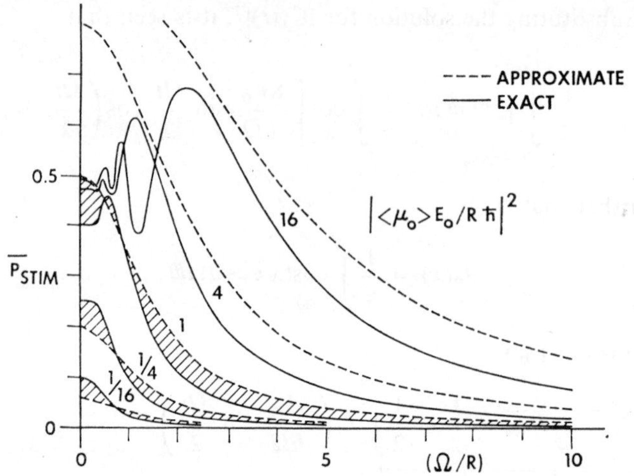

FIGURE 2-14. Comparison of exact calculation of P_{stim} in Eq. (2.152) with the additive approximation [Eq. (2.153)] for atoms at the center of the line in the presence of a standing wave. The shaded regions cancel in the integration over the velocity distribution at low values of the field.

As may be seen from Fig. 2-14, Eqs. (2.152) and (2.153) for the exact and approximate stimulated emission probabilities differ significantly. The difference is especially striking at extremely high values for the saturation parameter where the exact expression exhibits a large interference effect for $\Omega \approx R$. For low values of the saturation parameter, however, *the integrals of the two expressions over the velocity distribution are closely the same.* Here the curves cross at $\Omega \approx R$ and the shaded areas approximately cancel in the difference between the two integrals. This averaging effect of the velocity distribution is illustrated quantitatively in Fig. 2-15 where the results of a numerical integration over the velocity distribution of the two functions are shown for different ratios of the Doppler width to the zero-field Lorentz width. These results are presented in such a way as to permit a quick determination of the total fractional error in the population hole depth when the laser is tuned to line center which results from the additive approximation. Note that this error goes precisely to zero for an infinite Doppler width (compared to the Lorentz width) when the field goes to zero. As the field increases, the additive approximation overestimates the hole depth by an amount which at first increases linearly with the intensity in the mode, but rapidly saturates for values of the saturation parameter much in excess of ≈ 1. At the latter point, the error varies from about 20–30 percent as the ratio of the Doppler width to the Lorentz width increases from about four to infinity. At infinite Doppler width, the error saturates at about 50 percent for very large fields. The improved accuracy of the additive approximation at high fields which results from decreasing the ratio of Doppler width to Lorentz width occurs because the region of biggest error (at large Ω/R in Fig. 2-14) is being selectively suppressed.

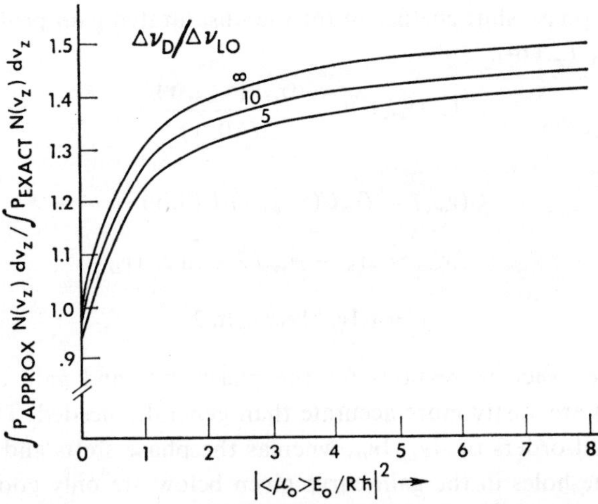

FIGURE 2-15. Comparison of the additive approximation and the exact value for P_{stim} when the expressions in Eqs. (2.152) and Eq. (2.153) are integrated over the velocity distribution.

2.22. Single Mode Equations and Hysteresis Effects

In the low field limit and through terms of order $(\Delta\nu_L/\Delta\nu_D)$, the phase shifts may be written in sufficiently compact form to warrant exact solution of Eq. (2.148) for the oscillator frequency through numerical evaluation of Eqs. (2.114), (2.142) and (2.149). Although Eq. (2.148) is an extremely non-linear function of the oscillator frequency, it is very easy to solve Eq. (2.148) for the cavity frequency using ν_{osc} as an independent variable in a computer program which evaluates the phase shift integrals to some prescribed accuracy. Adopting this procedure gets around the limitations imposed by the approximation $\Delta\nu_c \ll \Delta\nu_L$ and the restriction to frequencies near the Doppler center imposed by the expansion technique employed in Eq. (2.150). In addition, the exact solution of Eq. (2.148) shows up an interesting possible hysteresis effect which was not discussed by Lamb (1964), but should be present in his equations as well.

Re-writing Eq. (2.148), the oscillator and cavity frequencies are related by the transcendental equation

$$\nu_{\text{cav}} = \nu_{\text{osc}} + \frac{\Delta\nu_{\text{cav}}}{(f/L_G)} [\Delta\phi_g(\nu_{\text{osc}}) + \Delta\phi_{\text{hole}}(\nu_{\text{osc}})] \tag{2.154}$$

where L_G is the length of the active medium.

The phase shift coefficient for the unsaturated gain profile is given exactly by Eq. (2.114),

$$\Delta\phi_g(\nu_{\text{osc}}) = -\frac{G_m V(x_{\text{osc}}, y)}{2U(0, y)}$$

where‡

$$\begin{aligned} G(\nu_{\text{osc}}) &= G_m U(x_{\text{osc}}, y)/U(0, y) \\ x_{\text{osc}} &= (\nu_m - \nu_{\text{osc}})\, 2\sqrt{\ln 2}/\Delta\nu_D \\ y &= (\Delta\nu_L/\Delta\nu_D)\sqrt{\ln 2}. \end{aligned} \tag{2.155}$$

Note that the exact expressions for the phase-shift and gain coefficients in Eq. (2.155) are vastly more accurate than generally needed. These terms are good to all orders of $\Delta\nu_L/\Delta\nu_D$, whereas the phase shifts and gain coefficients for the holes in the gain curve given below are only good through the first order in $\Delta\nu_L/\Delta\nu_D$. Although the functions $U(x, y)$ and $V(x, y)$ can be run off as a subroutine in a computer program [see the discussion by Faddeyeva and Terent'ev (1961) regarding useful methods for computing these functions], there is a tendency to spend a disproportionate amount of time computing these functions in respect to their intrinsic importance to the present problem. Because the terms arising from the holes are generally much more important above threshold, it is usually adequate to replace the exact expressions in Eq. (2.155) by their values in the limit $y = 0$, previously discussed in connection with Eqs. (2.115)–(2.117). That is

$$\begin{aligned} G(\nu_{\text{osc}}) &\approx G_m U(x_{\text{osc}}, 0) = G_m \exp(-x_{\text{osc}}^2) \\ \Delta\phi(\nu_{\text{osc}}) &\approx -\frac{G(\nu_{\text{osc}})}{\sqrt{\pi}} \int_0^{x_{\text{osc}}} e^{t^2}\, dt \approx -0.282 G_m \sin\left(\frac{\nu_m - \nu_{\text{osc}}}{0.3\Delta\nu_D}\right) \end{aligned} \tag{2.155b}$$

where the sine approximation holds near the center of the line [see discussion connected with Eq. (2.117)].

‡ The functions $U(x, y)$ and $V(x, y)$ are discussed in Sec. 2.7.

The most important term in Eq. (2.154) well-above threshold arises from the "mirror image" gain hole at $2\nu_m - \nu_{osc}$. As shown in Eqs. (2.142) and (2.149), the phase shift coefficient from this gain hole is

$$\Delta\phi_{\text{hole}}(\nu_{osc}) = d_{osc}\left(\frac{\nu_m - \nu_{osc}}{\Delta\nu_L}\right)\left[1 + 4\left(\frac{\nu_m - \nu_{osc}}{\Delta\nu_L}\right)^2\right]^{-1} \tag{2.156}$$

where the depth of the gain hole is

$$d_{osc} = \frac{G(\nu_{osc}) - f/L_G}{1 + \left[1 + 4\left(\frac{\nu_m - \nu_{osc}}{\Delta\nu_L}\right)^2\right]^{-1}}$$

and $d_{osc} > 0$ for the oscillator to be above threshold. The hole depth equation contains the Lamb dip, as discussed in Sec. 2.20. As discussed in connection with Eq. (2.136), the Lorentzian approximation for the hole shapes ignores terms of order $(\Delta\nu_L/\Delta\nu_D)^2$.

Using approximation (2.155b), a complete frequency tuning characteristic [solution to Eq. (2.154)] can be run off in a few seconds on almost any rudimentary sort of computer or programmable desk calculator. The power tuning characteristic is also obtained in such a solution because the total power flux out of the oscillator at steady-state from both running wave interactions is

$$P_{\text{out}} = 2P_{\text{total}}(\nu_{osc})\, L_G = \left(\frac{\hbar^2 c R_1 R_2 L_G}{2\pi\langle\mu_0\rangle^2}\right) d_{osc} \equiv C_0\, d_{osc} \tag{2.157}$$

from Eq. (2.136b) where L_G is the length of the gain section. As discussed in Sec. 1.9 only a fraction T/f of the total power is usefully available in the beam, where T is the average coupling loss per pass (e.g., half the "dumping" mirror transmission coefficient when a high reflectance mirror is used at one end) because the total fractional loss f includes dissipative, as well as coupling, losses. The "circulating" power flux in the cavity is larger than Eq. (2.157) by a factor l/f due to Q-multiplication.

As discussed in connection with Eq. (2.150), it is really the phase shift produced by the "mirror image" hole which produces the most interesting effects near line center. Hole repulsion effects due to that term were seen to be responsible for the power dependent "pushing" effect near line center described by Lamb (1964). If we relax the restriction, $\Delta\nu_{cav} \ll \Delta\nu_L \ll \Delta\nu_D$, used in obtaining Eq. (2.150), still more striking effects can occur.

It is possible in principle to operate a high gain single mode gas laser far above threshold in a lossy cavity for which

$$\Delta\nu_{\text{cav}} \approx \Delta\nu_L \ll \Delta\nu_D. \tag{2.158}$$

In this limit the oscillation frequency is relatively unaffected by the original Doppler line and yet can be violently repelled by the "mirror image" hole near line center. This effect is not of much practical value as applied to the *normal* single-mode laser, but it does give useful insight regarding the nature of the self-stabilization process discussed below which can occur in the presence of a saturable absorber. Sufficiently far above threshold, the solutions for the oscillation frequency become triple-valued when condition (2.158) is satisfied. An example of this phenomenom is given in Fig. 2-16 for values of the linewidth parameters which were arbitrarily chosen to exaggerate the effect. The curve labelled "Threshold" represents solutions to Eqs. (2.154)–(2.156) for G_m just slightly greater than f. The curve labelled "10 × Threshold" corresponds to a computer solution to the same equations with $G_m = 10f$. Eqs. (2.154)–(2.156) do not hold accurately that far above threshold because of power broadening effects.

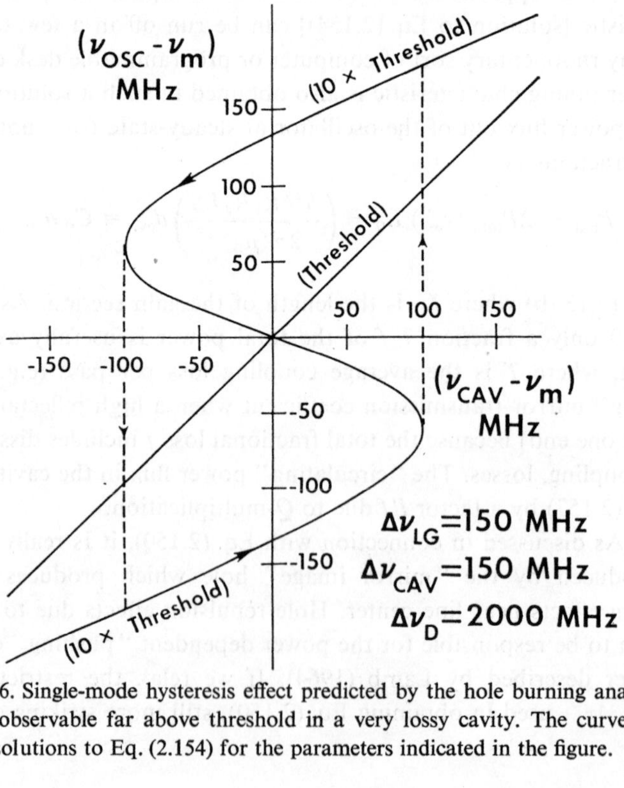

FIGURE 2-16. Single-mode hysteresis effect predicted by the hole burning analysis which should be observable far above threshold in a very lossy cavity. The curves represent numerical solutions to Eq. (2.154) for the parameters indicated in the figure.

As indicated in Fig. 2-16 the laser clearly would not tend to oscillate on the characteristic with negative slope at the center of the figure, unless one went to some effort to set the cavity frequency to line center before turning up the gain. Normally, as the laser cavity frequency was tuned towards line center from the left on the "10 × Threshold" curve, the oscillator frequency would lag behind the cavity frequency until it reached a point $\approx \Delta\nu_L/2$ below line center, at which it would suddenly jump along the vertical dashed line to a point $\approx \Delta\nu_L$ above line center. A similar discontinuous tuning jump would occur in the opposite direction as indicated by the dashed vertical arrow pointing downwards at the left side of Fig. 2-26. The phenomenon would exhibit itself by removing a piece from the central region of the tuning dip curve. Normally this hysteresis effect would be hard to observe because of the tendency of the laser to go into multimode oscillation far above threshold without the use of additional mode suppression techniques such as the titled-etalon method described in Sec. 1.10.

Strictly-speaking, the single-mode equations given above are only valid in the absence of power broadening. As implied by the discussion in Sec. 2.27, the neglect of power broadening is equivalent to the assumption that $2d_{osc}L_G \ll f$, or $G_mL_G - f \ll 0.5f$. Consequently the figures shown here which were computed for values of the gain far above threshold are intended primarily as a basis for qualitative discussion.

2.23. The Isotope Effect in Single Mode

As discussed by McFarlane, Bennett and Lamb (1963), the initial attempts to observe the tuning dip on the 1.1523 micron transition of neon in the helium-neon laser were complicated by the presence of isotope shifts intermediate between the Lorentz and Doppler widths in naturally occurring neon (Ne^{20} 90.92 percent, Ne^{21} 0.26 percent and Ne^{22} 8.82 percent). The initial measurements exhibited a pronounced asymmetry in the tuning curve near line center, but clear resolution of the dip required isotopically pure neon samples in the particular laser used.

Measurement of isotope shifts on a number of neon lines in the helium-neon laser have since been reported and are summarized in Table 2-2. The common isotopes are Ne^{20} and Ne^{22}, and in all cases reported the Ne^{22} resonance falls on the high-frequency side. The agreement between the separate measurements for the isotope shift on the 1.1523 micron line is remarkably good and the small shift involved explains the difficulties met in clearly resolving the Lamb dip on this transition in normal neon near oscillation threshold.

TABLE 2-2. Ne^{22}–Ne^{20} isotope shift measurements on neon transitions in the helium-neon laser. (Frequency shifts in Mc/sec.)

Line (in microns)			Reference
1.1523	0.6328	0.6096	
261 ± 3	—	—	Szoke and Javan (1963)
257 ± 15	1030 ± 50	—	Ballik (1964; 1971)
—	875 ± 12	—	Cordover, Jaseja and Javan (1965)
257 ± 8	—	1706 ± 30	Cordover, Bonczyk and Javan (1967 a)

The isotope effect provides an entertaining illustration of the hole-burning model. Consider the case illustrated schematically in Fig. 2-17 in which we assume the total small signal gain curve is composed of two separate Doppler distributions with maxima at ν_{m1} and ν_{m2}, corresponding to an isotope shift $\delta\nu_1 = \nu_{m2} - \nu_{m1}$, and unsaturated gain coefficients, $G_1(\nu)$ and $G_2(\nu)$ given by

$$G_j(\nu) = G_{jm} \exp(-x_j^2)$$

where (2.158)

$$x_j = (\nu - \nu_{mj})\, 2\sqrt{\ln 2}/\Delta\nu_{Dj} \quad \text{for} \quad j = 1, 2.$$

From Eq. (2-52), the Doppler widths vary as

$$\Delta\nu_{D1}/\Delta\nu_{D2} = \sqrt{M_2/M_1}$$

where M_1, M_2 are the masses of the two isotopes.

As shown in Fig. 2-17, there are four different gain holes in this problem above threshold, two in each isotopic gain profile. Figure 2-17 represents the separate Doppler gain curves as a function of frequency that would be seen by a running wave travelling in one direction. The figure specifically represents the gain for a running wave going to the left ($-z$ direction) in the case where $\nu_{osc} > \nu_{m2} > \nu_{m1}$. Steady state is characterized by the requirement that the total gain in a single pass for this running wave be reduced to the loss at the oscillation frequency, ν_{osc}. Thus the shaded holes at ν_{osc} are burnt directly in the two gain curves by the running wave going to the left. As indicaded by parent hesesin the figure, the two non-shaded holes are burnt in the opopposite sides of the two Doppler distributions by the running wave travelling in the opposite (+z) direction. [A similar figure could, of course, be drawn representing the gain seen by the running in the opposite ($+z$) direction.]

Assuming that the individual Doppler distributions are symmetric about their respective maxima, the symmetrically-placed holes in a given isotope gain profile are of equal depth. We shall let the gain-hole depths for one running wave interaction in the two isotopic species be d_1 and d_2, as indicated in Fig. 2-17. Using the additive approximation discussed in

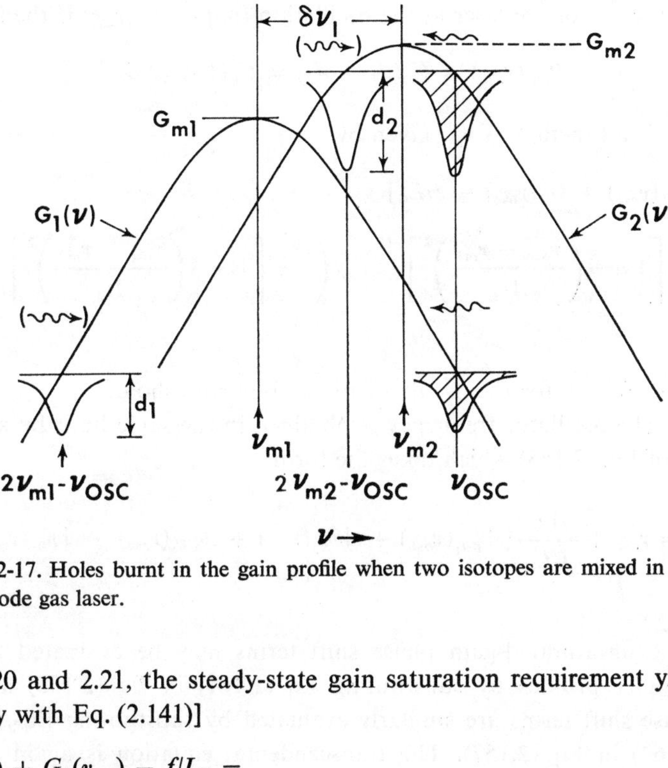

FIGURE 2-17. Holes burnt in the gain profile when two isotopes are mixed in the same single-mode gas laser.

Sec. 2.20 and 2.21, the steady-state gain saturation requirement yields [in analogy with Eq. (2.141)]

$$G_1(\nu_{\rm osc}) + G_2(\nu_{\rm osc}) - f/L_G =$$

$$d_1\left\{1 + \left[1 + 4\left(\frac{\nu_{\rm osc} - \nu_{m1}}{\Delta\nu_L}\right)^2\right]^{-1}\right\} + d_2\left\{1 + \left[1 + 4\left(\frac{\nu_{\rm osc} - \nu_{m2}}{\Delta\nu_L}\right)^2\right]^{-1}\right\}. \tag{2.159}$$

where L_G is the length of the amplifying medium. From the discussion of Eq. (2.136) in the Doppler broadened limit, the ratio of the two hole depths due to the individual running wave interactions is now a function of frequency given by

$$\gamma \equiv \frac{d_2}{d_1} = \frac{G_2(\nu_{\rm osc})}{G_1(\nu_{\rm osc})} = (G_{m2}/G_{m1}) \exp\left(x_{1\rm osc}^2 - x_{2\rm osc}^2\right) \tag{2.160}$$

where G_{m2}/G_{m1} is equal to the ratio of the isotopic abundance and

$$\begin{aligned} x_{1\text{osc}} &= (\nu_{\text{osc}} - \nu_{m1})\, 2\sqrt{\ln 2}\,/\Delta\nu_{D1} \\ x_{2\text{osc}} &= (\nu_{\text{osc}} - \nu_{m2})\, 2\sqrt{\ln 2}\,/\Delta\nu_{D2}. \end{aligned} \tag{2.161}$$

Assuming the same constant C_0 in Eq. (2.157) holds for both isotopes, the total power out of the laser at the oscillation frequency (ν_{osc}) is then

$$P_{\text{out}}(\nu_{\text{osc}}) = C_0(d_1 + d_2) = C_0(1 + \gamma)\, d_1 \tag{2.163}$$

where d_1 is a function of ν_{osc} given by

$$d_1 = [G_1(\nu_{\text{osc}}) + G_2(\nu_{\text{osc}}) - f/L_G] \times \left\{1 + \left[1 + 4\left(\frac{\nu_{\text{osc}} - \nu_{m1}}{\Delta\nu_L}\right)^2\right]^{-1} + \gamma\left(1 + \left[1 + 4\left(\frac{\nu_{\text{osc}} - \nu_{m2}}{\Delta\nu_L}\right)^2\right]^{-1}\right)\right\}^{-1} \tag{2.164}$$

and where $d_1 > 0$ for the oscillator to be above threshold.

The oscillator frequency is obtained in the same limit by a modification of Eq. (2.148) which takes the form

$$\nu_{\text{cav}} = \nu_{\text{osc}} + \frac{\Delta\nu_c}{f/L_G}\,[\Delta\phi_{g_1}(\nu_{\text{osc}}) + \Delta\phi_{g_2}(\nu_{\text{osc}}) + \Delta\phi_{h_1}(\nu_{\text{osc}}) + \Delta\phi_{h_2}(\nu_{\text{osc}})] \tag{2.165}$$

where the unsaturated-gain phase shift terms may be evaluated for the two Doppler profiles by substituting Eq. (2.161) in Eq. (2.155) and the hole phase shift terms are similarly evaluated by substituting Eqs. (2.160) and (2.164) in Eq. (2.157). The transcendental equation is again readily solved by treating ν_{osc} as the independent variable in a numerical computation of the type used to determine Fig. 2-16 in the normal, single-mode case. Hysteresis effects can again be obtained far above threshold in the limit that condition (2.158) is satisfied and are of a type which is similar, but more complex, than that illustrated in Fig. 2-16 for the normal single isotope case. The frequency tuning characteristics for isotopic mixtures far above threshold and in the limit that condition (2.158) is satisfied generally tend to *decrease* rather than improve the long term frequency stability of the laser in respect to drift of the cavity frequency. That is, one tends to get the opposite of the self-stabilization effect discussed below when a saturable absorber is placed within the cavity.

The form of the power tuning curve is of more interest in the case of isotopic mixtures. For example, the form of Eq. (2.163) is shown in Fig. 2-18 for equal concentrations of two isotopes whose mass ratio is 22/20. The line width parameters roughly correspond to the data in Table 2-2 for the 0.6328 micron neon transition. The slight asymmetry is due to the dependence of the Doppler widths on mass.

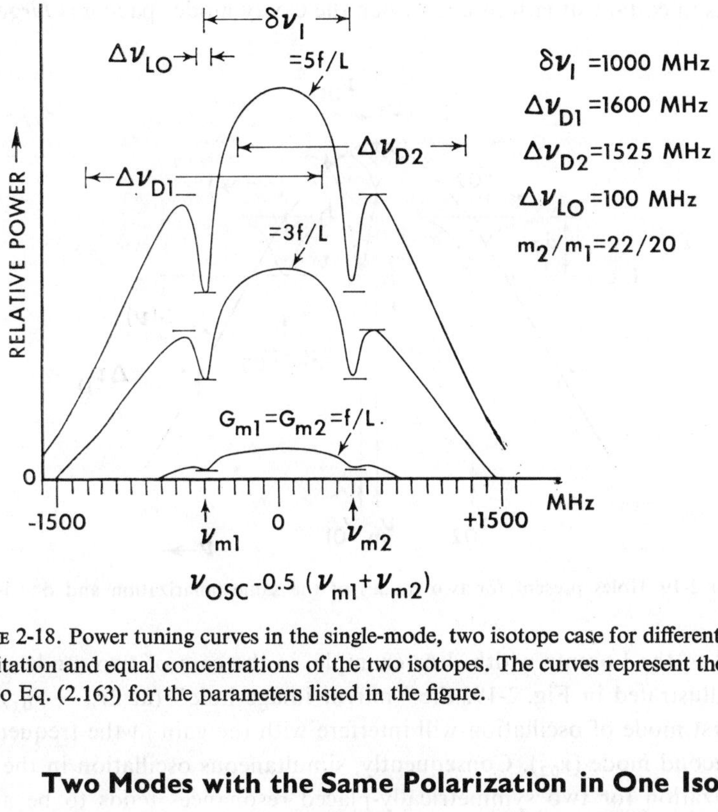

FIGURE 2-18. Power tuning curves in the single-mode, two isotope case for different levels of excitation and equal concentrations of the two isotopes. The curves represent the solutions to Eq. (2.163) for the parameters listed in the figure.

2.24. Two Modes with the Same Polarization in One Isotope

In the two-mode single-isotope case there are four gain holes burnt in the Doppler profile above threshold for the second mode. The four holes arise from the pairs of running wave interactions at the two different frequencies. A representative situation in which the two center holes are just resolved is shown in Fig. 2-19. This figure corresponds to the gain coefficient that would be seen as a function of frequency by a running wave travelling to the left ($-z$ direction) for the case where $\nu_{02} < \nu_m < \nu_{01}$, in which we adopt the convention that ν_{01} is the oscillation frequency for the first mode and ν_{02} is the oscillation frequency for the second. As the cavity resonance is tuned to higher values, the two shaded holes in Fig. 2-19 move to the right and the unshaded holes move to the left.

As is fairly obvious (Bennett, 1962a), simultaneous oscillation on two cavity modes of the same radial field distribution and spaced by less than a Lorentz width tends to be an unstable situation and the mode which initially had the most power will tend to suppress the other. However, the existence of the "mirror image holes" can produce similar interference effects in certain situations even when the cavity mode spacing is *large* compared to the Lorentz width. For example, in the case of two mode oscillation illustrated in Fig. 2-19, the "mirror image hole" (at $2\nu_m - \nu_{01}$) from the first mode of oscillation will interfere with the gain at the frequency of the second mode (ν_{02}). Consequently, simultaneous oscillation in the same polarization for two symmetrically-placed resonances tends to be an unstable condition and the stronger of the two modes will tend to put the weaker one out. Hence the behavior of the device will depend on its past history and one gets hysteresis effects such as those discussed below in connection with Fig. 2-20. [These hysteresis effects correspond to similar effects which would occur in the Lamb (1964) theory as a result of transitions from "weak" to "strong" coupling when the cavity resonances are tuned through the line.] For similar reasons, any symmetric placement of gain holes in multimode oscillation tends to be subject to hysteresis effects and one or more of the oscillator modes may "go out" or "jump" as the cavity length is

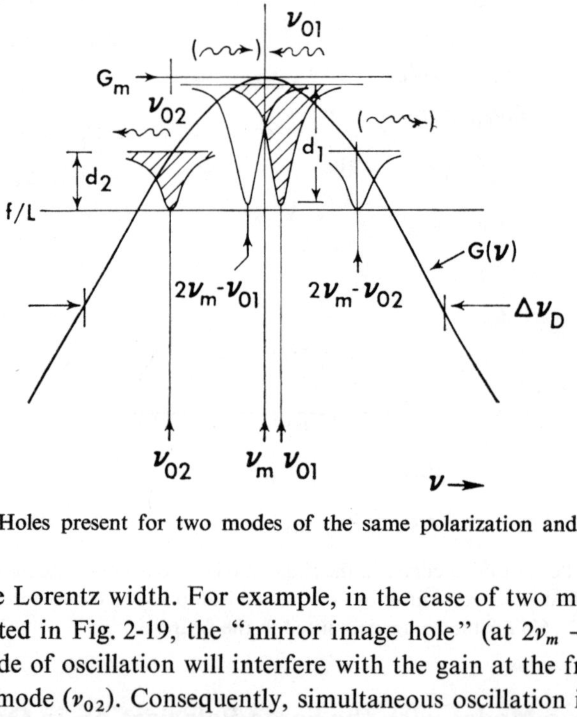

FIGURE 2-19. Holes present for two modes of the same polarization and one isotope.

adjusted to the symmetric condition. If the restriction on polarization is relaxed, level-degeneracy effects can permit simultaneous oscillation on orthogonal modes of polarization for a symmetric placement of two modes. However, level degeneracy effects are beyond the scope of the present two-level discussion.

The steady-state gain saturation requirement must be satisfied separately for both oscillation frequencies and each of the four running wave interactions contribute to the gain reduction at these two frequencies. Hence there are two separate gain saturation equations involving the four gain holes. Letting d_1 and d_2 characterize the gain hole depths due to single running waves at frequencies ν_{01} and ν_{02} when the gain holes are all well-resolved, and employing the additive approximation discussed in Sec. 2.20 and 2.21 when the holes overlap, steady-state is characterized in the two mode case by the equations

$$G(\nu_{01}) - f/L_G = d_1\left\{1 + \left[1 + 4\left(\frac{\nu_m - \nu_{01}}{\Delta\nu_L}\right)^2\right]^{-1}\right\}$$
$$+ d_2\left\{\left[1 + \left(\frac{\nu_{01} - \nu_{02}}{\Delta\nu_L}\right)^2\right]^{-1} + \left[1 + \left(\frac{2\nu_m - \nu_{02} - \nu_{01}}{\Delta\nu_L}\right)^2\right]^{-1}\right\} \tag{2.166a}$$

and

$$G(\nu_{02}) - f/L_G = d_1\left\{\left[1 + \left(\frac{\nu_{01} - \nu_{02}}{\Delta\nu_L}\right)^2\right]^{-1}\right.$$
$$\left. + \left[1 + \left(\frac{2\nu_m - \nu_{01} - \nu_{02}}{\Delta\nu_L}\right)^2\right]^{-1}\right\} + d_2\left\{1 + \left[1 + 4\left(\frac{\nu_m - \nu_{02}}{\Delta\nu_L}\right)^2\right]^{-1}\right\}. \tag{2.166b}$$

Note that Eqs. (2.166) have been written merely by application of Eq. (2.139) to the four gain holes shown in Fig. 2-19.

The oscillation frequencies are determined by rewriting Eq. (2.148) for the two modes:

$$\nu_{01} = \nu_{c1} - \frac{\Delta\nu_c}{f/L_G}\{\Delta\phi_g(\nu_{01}) + \Delta\phi_{h1}(\nu_{01}) + \Delta\phi_{h2}(\nu_{01})\} \tag{2.167a}$$

and

$$\nu_{02} = \nu_{c2} - \frac{\Delta\nu_c}{f/L_G}\{\Delta\phi_g(\nu_{02}) + \Delta\phi_{h1}(\nu_{02}) + \Delta\phi_{h2}(\nu_{02})\}. \tag{2.167b}$$

Here, the gain hole phase shift terms are given from Eqs. (2.149) by

$$\Delta\phi_{h1}(\nu_{01}) = \frac{d_1(\nu_m - \nu_{01})/\Delta\nu_L}{1 + [2(\nu_m - \nu_{01})/\Delta\nu_L]^2} \tag{2.168a}$$

$$\Delta\phi_{h2}(\nu_{01}) = d_2 \left\{ \frac{(\nu_{02} - \nu_{01})/2\Delta\nu_L}{1 + [(\nu_{02} - \nu_{01})/\Delta\nu_L]^2} + \frac{(2\nu_m - \nu_{02} - \nu_{01})/2\Delta\nu_L}{1 + [(2\nu_m - \nu_{01} - \nu_{02})/\Delta\nu_L]^2} \right\} \tag{2.168 b}$$

$$\Delta\phi_{h1}(\nu_{02}) = d_1 \left\{ \frac{(2\nu_m - \nu_{01} - \nu_{02})/2\Delta\nu_L}{1 + [(2\nu_m - \nu_{01} - \nu_{02})/\Delta\nu_L]^2} + \frac{(\nu_{01} - \nu_{02})/2\Delta\nu_L}{1 + [(\nu_{01} - \nu_{02})/\Delta\nu_L]^2} \right\} \tag{2.168 c}$$

$$\Delta\phi_{h2}(\nu_{02}) = \frac{d_2(\nu_m - \nu_{02})/\Delta\nu_L}{1 + [2(\nu_m - \nu_{02})/\Delta\nu_L]^2} \tag{2.168 d}$$

where $\Delta\phi_{h2}(\nu_{01})$ is the phase shift at the oscillation frequency ν_{01} produced by the holes burnt in the gain curve by the running waves at oscillation frequency ν_{02}, etc. The above equations have, of course, been written in the Doppler-broadened limit ($\Delta\nu_L \ll \Delta\nu_D$) and power broadening of the Lorentz widths has been ignored. The gaussian phase shift terms, $\Delta\phi_g(\nu)$, are given by Eqs. (2.155) or the approximation (2.155b).

These equations may also be solved to a prescribed accuracy by numerical methods. However, the coupling between the two oscillator frequencies which occurs through gain-hole phase-shift terms makes the problem more difficult than in the single mode case.

In the limit that

$$\Delta\nu_c \ll \Delta\nu_L \ll \Delta\nu_D \tag{2.169}$$

(where $\Delta\nu_c$ is the cavity resonance width and is assumed to be the same for both modes), the equations may be solved simply by letting

$$\nu_{01} \approx \nu_{c1} \quad \text{and} \quad \nu_{02} \approx \nu_{c2} \tag{2.170}$$

in the bracketed terms on the ride hand side of Eqs. (2.167), where the two cavity resonance frequencies are related by

$$\nu_{c2} = \nu_{c1} \pm c/2L. \tag{2.171}$$

The sign in Eq. (2.171) is generally determined by the requirement that both modes be above threshold for oscillation, or

$$d_1, d_2 > 0. \tag{2.172}$$

The power out in each mode is $P_j = C_0 d_j$ ($j = 1, 2$) from Eq. (2.157). A "negative" hole depth in the solution of Eqs. (2.166) either means that both modes are *not* above threshold (i.e., only the single-mode solution is stable), or that the stable solution for the weaker cavity mode has just jumped by ($2c/2L$) to the opposite of side the Doppler line. (The single-mode solution is discussed in Sec. 2.22.)

With the above precautions, one can solve Eqs. (2.166) through (2.168) in the following manner. First, we choose a value for one cavity resonance frequency, ν_{c1}. Assuming that $\nu_{c1} > \nu_m$ and that ν_{01} is the dominant mode, $\nu_{c2} = \nu_{c1} - c/2L$. Next we use approximation (2.170) and solve Eqs. (2.166) for the two hole depths, d_1, d_2. We next check to see that these hole depths satisfy requirement (2.172). If they do not, we let $\nu_{c2} = \nu_{c1} + c/2L$ and repeat the solution to Eqs. (2.166). Once condition (2.172) is satisfied for a particular pair of hole depths, d_1, d_2, we then calculate the phase shift terms through solution of Eqs. (2.168). We next calculate the gaussian phase shifts using either Eq. (2.155) or approximation (2.155b). Note that for $c/2L \ll \Delta\nu_D$, the sine approximation in Eq. (2.117) is perfectly adequate for most purposes. We next insert these phase shifts in Eqs. (2.167) and obtain corrected values for the oscillation frequencies. The power in each mode is approximately proportional to the gain hole depths through Eq. (2.157). Hence at this point, if condition (2.169) is well-satisfied, one has a solution for both oscillator frequencies and the power in each mode.

If condition (2.169) is not well-satisfied, it is straightforward to do the above calculation reiteratively until satisfactory convergence is obtained. That is, one takes the corrected values for the oscillator frequency, solves Eqs. (2.166) for the corrected hole depths, evaluates the phase shift expressions for the new hole depths, and determines a still better pair of values for the oscillator frequencies. This process tends to converge very rapidly unless the cavity resonances are sitting close to an unstable region, in which case condition (2.172) may be violated and cause wild fluctuations in the reiterative values.

A representative computer solution of Eqs. (2.166)–(2.168) for the two-mode tuning characteristic showing the hysteresis effect described above is illustrated in Fig. 2-20. In the case illustrated, mode No. 1 remains dominant at all times for $-0.5(c/2L) < (\nu_{c1} - \nu_m) < +0.5(c/2L)$. (The relative power in the two oscillating modes is shown in the lower portion of the figure.) As the symmetric tuning points are approached, the "mirror image hole" for the dominant mode puts the weaker mode out altogether and for a brief range the behavior is governed by the single mode equations. As indicated by the frequency tuning curves at the top of Fig. 2-20, for $\nu_{01} \ll \nu_m$ (but above threshold) the second mode starts oscillating at a frequency $\nu_{02} \approx \nu_{01} + c/2L$ and follows the top curve until $(\nu_{c1} - \nu_m) \gtrsim 0.5\Delta\nu_L$. At that point the second mode jumps to the opposite side of the line where it remains until it is put out again by the dominant mode near the point $(\nu_{c1} - \nu_m) \approx +0.5(c/2L)$. The hysteresis effects (indicated by the arrows in Fig. 2-20) in the central region of the tuning curve are due entirely to gain suppression on the opposite side of the line from the weaker mode resonance (ν_{02}) by its own "mirror image hole" at $2\nu_m - \nu_{02}$ (see Fig. 2-20). The mode pushing and pulling effects which occur for the parameters chosen are too small to show up on the scale used in Fig. 2-20. These effects are nevertheless present and, as described in the single mode case, the "pushing" arises from the repulsion between adjacent gain holes.

It is of interest to follow the power and frequency development of one mode as its corresponding cavity resonance is tuned completely through the Doppler line. The latter has been illustrated in Fig. 2-21 for

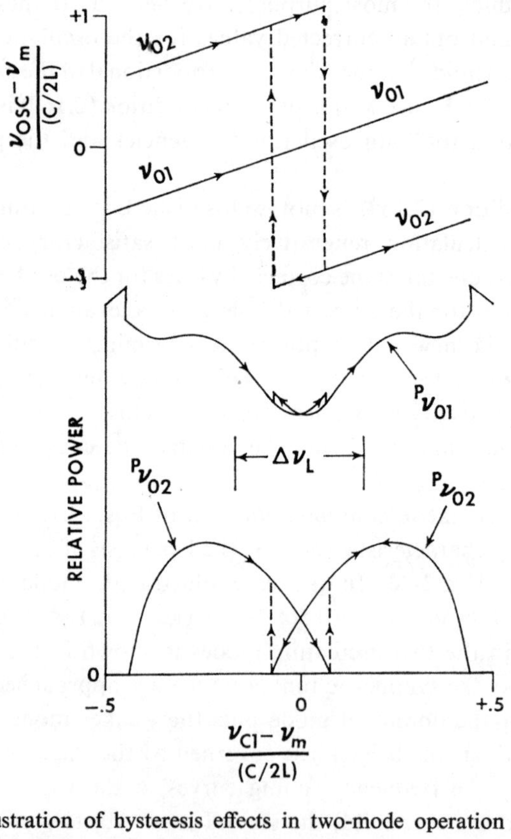

FIGURE 2-20. Illustration of hysteresis effects in two-mode operation produced by the hole burning effects illustrated schematically in Fig. 2-19. The upper part of the figure shows the frequency relationship of the two modes and the lower part of the figure shows the relative power tuning curves. The curves were computed from Eq. (2.166) and (2.167) for $G_m L_G/f = 1.5$, a cavity mode separation of 1/3 the Doppler width and a Doppler width 6 times the Lorentz width.

the same parameters used in Fig. 2-20. Here the cavity resonance first crosses threshold for oscillation at point A and undergoes the tortuous path indicated by the dashed line until it is finally extinguished at point B, as the cavity resonance frequency ν_{c1} continuously increases in frequency. The additional complexity associated with the short period of single-mode oscillation near the symmetric tuning points has been ignored in Fig. 2-21 for the sake of clarity. The principal hysteresis effects are indicated by the arrows in the figure.

Additional hysteresis effects of the type discussed in Sec. 2.22 for single-mode oscillation could in principle occur in the two-mode case

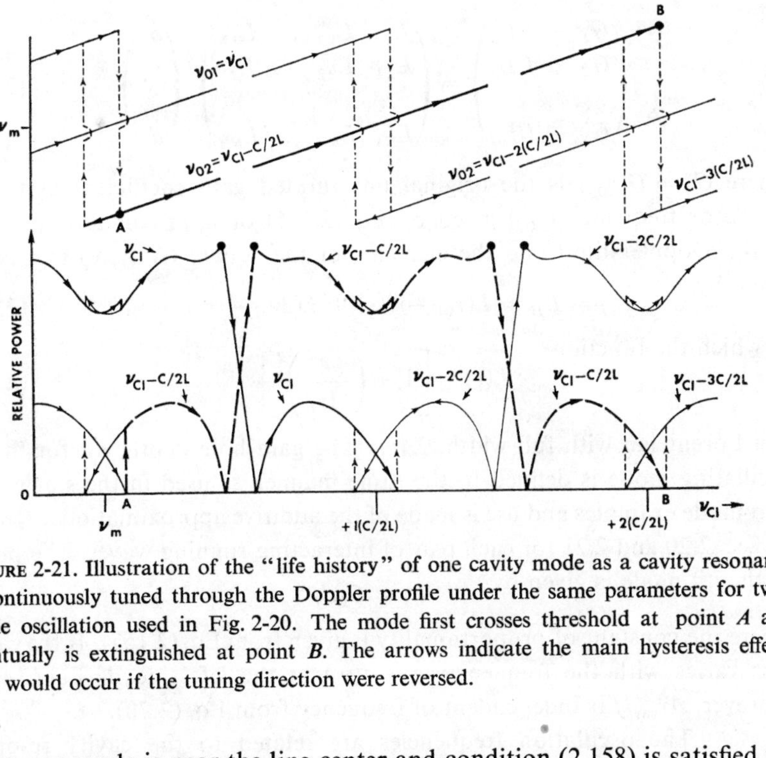

FIGURE 2-21. Illustration of the "life history" of one cavity mode as a cavity resonance is continuously tuned through the Doppler profile under the same parameters for two-mode oscillation used in Fig. 2-20. The mode first crosses threshold at point A and eventually is extinguished at point B. The arrows indicate the main hysteresis effects that would occur if the tuning direction were reversed.

when one mode is near the line center and condition (2.158) is satisfied far above threshold. However, to see that type of hysteresis effect in practice would require some additional mode suppression technique to keep more than two modes from going into oscillation. Hence as a practical matter it appears unlikely that the analogue of the single-mode hysteresis effect illustrated in Fig. 2-16 would ever occur in a normal laser cavity in multi-mode oscillation.

2.25. N-Modes with the Same Polarization

One advantage of the hole-burning model is that the same approximations used to describe the single- and two-mode case may be readily extended to handle N-modes of the same polarization in the same isotope. This extension is only useful when based entirely on reiterative solution by computer analysis. The latter is facilitated by setting up the equations in matrix form, in which case the transition from N to M modes can be accomplished merely by redimensioning the matrices in the middle of the program.

The gain saturation equations analogous to Eqs. (2.166) in the two-mode case may be written for N modes,

$$\begin{pmatrix} G_1 - f/L_G \\ G_2 - f/L_G \\ \cdot\ \cdot\ \cdot\ \cdot\ \cdot \\ G_N - f/L_G \end{pmatrix} = \begin{pmatrix} L_{11} & L_{12} & \dots & L_{1N} \\ L_{21} & L_{22} & \dots & L_{2N} \\ \cdot & \cdot & \cdot & \cdot \\ L_{N1} & L_{N2} & \dots & L_{NN} \end{pmatrix} \begin{pmatrix} d_1 \\ d_2 \\ \cdot\ \cdot \\ d_N \end{pmatrix} \tag{2.173}$$

where $G_j = G(\nu_{0j})$ is the original unsaturated gain coefficient at the jth oscillator frequency (ν_{0j}) given by Eq. (2.155) or approximation (2.155b) in the Doppler limit and the elements of the square ($N \times N$) matrix are given by

$$L_{ij} = L_{ji} = L(\nu_{0i} - \nu_{0j}) + L(2\nu_m - \nu_{0i} - \nu_{0j}) \tag{2.174}$$

in which the function

$$L(\nu) = \left[1 + \left(\frac{\nu}{\Delta\nu_L}\right)^2\right]^{-1} \tag{2.175}$$

is a Lorentzian with full width, $2\Delta\nu_L$. The gain hole depth, d_j, for the jth oscillating mode is defined in the same manner as used in the single- and two-mode examples and use is made of the additive approximation discussed in Sec. 2.20 and 2.21 for each pair of interacting running waves. The power in the jth mode is given by

$$P_j \approx C_0 d_j$$

where the constant of proportionality is given from Eq. (2.157). If the cavity loss varies with the frequency, $f = f(\nu_j)$ in the left side of Eq. (2.173); however, $\Delta\nu_{\text{cav}}/f$ is independent of frequency from Eq. (2.70).

The oscillation frequencies are related to the cavity resonant frequencies and total phase shift coefficients introduced by the medium through the extension of Eq. (2.148),

$$\begin{pmatrix} \nu_{01} \\ \nu_{02} \\ \cdot\ \cdot \\ \nu_{0N} \end{pmatrix}_{\text{osc}} = \begin{pmatrix} \nu_{c1} \\ \nu_{c2} \\ \cdot\ \cdot \\ \nu_{cN} \end{pmatrix}_{\text{cav}} - \left(\frac{\Delta\nu_{\text{cav}}}{f/L_G}\right) \begin{pmatrix} \Delta\phi_1 \\ \Delta\phi_2 \\ \cdot\ \cdot \\ \Delta\phi_N \end{pmatrix}_{\text{medium}}, \tag{2.176}$$

where the total phase shift coefficients may be written in terms of contributions from the original unsaturated gaussian gain curve and terms from holes,

$$\begin{pmatrix} \Delta\phi_1 \\ \Delta\phi_2 \\ \cdots \\ \Delta\phi_N \end{pmatrix} = \begin{pmatrix} \Delta\phi_{g1} \\ \Delta\phi_{g2} \\ \cdots \\ \Delta\phi_{gN} \end{pmatrix}_{\text{gain}} + \begin{pmatrix} h_1 \\ h_2 \\ \cdot\cdot \\ h_N \end{pmatrix}_{\text{holes}}, \tag{2.177}$$

Here, $\Delta\phi_{gj} = \Delta\phi_g(\nu_{0j})$, is the phase shift coefficient introduced at the jth oscillator frequency by the original line and is given exactly by Eq. (2.155), or approximately by Eq. (2.155b). The phase shifts from the $2N$ holes are included in the column matrix (h) given by

$$\begin{pmatrix} h_1 \\ h_2 \\ \cdot\cdot \\ h_N \end{pmatrix} = \begin{pmatrix} H_{11} & H_{12} & \cdots & H_{1N} \\ H_{21} & H_{22} & \cdots & H_{2N} \\ \cdot & \cdot & \cdot & \cdot \\ H_{N1} & H_{N2} & \cdots & H_{NN} \end{pmatrix} \begin{pmatrix} d_1 \\ d_2 \\ \cdot\cdot \\ d_N \end{pmatrix} \tag{2.178}$$

The elements of the matrix (H) are determined from Eq. (2.149a) and are given by

$$H_{ij} = \left(\frac{\nu_{0j} - \nu_{0i}}{2\Delta\nu_L}\right) L(\nu_{0j} - \nu_{0i}) + \left(\frac{2\nu_m - \nu_{0j} - \nu_{0i}}{2\Delta\nu_L}\right) L(2\nu_m - \nu_{0j} - \nu_{0i}) \tag{2.179}$$

where the function $L(\nu)$ is given by Eq. (2.175). The term, $H_{ij}d_j$, corresponds to the phase shift produced at ν_{0i} due to the holes at ν_{0j} and $2\nu_m - \nu_{0j}$, of common depth d_j.

These equations may be solved reiteratively in a manner similar to that discussed before in the two-mode case. First, we pick the cavity resonance frequency, ν_{c1}, closest to the line center (ν_m), and then, based on some initial estimate of the number (N) of oscillating modes involved, order the remaining resonances in terms of their successive proximity to ν_m. This ordering can be done through the algorithm,

$$\nu_{cj+1} = \nu_{cj} + j(c/2L) \quad \text{if} \quad \nu_{cj} < \nu_m,$$

and (2.180)

$$\nu_{cj+1} = \nu_{cj} - j(c/2L) \quad \text{if} \quad \nu_{cj} > \nu_m.$$

That is, the successive modes will tend to appear on opposite sides of the (initially-symmetric) line.

Next we let $\nu_{0j} = \nu_{cj}$ for $j = 1$ to N and calculate the matrix elements $G_j - f/L_G$. If $G_M < f/L_G$, we redimension the matrices so $N = M$.

Otherwise we go on and calculate the matrix elements L_{ij} and invert the matrix (L). The latter is generally non-singular except near the symmetric placement condition. There, the oscillator itself is very unstable and it is best to avoid the singular region by retuning ν_{c1}. We next calculate the matrix for the hole depths, using the matrix equation,

$$(d) = (L)^{-1}\,(G - f/L_G) \tag{2.181}$$

and examine all the hole depths to make sure they are positive. If $d_M < 0$, we then redimension $N = M$ and start over. The first time we solve Eq. (2.181), if no hole depths are negative, the gain at ν_{cN+1} should be checked in the presence of the holes to see if the matrices should be redimensioned for $N + 1$ modes. [This is probably easiest to do by looking for negative hole depths in Eq. (2.181) for N + 1 modes.]

After solving Eq. (2.181) and determining the correct choice for N, we substitute the value of H_{ij} from Eq. (2.179) in (2.178) and determine the column matrix (h). We next solve Eq. (2.177) for the total phase shifts and determine the corrected oscillator frequencies from Eq. (2.176). The latter are substituted in Eq. (2.173), we invert the matrix again and solve Eq. (2.181) for the hole depths, going through the same check list on the hole depths. This process is then continued reiteratively until satisfactory convergence on the mode parameters is obtained.

A similar set of matrix equations to those described above could also be written using the more accurate expression for the gain holes given in Eq. (2.136). As noted in the discussion of Eq. (2.136), the Lorentzian approximation for the gain holes neglects symmetric terms of order $(\Delta\nu_L/\Delta\nu_D)^2$ and asymmetric terms of order $(\Delta\nu_L/\Delta\nu_D)^3$ in the numerator of Eq. (2.136). However, as discussed in connection with Eq. (2.136), the single running wave hole depths at resonance have exactly the same dependence on the power gain coefficient as used in Eq. (2.140) in the Doppler limit. Hence all we have to do to modify the right hand side of Eq. (2.173) in the case that the Lorentz width is comparable to the Doppler width is to replace the Lorentzian function in Eq. (2.175) by a gain hole function proportional to the expression in Eq. (2.136) and which is normalized to unity at the resonance.

The problem gets significantly more complicated in evaluating the phase shift coefficients from the various gain hole terms which determine the matrix elements, H_{ij}, in Eq. (2.178). Equation (2.179), based on the assumption of Lorentzian gain holes, is no longer valid for $\Delta\nu_L \approx \Delta\nu_D$. Because the gain hole has an asymmetric term, there will now be a residual phase shift at the center of the hole due to itself. Although it may be possible to express the phase shifts for the non-Lorentzian gain hole function of Eq. (2.136) in terms of some other tabulated functions (perhaps even $w(z)$ itself), it seems more useful at this point to have a subroutine run off the phase shifts for each gain hole function through numerical integration of the Kramers–Kronig relations [Eq. (2.94)]. The physical interpretation of the two terms that would replace those originally in Eq. (2.179) remains unchanged. That is, the replacement term for H_{ij} would still correspond to the phase shift contributions at ν_{0i} due to the gain holes at ν_{0j} and $2\nu_m - \nu_{0j}$, produced by the pair of running waves in the jth oscillating mode. However, the first of these terms would no longer be zero for $j = i$ and these terms would be a function of $\Delta\nu_L/\Delta\nu_D$. Eventually, of course, the averaging effect of the velocity distribution which justifies the additive approximation (see Fig. 2-15) in the first place will fail. Hence a modification to include non-Lorentzian hole widths does not appear to be of practical value. Also the effects of power broadening have been neglected and are intrinsically more important.

2.26. Comparison with the Lamb Theory

Because the present method of calculation based on the hole burning model (Bennett, 1962a; 1962b; 1963) represents a totally different and independent method of approach to the problem, it is of considerable interest to compare the results obtained with those of the third-order Lamb theory (Lamb, 1964). Obviously it is only meaningful to compare the present results with the steady-state solutions to Lamb's equations. As previously noted (Bennett, 1963), the single-mode solutions for the power and frequency characteristics are exactly the same in the two methods of analysis, hence a comparison of the multi-mode results is of primary concern here.

The steady-state equations for Lamb's theory in the case of three modes take the form

$$\begin{aligned} \alpha_1 &= \beta_1 E_1^2 + \theta_{12} E_2^2 + \theta_{13} E_3^2 + \text{terms}\,(\eta, \xi, \psi, E_1^{-1} E_2^2 E_3) \\ \alpha_2 &= \theta_{21} E_1^2 + \beta_2 E_2^2 + \theta_{23} E_3^2 + \text{terms}\,(\eta, \xi, \psi, E_1 E_2 E_3) \\ \alpha_3 &= \theta_{31} E_1^2 + \theta_{32} E_2^2 + \beta_3 E_3^2 + \text{terms}\,(\eta, \xi, \psi, E_1 E_2^2 E_3^{-1}) \end{aligned} \tag{2.182}$$

and

$$\begin{aligned} \nu_1 &= \Omega_1 + \sigma_1 + \varrho_1 E_1^2 + \tau_{12} E_2^2 + \tau_{13} E_3^2 + \text{terms}\,(\eta, \xi, \psi, E_1^{-1} E_2^2 E_3) \\ \nu_2 &= \Omega_2 + \sigma_2 + \tau_{21} E_1^2 + \varrho_2 E_2^2 + \tau_{23} E_3^2 + \text{terms}\,(\eta, \xi, \psi, E_1 E_3) \\ \nu_3 &= \Omega_3 + \sigma_3 + \tau_{31} E_1^2 + \tau_{32} E_2^2 + \varrho_3 E_3^2 + \text{terms}\,(\eta, \xi, \psi, E_1 E_2^2 E_3^{-1}). \end{aligned} \tag{2.183}$$

The final terms involving η, ξ, ψ on the right hand side of each equation have not been stated in detail because they are totally neglected from the beginning in the hole burning model. These terms contribute to pulsations and combination tones in the Lamb theory and can be important under very special conditions of cavity tuning and excitation. For example, they lead to the frequency-locking effect observed in the first helium-neon laser in three mode operation when the beat frequencies between adjacent modes were tuned within $\approx$ 1 kHz of each other. These terms all depend strongly on spatial Fourier components of the excitation distribution along the laser axis. With certain distributions of the excitation, the locking effect can apparently vanish altogether. To the author's knowledge, it has never been observed to occur in a laser constrained to oscillate in one polarization. In any case, the comparison with the hole-burning model should be made without the terms involving η, ξ, ψ because those terms average out to zero except in very special cases. [An experimental study of pulsation effects in a helium-neon laser has been given by Fork, Hargrove and Pollack (1964).]

The remaining terms in Eqs. (2.182) and (2.183) all have a specific analogue in the hole burning equations discussed in Sec. 2.25. Equation (2.182) is identical in form to Eq. (2.173). The coefficients α_j correspond to the elements $G_j - f/L_G$ of the column matrix on the left of Eq. (2.173). The coefficients β_j correspond to the diagonal elements L_{jj} of the square matrix on the right side of Eq. (2.173), and the coefficients θ_{ij} correspond to the off-diagonal elements L_{ij} of this same matrix. Finally the hole depths, d_j in the column matrix on the right side of Eq. (2.173) are each proportional to E_j^2 through Eq. (2.136b).

Equations (2.183) of the Lamb theory (without the "pulsation" terms) are of the same form as Eqs. (2.176) through (2.178) of the hole burning model. The terms Ω_j in Eq. (2.183) equal $2\pi\nu_{cj}$ in Eq. (2.176); the terms σ_j in Eq. (2.183) equal c times $\Delta\phi_{gj}$ in Eq. (2.177); the terms ϱ_j in Eq. (2.183) correspond to the diagonal elements H_{jj} of the square matrix on the right side of Eq. (2.178) and the terms τ_{ij} in Eq. (2.183) correspond to the off-diagonal matrix elements H_{ij} in Eq. (2.178).

By expressing the steady-state Eqs. (2.182) in terms of gain coefficients, common factors of the type $\nu\mathscr{P}^2/(\varepsilon_0\hbar Ku)$ cancel out of Lamb's equations. The "relative excitation" $(\bar{N}/\bar{N}_T)$ in Lamb's theory is also more conveniently expressed for experimental purposes in terms of the quantity, $G_m L_G/f$, used in the present work which represents the single-pass fractional energy gain $(G_m L_G)$ at the maximum of the unsaturated line divided by the average fractional energy loss (f) per pass. The latter quantities can usually be measured directly, whereas the average excitation (or effective inversion density) and cavity Q are harder to determine directly.

The third-order theory involves the neglect of the field-dependent term in the Lorentz width [Eq. (2.124)] and in the denominator of Eq. (2.130), in which R_1, R_2 correspond to γ_a, γ_b and $\langle\mu_0\rangle^2$ corresponds to $\mathscr{P}^2$ in the Lamb theory. Relating the effective inversion density to the gain coefficient involves expressing the quantity Ku of the Lamb theory in terms of the Doppler width and the use of Eqs. (2.53) or (2.60).

Pursuing this comparison in more detail in the three-mode case and allowing for the differences in notation, it is seen that the terms α_j, β_j, ν_j, Ω_j, σ_j, and ϱ_j are in precise agreement between the two methods of calculation (based on the above identification), as are the dominant frequency-dependent terms in θ_{ij} and τ_{ij}. The dominant terms in these last two quantities correspond to the second terms in Eqs. (2.174) and (2.179), respectively. The first terms in Eqs. (2.174) and (2.179) go over to their counter parts in the Lamb theory $[4\gamma_{ba}^2 \Delta^{-2} = \Delta\nu_L^2/(c/2L)^2]$ in the limit that $\Delta\nu_L \ll c/2L$. The latter difference is certainly not a significant one and probably arises merely through expansion of the terms to order $1/\Delta^2$ in the Lamb theory. [In making the comparison it is, of course, necessary to note the similarity between the function $w(z)$ tabulated by Faddeyeva and Terent'ev (1961) and the plasma dispersion function tabulated by Fried and Conte (1961). An explicit relationship between the output power and the gain coefficient hole depths also involves the use of the constant of proportionality in Eq. (2.157).]

There are some real differences between the terms θ_{ij}, τ_{ij} in the Lamb theory and their counterparts in the hole burning equations which should be noted. These consist entirely of terms arising from spatial Fourier projections along the laser axis of the effective inversion density. These terms are of order $\left(\dfrac{N_\mu \gamma_a \gamma_b}{\bar{N}\Delta^2}\right)$ in Lamb's notation, in respect to the dominant Lorentzian terms, of relative order unity. These terms are generally expected to be small, but are usually of unknown magnitude. Lamb notes that these terms could be enhanced by very non-uniform axial distributions of the inversion density. Such terms were neglected in the hole-burning model from the start on the assumption that they would average out to zero. Although their effect is identically zero in the single-mode case, they clearly introduce a small, but real difference in the multi-mode equations.

Fork and Pollack (1965) performed numerical solutions of Lamb's two-mode equations to illustrate the importance of the mode competition effects which arise from the θ, τ coefficients. Their results were quite similar to the numerical solutions to the two-mode problem based on the hole burning model shown in Figs. 2-20 and 2-21. Although Fork and Pollack discuss the problem qualitatively in terms of hole burning effects, they apparently were unaware that analagous terms are present in the hole-burning model.

2.27. Approximate Semi-Empirical Form for Power Broadening

A mutual inadequacy of the third-order Lamb theory *and* of the hole-burning equations stated in Sec. 2.25 is that they do not include the effects of power broadening. Even with the initial helium-neon laser it was possible to make the power-dependent term greater than unity in expression (2.124) for the Lorentz width. The effects of this term don't appear until 5th order in the Lamb theory (see, for example, Vehara and Shimoda, 1965) and one can expect that the 7th-, 9th-, etc., order terms aren't far behind for values of the saturation parameter close to unity. The complexity of the 3rd-order terms is sufficiently formidable to discourage most mortals from undertaking a thorough study of the higher-order terms in the multi-mode Lamb theory. At some level in the development, it would seem just as enlightening to integrate the equations of motion for the density matrix numerically from the start for each set of laser parameters [e.g., see Holt, 1970].

Modifying the hole-burning model to allow for power broadening is not a trivial matter either. If one is willing to tolerate errors of the magnitude implied by Fig. 2-15, the additive approximation could also be employed in the region where significant power broadening of the Lorentz widths occurs.

In order to see how to modify the hole-burning equations for the effects of power broadening, it is necessary to return to the exact formula in Eq. (2.135) for the shape of the gain hole burned in the Doppler line by a single running wave. One could in principle modify the equations in Sec. 2.25 for gain holes which are computed in each instance from the exact expression in Eq. (2.135). However, that process becomes fairly unpleasant even for computer analysis of the problem; for example, the hole depths are related to the power gain coefficient by the rather messy transcendental expression in Eq. (2.135b). It seems more appropriate to consider a semi-empirical representation of these expressions instead.

If we expand the terms in Eq. (2.135) for $\Delta\nu_L < \Delta\nu_D$ (or $y_1 < 1$), the resulting expression simplifies considerably in two limiting cases: a) $y_1 = y_2$. This corresponds to the limit of no power broadening and the expressions obtained from it were used in Sec. 2.25. b) $y_1 \gg y_2$. This corresponds to large power broadening.

Neglecting terms of order $y_1{}^2$ and expanding the remaining terms in Eq. (2.135) in the limit $y_1 \gg y_2$ yields the result for the large power broadening case that the gain hole is given by

$$\Delta G(\nu) \approx d'\left[1 + \left(\frac{\nu - \nu_0}{\Delta\nu_L/2}\right)^2\right]^{-1} \tag{2.184}$$

where

$$d' \approx \frac{2\pi\langle\mu_0\rangle^2 D_m U(x_1, 0)}{\hbar\lambda} \approx \frac{8\pi\langle\mu_0\rangle^2 P_{\text{total}}(\nu_0)}{\hbar^2 c R_1 R_2}\left(\frac{\Delta\nu_{LO}}{\Delta\nu_L}\right) \tag{2.185}$$

from Eq. (2.132).

The total intensity flux in the laser cavity is now

$$I = 2P_{\text{total}}(\nu_0)\, L_G/f = \frac{L_G\hbar^2 c R_1 R_2}{4\pi f\langle\mu_0\rangle^2}\left(\frac{\Delta\nu_L}{\Delta\nu_{LO}}\right)$$

Hence the power-broadened Lorentz width is given from Eqs. (2.124), (2.39) by

$$\Delta\nu_L = \Delta\nu_{LO}\sqrt{1 + K_1(\Delta\nu_L/\Delta\nu_{LO})\, d'}$$

where $K_1 \approx 2L_G/f$ and L_G equals the length of the gain tube.

Hence

$$\Delta\nu_L \approx \Delta\nu_{LO}\left[\sqrt{1 + K_1 d'/4} + K_1 d'/2\right] \tag{2.186}$$

in the present limit. [The above results may also be obtained by taking $D(\nu') \approx D(\nu_0)$ out of the integral in Eq. (2.134) and by noting that the sharper Lorentzian acts like a δ-function].

Comparing these equations with Eqs. (2.139) and (2.140) holding in the absence of power broadening, it is seen that two separate factors of ($\frac{1}{2}$) enter in the transition from no power broadening to large power broadening [one in the Lorentz width in Eq. (2.184) and one in Eq. (2.185) relating the power gain coefficient to the gain hole depth]. The intermediate behavior can be determined through numerical evaluation of Eq. (2.135) and is at least roughly monotonic in its dependence on the gain hole depth. Consequently there is justification for defining a semi-empirical factor, Γ, which varies monotonically from 1 to (1/2) as the hole depth (and hence power broadening effect) increases from zero to large values. One function that satisfies the desired requirement is (for the jth hole depth)

$$\Gamma_j \equiv \frac{1 + e^{-K_2 d_j}}{2} \tag{2.187}$$

where K_2 is presumably of about the same magnitude as K_1 ($\approx 2L_G/f$). Hence, if we assume the jth gain hole shape is given by

$$\Delta G_j(\nu) = d_j\left[1 + \left(\frac{\nu - \nu_{0j}}{\Delta\nu_{\mathrm{Leff}_j}}\right)^2\right]^{-1} \tag{2.188}$$

where

$$\Delta\nu_{\mathrm{Leff}_j} = \Gamma_j \Delta\nu_L \tag{2.189}$$

using the expression for $\Delta\nu_L$ in Eq. (2.186) with $d' = d_j$, we get a continuous transition between the two limiting cases discussed above in which the basic equations determining the gain hole depths and oscillation frequencies are of exactly the same form as those discussed in the multi-mode case in Sec. 2.23. The only differences are that the effective values of the Lorentz width given by Eq. (2.189) are used to replace Eqs. (2.174) and (2.175) with

$$L_{ij} = L_j(\nu_{0i} - \nu_{0j}) + L_j(2\nu_m - \nu_{0i} - \nu_{0j}) \tag{2.190}$$

where the function is now

$$L_j(\nu) = \left[1 + \left(\frac{\nu}{\Delta\nu_{\mathrm{Leff}_j}}\right)^2\right]^{-1} \tag{2.191}$$

and the matrix elements in Eq. (2.179) are replaced by

$$H_{ij} = \left(\frac{\nu_{0j} - \nu_{0i}}{2\Delta\nu_{\mathrm{Leff}_j}}\right) L_j(\nu_{0j} - \nu_{0i}) + \left(\frac{2\nu_m - \nu_{0j} - \nu_{0i}}{2\Delta\nu_{\mathrm{Leff}_j}}\right) L_j(2\nu_m - \nu_{0j} - \nu_{0i}). \tag{2.192}$$

Also, the effects of exponential growth on the gain saturation condition must be included because the gain coefficients may now be large; the latter modification may be accomplished simplify by replacing "f" on the left side of Eq. (2.173) by the quantity, "$-\log_e(1 - f)$". The power in the jth mode is determined from the hole depths through the relation,

$$P_{0j} = C_0 \Gamma_j \left(\frac{\Delta\nu_L}{\Delta\nu_{L0}}\right) d_j \tag{2.193}$$

where C_0 is given in Eq. (2.157). Actually solving Eqs. (2.173) with the above substitutions in the N-mode case is a much more formidable problem than it was without the power broadening terms present because the modified version of Eq. (2.173) is transcendental in the N hole depths. In principle this transcendental equation could be solved reiteratively for each set of frequencies by starting with an estimate of the hole depths based on the non-power-broadened form.

2.28. Collision Broadening

As discussed in connection with Eq. (2.20) and (2.40) both radiative *and* collision processes contribute to the total phase interruption rates for the upper and lower states of the laser transition.

Total lifetimes may be measured through multi-channel delayed coincidence analysis of the transient decay of states which are sharply- and selectively-excited at some instant in time. For example, using short pulses of threshold energy electrons in an apparatus illustrated schematically in Fig. 2-22, Bennett and Kindlmann (1966) were able to determine total lifetimes in excess of about 3 nanoseconds within limiting errors of about one percent. In many instances where the excited state involved can radiate in the vacuum ultra-violet on a transition to the atom ground state, the total radiative decay rate is much shorter than $(3\text{ nsec})^{-1}$ and immeasurable

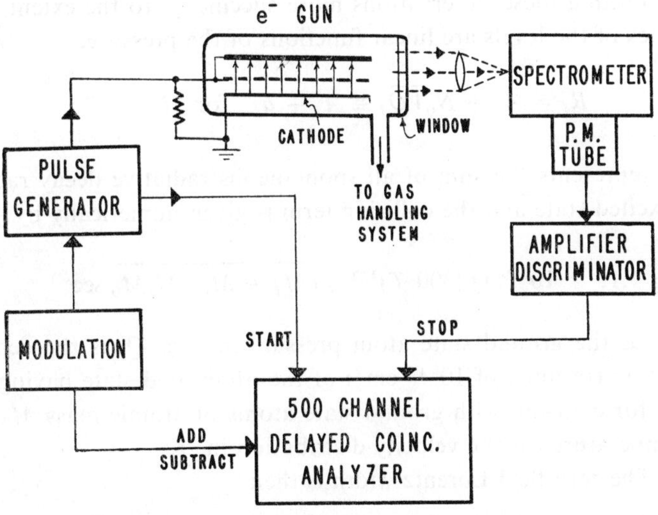

FIGURE 2-22. Schematic diagram of the method used by Bennett and Kindlmann (1966) to measure excited state lifetimes. The pulse amplitude applied to the electron gun is modulated at a slow rate about the threshold value for excitation to eliminate errors from radiative cascade.

in practice by the electron impact method. [The basic limit in response time in these experiments is generally due to the drift time for the threshold energy electrons to go through the excitation region.] It seems probable that the use of mode-locked laser pulses as a selective and extremely-short source of initial excitation may eventually provide a general method of measuring subnanosecond lifetimes with comparable precision (see, for example, Arrathoon and Sealer, 1971). However, in most experiments to date the delayed-coincidence method has only been able to yield precise values of the radiative lifetimes on one (the longest-lived) of the two states involved in the more important gas laser transitions in the near-visible spectrum. (For the rapidly-decaying states, the "resonance-trapped" lifetimes, which determine the actual excited state densities, are easily obtained in practice). The delayed-coincidence experiments also provide values for the total inelastic collision destruction cross sections. Except in special instances involving very closely-spaced energy levels, these cross sections are seldom in excess of $\approx 10^{-15}$ cm^2 (see the review by Bennett, Kindlmann and Mercer, 1965). Consequently the total contribution of inelastic collisions to the Lorentz width for a given transition is typically in the order of a Mhz/torr.

Putting these observations more succinctly, to the extent that the decay rates of the levels are linear functions of the pressure,

$$R_j = A_j + N_0 \overline{VQ}_j \equiv A_j + a_j \quad \text{for} \quad j = 1, 2. \qquad (2.194)$$

Here A_j represents the sum of all spontaneous radiative decay rates from the jth excited state and the collision term is given numerically by

$$a_j = 1.15 \times 10^7 P_0 Q_j (300/T)^{1/2} \sqrt{(M_j + M_0)/M_j M_0} \ \text{sec}^{-1} \qquad (2.195)$$

where P_0 is the ground state atom pressure in torr, Q_j is the destructive cross section (in units of 10^{-15} cm^2) of the jth excited state having atomic mass M_j for collision with ground state atoms of atomic mass M_0, and T is the temperature for the velocity distribution in °K.

The zero-field Lorentz width is then

$$\Delta\nu_{L0} = \left(\frac{A_1 + A_2}{2\pi}\right) + \left(\frac{a_1 + a_2}{2\pi}\right) \equiv \Delta\nu_{\text{nat}} + BP_0 \qquad (2.196)$$

where the first term on the right is the "natural width". The term $B = 1.2$ MHz/torr at 300 °K for $M_j = M_0 = 20$ when both upper and lower states have destructive cross sections of 10^{-15} cm^2. Measurements of line broadening on laser transitions in the near visible range have exhibited values of the coefficient "B" in Eq. (2.196) which are *typically* ≥ 50 MHz-torr^{-1}, hence corresponding to individual excited state cross sections

$\gtrsim 5 \times 10^{-14}\,\text{cm}^2$ thereby ruling out the involvement of inelastic atom-atom collisions to any significant extent in most cases.

It seems clear from various points of view that elastic collisions in which the atoms' resonance frequency is Doppler-shifted by an amount much larger than the "natural width" will constitute excited state destruction in essentially the same sense as the decay processes introduced phenomenologically in Eq. (2.20). "Large-angle" elastic scattering in which the resonance is shifted by amounts comparable to $\Delta\nu_D$ (the full width of the gaussian line) are obviously sources of excited state destruction in the above sense. However, average "large-angle" elastic scattering cross sections (as determined, for example, from excited metastable state diffusion experiments) are also only in the order of 2 or $3 \times 10^{-15}\,\text{cm}^2$. Hence, "large angle" elastic scattering has about the same relative unimportance as inelastic scattering in explaining the general magnitude of the observed pressure broadening coefficients in Eq. (2.196).

From a semi-classical point of view, one might expect that small-angle elastic collisions would play a more important role in laser line broadening experiments due to the long range interaction between pairs of atoms (i.e. Van der Waals forces). For example, if one assumes an interaction potential of the form

$$W(r) = W_0/r^n \tag{2.197}$$

one can do a classical calculation of the cross section for an atom to be scattered through an angle such that the atom's resonance frequency is doppler-shifted by an amount equal to the natural width for the transition when viewed along the laser axis. Such a calculation does indeed yield line-broadening cross sections $\approx 10^{-14}\,\text{cm}^2$ for dipole–dipole interactions ($n = 6$) on near-visible transitions.

From a quantum mechanical point of view, the small angle scattering problem also leads to broadening in the stimulated emission case. However, the interpretation is rather different. As noted by Berman and Lamb (1970), it is important to remember that the atom is in a mixed state in the presence of the laser field and that the separate upper and lower laser states will generally have quite different interaction coefficients with a neighboring atom. The separate scattering trajectories for atoms in the upper and lower state would thus be quite different and a wave packet describing the atom in a mixed state would break up during the collision if one of the scattering trajectories were adequate to produce a doppler shift comparable to the natural width. In this case the mixed state would be destroyed during the collision and, although broadening would occur, the mechanism isn't quite that pictured in the semi-classical model above. This argument also appears to rule out the concept of "soft" collisions in which the phase of the mixed state is preserved in small angle scattering through doppler shifts comparable to the natural width.

From either the classical or quantum mechanical point of view, some evidence of the Dicke (1953) line narrowing effect would presumably be observable in spontaneous emission experiments on the laser transition if small-angle scattering were the only source of pressure broadening in the laser tuning dip experiments. Although the line broadening experiments are very difficult and prone to many sources of systematic error, both the single-mode tuning dip experiments and the spontaneous emission experiments now appear to be in fairly good quantitative agreement on the neutral atom laser transitions that have been studied so far (primarily, the helium-neon system). Consequently, the main source of line broadening in these experiments does not appear to be due directly to elastic scattering, but rather to phase interruption of the transition due to a change in the internal electronic energy levels produced by the atom-atom interaction in grazing collisions.

The magnitude of the line broadening effect can be estimated in the manner of Weisskopf (1932) by determining an impact parameter, b, which characterizes a grazing collision in which a phase change of ≈ 1 radian occurs as a result of the interaction between atoms. If the relative velocity is in the x direction

$$r^2 = x^2 + b^2 \tag{2.198}$$

where r is the distance to the perturbing atom, the total phase change in one collision for an atom in one level is

$$\delta\phi \approx \frac{1}{V\hbar} \int_{-\infty}^{+\infty} \frac{W_0\, dx}{r^n} \tag{2.199}$$

where we have assumed an interaction of type (2.197) and V is the constant relative velocity. Substituting Eq. (2.198) and demanding that the phase shift be ≈ 1 radian during the collision yields an equation determining the impact parameter,

$$\frac{W_0}{V\hbar b^{n-1}} \int_{-\infty}^{\infty} \frac{d\xi}{(1 + \xi^2)^{n/2}} \equiv 1. \tag{2.200}$$

The integral can be expressed in terms of gamma functions (whose numerical values are ≈ 1). Hence the cross section for broadening is roughly

$$Q_j \approx \pi b^2 \approx \pi \left[\frac{4W_0\sqrt{\pi}}{V\hbar} \frac{\Gamma\left(\frac{n+1}{2}\right)}{(n-1)\Gamma\left(\frac{n}{2}\right)} \right]^{2/(n-1)} \tag{2.201}$$

The general magnitude indicated by Eq. (2.201) is adequate to explain the observed broadening if realistic values of the interaction coefficient between excited states and the perturber are assumed. Some other quantitative features of Eq. (2.201) are at least in approximate agreement with experiment. For example, the dependence of the cross section on relative velocity in Eq. (2.201) implies a specific temperature dependence of the average collision rate per atom:
for resonant interactions between identical atoms ($n = 3$),

$$VQ = \text{constant}; \tag{2.202}$$

for the normal Van der Waals, or dipole-dipole interaction between different atoms ($n = 6$),

$$VQ \propto T^{0.3} \tag{2.203}$$

Although a significant role or resonant interactions on the broadening on laser transitions has not been established, both stimulated emission and spontaneous emission experiments have been reported in the helium-neon system which are compatible with the temperature dependence in Eq. (2.203).

Although not specifically considered by Weisskopf, the atomic interaction should also lead both to small frequency shifts and to line asymmetries. The absolute frequency shifts occur simply because the net effect of the atomic interaction is generally different for the two levels involved in the transition. For example, if the interaction is attractive and largest for the upper state of the transition, one expects a red shift. Slight line asymmetries should also arise from the pressure shift in the case where the collision rate (VQ) is speed-dependent. In that case, the collision rate is greatest in the wings of the line and least in the center. Since the frequency shift is in the same direction, one should see an asymmetry in the Doppler profile which is proportionate to the shift. Extensions of the spontaneous emission line broadening theory to include both shifts and asymmetries have been given by Lindholm (1945), Foley (1946), and others.

Although pressure shifts $\approx$ 10 to 20 MHz per torr have clearly been observed in laser frequency stabilization experiments, the experimental evidence for line asymmetries is much more questionable. Here one should note that it is the relative velocity between the radiating atom and the perturber which determines the collision rate, whereas it is the velocity of the radiating atom in the laboratory reference frame which determines the Doppler profile. Thus, in a case such as the helium-neon system where a heavy radiator is colliding with a light perturber, very small asymmetries over the Doppler profile are to be expected. One presumably could enhance the asymmetry by observing a light radiating atom colliding with a heavy perturber. However, the available data has mainly been taken in the helium-neon system under conditions in which either sources of systematic error would be expected to produce slight asymmetries (as in most laser tuning dip experiments), or where the inherent sensitivity is not adequate to detect the asymmetry (as in most spontaneous emission profile experiments).

Most tuning dip and spontaneous emission profile experiments have been analyzed by merely adding a pressure-dependent collision term to the Lorentz width (or to γ_{ab} in the Lamb theory). The agreement with experiment in recent work of this type has been remarkably good and it is not clear that the results are sufficiently free from systematic error to warrant analysis with the more cumbersome expressions which result when speed-dependent broadening is included.

The first pressure-dependent studies of the tuning dip in single-mode oscillation were made on the 1.15 micron transition of neon by Szöke and Javan (1963, 1966). Szöke and Javan encountered significant line asymmetries in their work and found it advantageous to analyze their data using two pressure-dependent linewidth parameters based on the concept of "hard" and "soft" collisions. Cordover and Bonczyk (1969) reported similar measurements on the 6328 Å neon transition using the same method of analysis. Fork and Pollack (1965) reported studies of the 6328 Å neon line under conditions of two-mode excitation which indicated large pressure-dependent asymmetries which they included in a modification of the Lamb theory. However, any two (longitudinal)-mode experiment would be especially subject to apparent asymmetries arising from the excitation of higher-order transverse modes. Smith (1966) studied the 6328 Å line in single-mode while monitoring the beat spectrum for the presence of the first higher order transverse mode. Smith notes that a good fit of his single-mode data can be made to the original form of the third-order Lamb theory if only one pressure-dependent term is added to the Lorentz width. Smith also checked the saturation parameter directly for this transition to minimize the effects of power broadening on linewidth measurement. Data by Lisitzin and Chebotayev (1968; 1969) also could be fit with single pressure-dependent linewidth parameters, but additionally show small pressure-dependent shifts in the line center. Mikhnenko *et al.* (1970) have found both a linear pressure dependence for the Lorentz width in tuning dip experiments on the 0.63 micron helium-neon laser transition and a temperature dependence of the product VQ which varied as $T^{(0.33\pm.04)}$ and is compatible with a simple model based on Van der Waals interactions. Measurements of Lorentz widths have also been made in single-frequency running wave experiments by several authors (Shank and Schwarz, 1968; Hänsch and Toschek, 1969; Smith and Hänsch, 1971) in which the Lorentz width parameter can be fit to a constant plus a linear pressure-dependent term.

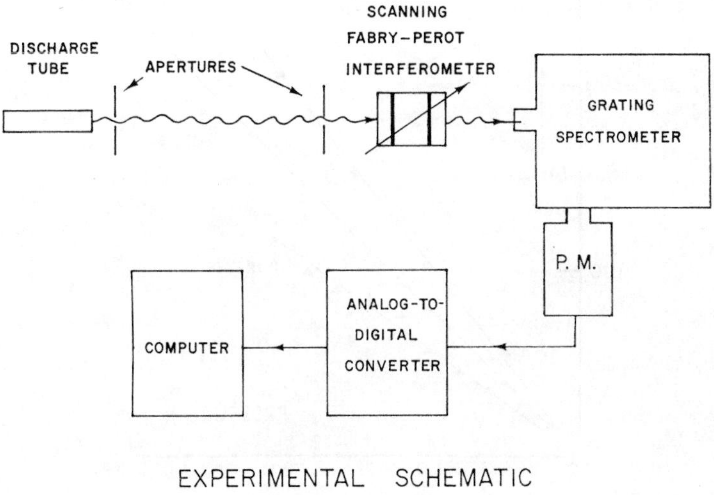

FIGURE 2-23. Schematic diagram of method used by Bennett, Knutson and Sze (1969) to analyze spontaneous emission profiles.

Evidence for "cross relaxation" effects due to the presence of neon has been reported by Beterov *et al.* (1970) and by Smith and Hänsch (1971). By "cross relaxation" is meant a process which redistributes the atoms in the velocity distribution by amounts comparable to the full Doppler width. Beterov *et al.* specifically consider the trapping of resonance radiation and Smith and Hänsch explain their data in terms of "hard" collisions (large-angle elastic scattering). It is also clear that resonant $1/r^3$ interactions between identical atoms could lead to the exchange of excitation in collisions between atoms which are widely-separated over the Doppler profile. The data of Smith and Hänsch appears to require an additive correction ≈ 20 percent of the usual Lorentzian term at ≈ 3 torr of Ne^{20}. They also report a much smaller effect in neon-helium collisions.

The initial measurements of line broadening on the 6328 Å line of neon from spontaneous emission analysis by Bennett *et al.* (1967b)

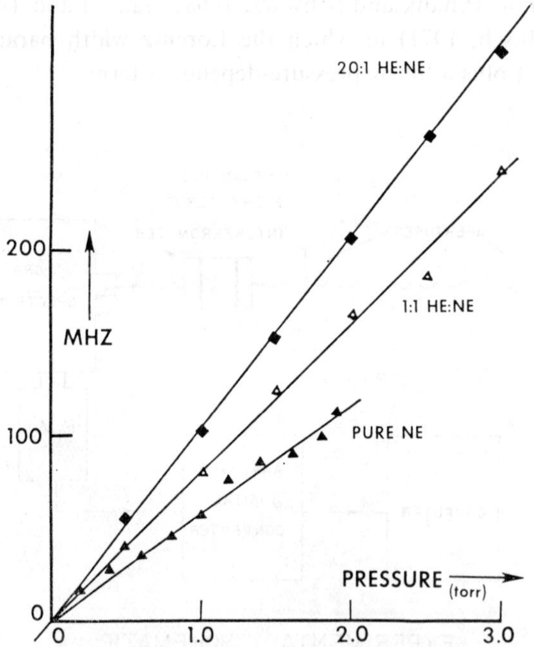

FIGURE 2-24. Pressure-broadened increase in the full Lorentz width of the 6328 A° neon laser transition determined by Knutson and Bennett (1971).

warrant additional comment. These data were based on scanning Fabry–Perot analysis using a graphical technique developed by Ballik (1966) to unfold the Voigt profile and separate the distortion due to the Fabry–Perot. As initially developed by Ballik, the method makes use of three data points across the line and works reasonably-well so long as the Lorentzian tails of neighboring transitions are well-isolated without significant distortion of the line under observation. We have since developed much more precise methods of spontaneous emission analysis in which four or five hundred points are taken across a little more than one free spectral range of the scanning Fabry–Perot and subjected to a least-squares fit to an exact representation of the Voigt profile as distorted by the Fabry–Perot (Bennett, Knutson and Sze, 1969; Knutson and Bennett, 1971; Sze and Bennett, 1972). The method is illustrated schematically in Fig. 2-23. Using this method to re-examine the 6328 Å transition we found that a large systematic error was involved in the initial determination of the neon-neon broadening term due to scattered light from a neighboring transition which is more strongly-excited in pure neon. (The 6328 Å line is extremely weak without enhancement by the $2\,{}^1S$ helium metastable.) Lorentz broadening data taken from spontaneous emission studies of the 6328 Å line are shown in Fig. 2-24 (Knutson and Bennett, 1971). As may be seen from Fig. 2-24, the broadening from neon-neon collisions is significantly smaller than that from neon-helium collisions and roughly in proportion to the mass dependence in Eq. (2.195). The product VQ varied as $T^{(0.33\pm.03)}$ in the case of neon-

Table 2–3.

Lorentz broadening parameters observed in spontaneous emission by Knutson and Bennett (1971; further work to be published). Full-width parameters and standard deviations (in parentheses) are given in MHz per 3.25×10^{16} atoms/cm^3 at 295°K.

Ne I Line (in Å units)	Ne^{20} - Ne^{20}	Ne^{20} - He^4
5434	81 (13)	130 (7)
5939	- - -	133 (8)
6046	106 (10)	129 (7)
6118	78 (12)	134 (7)
6328	94 (10)	126 (5)
6351	91 (14)	133 (7)

helium collisons. Studies of the line symmetry in spontaneous emission have also been made by Knutson and Bennett (1971) which exhibit no detectable asymmetries on any of the neon laser transitions observed within limits of sensitivity $\approx$0.1 percent of the maximum intensity for pressures $\approx$1 torr.

Summarizing the discussion of pressure broadening measurements on neutral-atom laser transitions in the near-visible spectrum, both the stimulated emission and spontaneous emission experiments seem largely-explained through the simple addition of a linear pressure-dependent term to the Lorentz width which is velocity independent. Some of the experiments indicate small pressure shifts of the line center, but the introduction of pressure-dependent asymmetries does not seem warranted by the experimental data. The existence of small cross-relaxation effects is also indicated by some experiments.

The practical advantages of the velocity-independent broadening approximation should not be lightly discarded. Any complete theory which incorporates significant distortions of the velocity distribution and includes line-asymmetry effects appears at this point to be horrendously-complicated. A representative feeling for the complexity involved can be obtained from the collision theories for laser oscillation developed by Rautian and Sobel'-man (1966), Rautian (1966), Gyorffy, Borenstein and Lamb (1968), Berman and Lamb (1970; 1971), and Borenstein and Lamb (1972).

As discussed by Rautian (1966), non-symmetric line shapes will in general result if the collision processes which modify the Doppler profile and produce Lorentz broadening are statistically correlated. If these collision mechanisms are uncorrelated, both mechanisms will tend to produce symmetric line shapes (even when the line center is shifted). At the moment, the experimental evidence with the neutral atomic gas lasers does not appear sufficiently free of systematic error to warrant abandoning the latter (much simpler) approximation.

The situation in the noble gas ion lasers is rather different and worthy of separate discussion. Here, direct measurements of the applied *dc* electric field in the direction of the laser axis imply significant amounts of acceleration within the radiative lifetime of the upper laser state (typically $\approx$ 10 nsecs in fields $\approx$ 5 volts/cm for the *cw* argon ion laser). Consequently, Doppler shifts comparable to the natural width can be anticipated in some instances due to the fact that the radiating ion is also accelerating. In that instance a non-Lorentzian response function of the individual ions is to be expected in the *absence* of collisions. The line shape can be estimated classically using the method of Bennett, Ballik and Mercer (1966).

Consider an excited ion formed with initial velocity, V_0, in the direction of the applied field and subject to a mean random phase interruption rate, R. This ion travels a distance $V_0 t + (eE/2m)\, t^2$ along the field in time t. The Doppler-shifted resonant frequency seen by a fixed observer looking at the receding ion, averaged over time t, will be $\omega_0[1 - (V_0/c) - (eE/2mc)\, t]$ where ω_0 is the resonant frequency of the stationary ion. It follows that the average time-dependent optical field seen by the observer is of the form

$$\langle E(t)\rangle = E_0 \exp\{i\omega_0[1 - (V_0/c) - (eE/2mc)\,t]\,t - Rt/2\}. \qquad (2.204)$$

The spectral intensity distribution, $I(\omega) = |E_\omega|^2$ is obtained from the Fourier transform of Eq. (2.204) and is given by

$$I(\omega) = (\pi/2\xi)\, I_0 |w(z)|^2 \qquad (2.205)$$

where the function $w(z)$ is tabulated by Faddeyeva and Terent'ev (1961) and is given in terms of the complex argument

$$z = \frac{(1 - \Omega) + i(1 + \Omega)}{2\sqrt{\xi}}, \qquad (2.206)$$

Here

$$\Omega \equiv 2(\omega - \omega_0')/R \quad \text{and} \quad \xi \equiv 4\omega_0 eE_{dc}/mcR^2. \qquad (2.207)$$

The constant Doppler shift (V_0/λ) has been absorbed in the resonant frequency (ω_0') and the full Lorentz width is $R/2\pi$. (By analogy with the transition from classical to quantum mechanics in the non-accelerated case, it is to be expected that the value of R to be used in Eq. (2.207) should correspond to the sum of the lower- and upper-state decay rates). The exact form of the line shape obtained from Eq. (2.205) is shown plotted in Fig. 2-25 for various values of the acceleration parameter, ξ. For $\xi = 0$, the expression reduces to original Lorentzian line shape and becomes increasingly-asymmetric with increasing values of ξ.

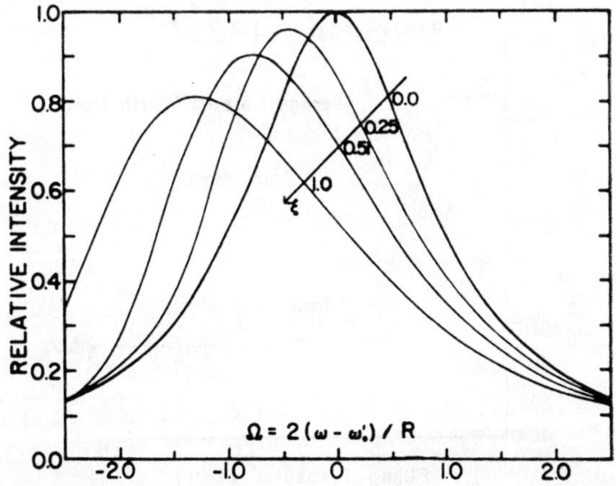

FIGURE 2-25. Line shape for a radiating, accelerating ion [Eq. (2.205)] (from Bennett, Ballik and Mercer, 1966).

Nevertheless, in most cases extremely symmetric spontaneous emission profiles are observed in the axial direction which fit the functional form of a Voigt profile (Gaussian distribution of Lorentzians) extremely well (see, Bennett *et al.*, 1966; Sze *et al.*, 1972). The reason for the excellent fit is to a large extent due to the importance of collision effects at pressures which are low enough to yield significant values of the acceleration parameter. The collisions of importance are, of course, not ion-neutral collisions. They are instead comprised of ion-ion and ion-electron collisions. The restoration of the Voigt profile is not as dramatically-dependent on collisions as initially concluded by Bennett, Ballik and Mercer (1966) in that the natural widths (typically $\approx$ 500 MHz) on the argon ion transitions are now known to be about five times larger than had been initially calculated (Statz *et al.*, 1965; 1968; also see, Sze, Antropov and Bennett, 1972).

An example of the dependence of observed Lorentz width on current and filling pressure on one argon ion laser transition is shown in Fig. 2-26 (Sze and Bennett, 1972). At high pressures, the Lorentz width extracted from the Voigt profile analysis closely approaches the natural width. At low filling pressures a strong and complex current dependence is encountered which partly arises from changes in the basic character of the discharge. Average Doppler widths (expressed as "Ion temperatures") for the same conditions are illustrated in Fig. 2-27.

Analysis of the broadening mechanisms in the ion laser experiments is hampered by lack of information regarding the average electron-ion density. Available data indicates that the ion density varies from $\approx 3 \times 10^{13}$

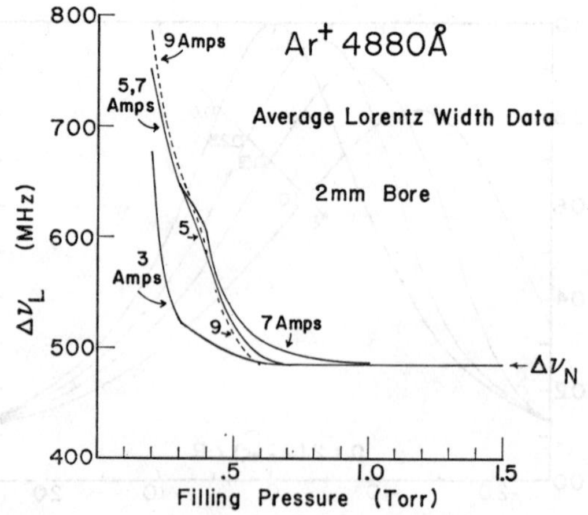

FIGURE 2-26. Lorentz width observed in the axial direction for the 4880 Å argon ion laser transition as a function of discharge parameters (from Sze and Bennett, 1972).

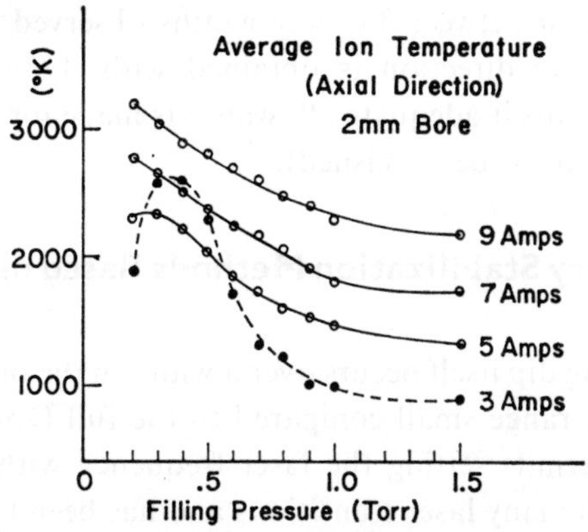

FIGURE 2-27. Average ion temperatures extracted from Doppler widths observed in the axial direction for the same argon discharge parameters involved in Fig. 2-26 (from Sze and Bennett, 1972).

to perhaps 3×10^{14} cm^{-3} for the data in Fig. 2-26 and the ion and electron temperatures involved are strongly-affected by the electron-ion interaction and the establishment of strong electric fields the in radial direction. Estimates of Lorentz broadening due to small-angle ion-ion scattering give values of the increased Lorentz width which are not as large as those observed. Further, from the $1/V^4$-dependence of the Rutherford scattering cross section, it would be expected that the total smallangle ion-ion scattering rate would vary as $n_{\mathrm{ion}}(T_{\mathrm{ion}})^{-3/2}$, a quantity which can be determined from the spectroscopic studies. Values of the latter product are actually observed to decrease slightly with increasing current, thereby ruling out a dominant involvement of small-angle ion-ion scattering in determining the collision broadened Lorentz widths in the argon ion laser transitions. However, the observed Lorentz widths vary with discharge parameters in about the same manner that would be expected from the Griem (1964) theory of line broadening. [The broadening in the Griem theory arises through electron impact by inducing transitions which are normally "forbidden", but which become allowed as a result of Stark mixing in the field of the ions.] Application of the Griem Stark broadening coefficients to the argon ion transitions yields agreement between the observed Lorentz broadening and independent measurements of the electron density using microwave techniques to within a factor of about two or three. (See discussion in Sze and Bennett, 1972). Details of the radial velocity distribution in this type of arc discharge are quite complex, but do not appear to affect the axially-observed profiles, except possibly by contributing to the Lorentz broadening. In this connection it is worth noting that reasonable agreement between Lorentz widths observed from spontaneous emission in the axial direction is obtained with those determined from tuning dip experiments if adequate allowance is made for power broadening (Wexler and Bennett, to be published).

2.29. Frequency Stabilization Methods Based on Hole–Burning Effects

The tuning dip itself occurs over a width in the order of the Lorentz width and hence a range small compared to the full Doppler width of the line. This effect permits setting the laser frequency within a few MHz of the line center with many laser transitions and has been the basis of a number of automatic stabilization schemes by Shimoda and Javan (1965) and others. In these systems the cavity length is periodically varied (or "dithered") and an error signal is generated from the second harmonic of the dither frequency in the laser output obtained from a phase-sensitive detector which uses the second harmonic of the dither signal as a reference. This type of scheme provides long-term stability which is typically good to several MHz and such systems are now commercially available for several laser transitions. However, a problem with this type of stabilization method arises in that the dither signal impresses a frequency modulation on the laser output due to the dither signal even at the stabilization point. Consequently, although the average frequency of the stabilized oscillator may be good to a few MHz, the instantaneous frequency typically undergoes a much larger excursion due to the dither signal itself. Further, since the laser output power goes through a quadratic minimum at the line center, the error signal becomes weakest precisely at the point where one is trying to stabilize the laser. Consequently, a rather larger dither is required to achieve stabilization.

It is obviously desirable to achieve stabilization using an error signal which goes linearly to zero at the stabilization point. Further, if a dither signal is to be applied, it is desirable to choose a system in which the variation in the laser frequency due to the dither signal goes to zero at the stabilization point. One such method for laser stabilization was proposed some time ago by the author (Bennett, 1963) and is based on the power-dependent repulsion effects which occur near line center as discussed in a previous section. From Eqs. (2.154)–(2.156) near the center of the profile, the change in oscillator frequency due to a change in single pass gain coefficient at line center by amount δG_m is

$$\delta\nu_{\text{osc}} \approx \left(\frac{c/2L}{2\pi\,\Delta\nu_L}\right)\left[1 - 1.88\,\frac{\Delta\nu_L}{\Delta\nu_D}\right](\nu_{\text{osc}} - \nu_m)\,\delta G_m L_G$$

in the limit $\Delta\nu_L \ll \Delta\nu_D$. In this expression L_G represents the cavity length occupied by the amplifying medium. Near the center of the line the expression is a linearly varying quantity which goes to zero as $\nu_{osc} \rightarrow \nu_m$. This quantity can be a much more sensitive indication of line center in a negative feedback stabilization scheme than the tuning dip itself, as illustrated by the two experimental curves taken by Bennett *et al.* (1964) with a 3.39-micron helium–neon laser shown in Fig. 2-28.

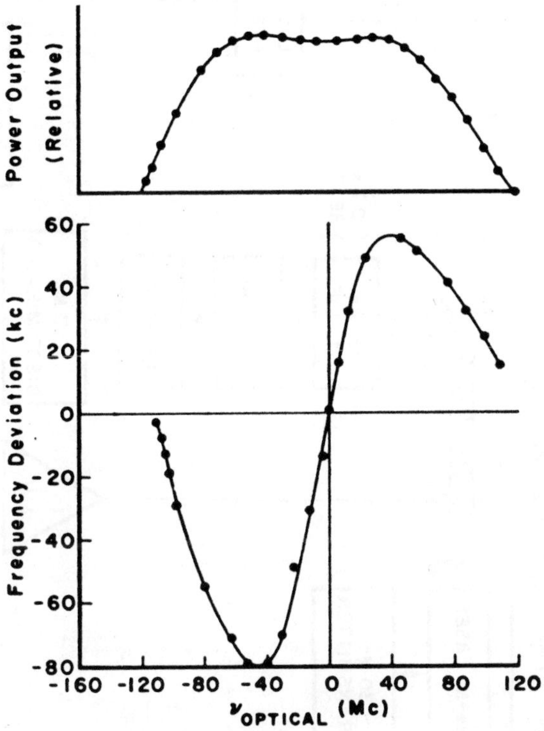

FIGURE 2-28. Tuning and oscillator dispersion characteristic ($\delta\nu_0/\delta G_m$) for the 3.39 micron He–Ne laser transition obtained by Bennett *et al.* (1964).

The method used to obtain the data in Fig. 2-28 is illustrated schematically in Fig. 2-29. Here a reference laser is locked through a slow servo loop to a frequency 10.7 MHz away from the laser being subject to the gain modulation at 20 kc/sec. The reference laser thus permits the shift in oscillator frequency of the first laser provided by the constant 20 kc/sec gain modulation as the first laser is slowly swept through the Doppler line by changing its cavity length. This effect was used by Bennett

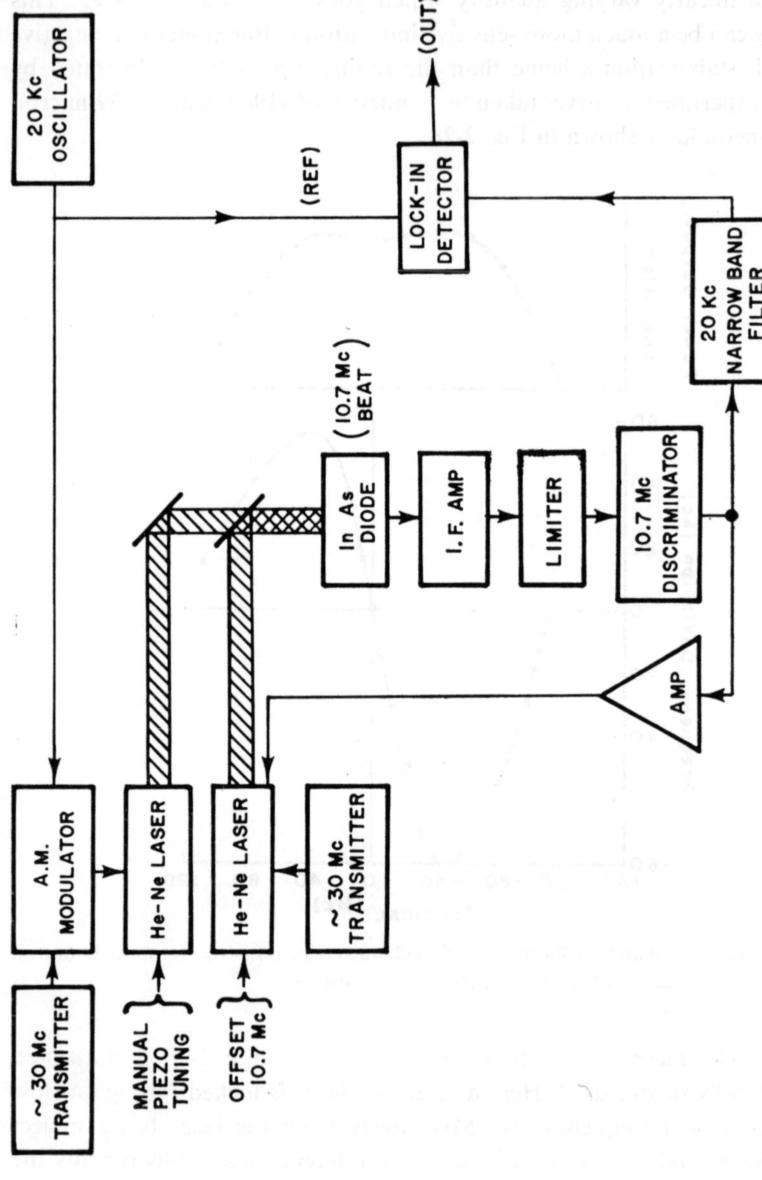

FIGURE 2-29. Schematic illustration of the method used by Bennett *et al.* (1964) to obtain the oscillator dispersion curve in Fig. 2-19.

et al. (1964) as the reference signal to obtain automatic frequency stabilization of a single-mode laser in respect to the center of the spontaneous emission profile. As illustrated in Fig. 2-30, it was found possible in practice to obtain long-term stabilization of the laser within ≈ 10 kc of the line center. However, there are two practical difficulties which were encountered with the method which should be mentioned:

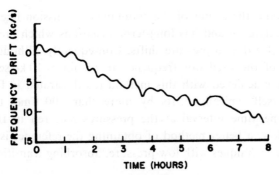

FIGURE 2-30. Drift in the oscillator frequency from line center determined for the stabilization experiment in Bennett *et al.* (1964).

First, the gain modulation (or "dither" signal) must be applied in such a way that it does not alter the refractive index of the medium except in its effect on the resonance line itself. For example, merely modulating the discharge in a gas laser is not desirable because the changing electron density affects the average refractive index of the medium and therefore would be equivalent to varying the cavity length in addition to the gain coefficient. As may be calculated from the expression for the refractive index of a plasma with electron density n_e, the shift in oscillator frequency due to a variation in the electron density by amount δn_e is

$$\delta\nu_{\text{osc}} \approx \frac{e^2}{2\pi m_e \nu}\left(\frac{L_G}{L}\right)\delta n_e$$

$$\approx 7.2 \times 10^{-7}\left(\frac{L_G}{L}\right)\delta n_e \ \text{Hz} \quad (\text{for } n_e \text{ in cm}^{-3})$$

at $\lambda = 3.39$ microns. Consequently, a stabilization method which dithers the discharge current in the active medium automatically stabilizes below the line center by an amount

$$\delta\nu \approx \frac{(2\Delta\nu_L e^2\lambda/mc^2)\,(\partial n_e/\partial G_m)}{1 - 1.88\Delta\nu_L/\Delta\nu_D}.$$

In the case of the 3.39-micron laser stabilization experiment reported in Bennett *et al.* (1964), this shift amounted to about 300 kHz. Although small compared to pressure shifts of the line center, it was still large compared to the inherent possibility of the method. In practice it is more effective to vary either the lower- or upper-state density by optical pumping with light generated in an externally-modulated discharge tube as done by Boyne *et al.* (1965).

Second, the center of the spontaneous emission line itself is not an absolute reference and has long-term variations which are typically in excess of a MHz due to pressure shifts. Consequently, although long-term stabilization of the oscillator frequency to within $\approx$ 10 kc/sec of the line center has been achieved with this method (as illustrated in Fig.2-30), the line center itself typically drifts by more than 100 times that amount over the same time interval at the pressures required in the laser. As discussed below, a better method of obtaining the reference is through the use of similar techniques with low-pressure, absorbing transitions.

2.30. Single-Mode with a Saturable Absorber

This problem is of considerable practical importance because of its relation to the frequency stabilization and mode isolation techniques discussed below. The problem also provides a nice illustration of the practical value of the hole-burning model in that one can write down the answers for the power and frequency tuning characteristics from inspection with as much accuracy as could be anticipated from the far more complex analysis that would be required to solve the same problem with the third-order Lamb theory. Before discussing the application of hole-burning effects in saturable absorbers to laser frequency stabilization, we will first develop some fairly general equations to describe the single-mode laser characteristics in the presence of a saturable absorber. We shall limit this discussion to gaseous absorbing transitions of a type similar to those used in a laser amplifier itself. The absorption is achieved by having more atoms in the lower, rather than the upper, energy level of the resonance transition.

The single-mode equations with a saturable absorber present are very similar to those discussed in Sec. 2.23 in connection with the isotope effect. The principal differences are that one "isotope" is now replaced by an absorber and that there can be large differences in the relaxation rates and electric dipole matrix elements. In addition, because the saturable absorber will in general be located in a different place from the gain section, focussing effects can result in different field intensities (power flux) along the cavity axis even when the amplifier and absorber transmit the same total power in the same running wave. In visualizing the problem, it will be helpful to consider the arrangement of holes in the gain and absorber profiles shown in Fig. 2-31. We will assume both gain and absorber line profiles

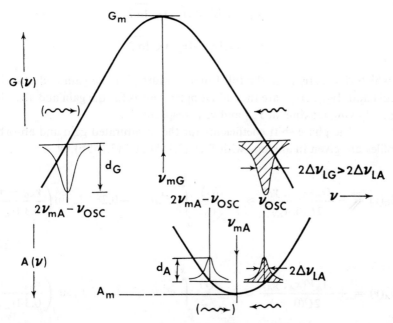

FIGURE 2-31. Holes burnt in the gain and absorption profile of a single-mode laser containing a saturable absorber inside the cavity.

are symmetric and that power broadening effects can be ignored. Quantities that involve the gain section will be described by the subscripts "G" and quantities that involve the absorber will be denoted by the subscript "A". Otherwise, the notation will be that previously used in the hole-burning analysis discussed in Sec. 2.22 through 2.25.

The small-signal (unsaturated) gain and absorption coefficients are given in analogy with Eq. (2.155) by

$$G(\nu) = \frac{G_m U(x_G, y_G)}{U(0, y_G)} \approx G_m \exp(-x_G^2)$$
$$A(\nu) = \frac{A_m U(x_A, y_A)}{U(0, y_A)} \approx A_m \exp(-x_A^2) \qquad (2.208)$$

where

$$x_G(\nu) = (\nu_{mG} - \nu)\, 2\sqrt{\ln 2}/\Delta\nu_{DG}$$

$$x_A(\nu) = (\nu_{mA} - \nu)\, 2\sqrt{\ln 2}/\Delta\nu_{DA}$$

$$y_G = (\Delta\nu_{LG}/\Delta\nu_{DG})\sqrt{\ln 2} \tag{2.209}$$

$$y_A = (\Delta\nu_{LA}/\Delta\nu_{DA})\sqrt{\ln 2}$$

in which $\Delta\nu_{LG}$, $\Delta\nu_{LA}$ are the full Lorentz widths for the gain and absorber media and $\Delta\nu_{DG}$, $\Delta\nu_{DA}$ are the full Doppler widths for the gain and absorber media having maxima at ν_{mG} and ν_{mA}, respectively.

The phase shift coefficients for the unsaturated gain and absorber profiles are given in analogy with Eqs. (2.155) and (2.155b) by

$$\Delta\phi_G(\nu) = -\frac{G_m V(x_G, y_G)}{2U(0, y_G)} \approx -\frac{G(\nu)}{\sqrt{\pi}} \int_0^{x_G} e^{t^2} dt \approx -0.282 G_m \sin\left(\frac{\nu_{mG} - \nu}{0.3\Delta\nu_{DG}}\right)$$

$$\Delta\phi_A(\nu) = +\frac{A_m V(x_A, y_A)}{2U(0, y_A)} \approx +\frac{A(\nu)}{\sqrt{\pi}} \int_0^{x_A} e^{t^2} dt \approx +0.282 A_m \sin\left(\frac{\nu_{mA} - \nu}{0.3\Delta\nu_{DA}}\right) \tag{2.210}$$

where the accuracy of the approximations has been discussed in Sec. 2.16.

Steady-state gain saturation in the single-mode oscillator is now described by

$$\begin{aligned} G(\nu_{\text{osc}})\, L_G - A(\nu_{\text{osc}})\, L_A - f &= d_G L_G \left\{1 + \left[1 + 4\left(\frac{\nu_{\text{osc}} - \nu_{mG}}{\Delta\nu_{LG}}\right)^2\right]^{-1}\right\} \\ &\quad - d_A L_A \left\{1 + \left[1 + 4\left(\frac{\nu_{\text{osc}} - \nu_{mA}}{\Delta\nu_{LA}}\right)^2\right]^{-1}\right\} \end{aligned} \tag{2.211}$$

where L_G is the length of the gain section and L_A is the length of the absorber (and average values are implied). Here, the quantities d_G and d_A represent the gain coefficient and absorption coefficient hole depths.

From Eqs. (2.136b), (2.130) and (2.60), it may be seen that the quantity

$$\alpha \equiv \frac{d_A L_A}{d_G L_G} = \frac{\left(\dfrac{\langle\mu_0\rangle^2}{R_1 R_2}\right)_A P_{\text{total}_A}(\nu_0)\, L_A}{\left(\dfrac{\langle\mu_0\rangle^2}{R_1 R_2}\right)_G P_{\text{total}_G}(\nu_0)\, L_G}$$

$$\approx \frac{\langle\mu_0\rangle_A^2\, D_{mA} U(x_{\text{osc}A}, y_A)\, L_A}{\langle\mu_0\rangle_G^2\, D_{mG} U(x_{\text{osc}G}, y_G)\, L_G} \approx \frac{\left(\dfrac{\langle\mu_0\rangle^2}{R_1 R_2}\right)_A A_m\, \Delta\nu_{DA} I_A L_A\, e^{-x^2_{\text{osc}A}}}{\left(\dfrac{\langle\mu_0\rangle^2}{R_1 R_2}\right)_G G_m\, \Delta\nu_{DG} I_G L_G\, e^{-x^2_{\text{osc}A}}} \tag{2.212}$$

is frequency-dependent through Eqs. (2.209) if $\nu_{mA} \neq \nu_{mG}$ and the two Doppler widths are not equal. The ratio $I_A/I_G \approx$ constant (even if not $= 1$) in the small gain case.

Substituting the definition of α in Eq. (2.211),

$$d_G L_G = [G(\nu_{\text{osc}})\, L_G - A(\nu_{\text{osc}})\, L_A - f] \times \left\{1 - \alpha + \left[1 + 4\left(\frac{\nu_{\text{osc}} - \nu_{mG}}{\Delta\nu_{LG}}\right)^2\right]^{-1} - \alpha\left[1 + 4\left(\frac{\nu_{\text{osc}} - \nu_{mA}}{\Delta\nu_{LA}}\right)^2\right]^{-1}\right\}^{-1}, \tag{2.213}$$

where the frequency dependence of α is given approximately by the last term in Eq. (2.212).

From Eq. (2.212) it is seen that the total power flux loss and gain coefficients for one running wave interaction are related by

$$P_{\text{total}_A}(\nu_0)\, L_A = \alpha' P_{\text{total}_G}(\nu_0)\, L_G \tag{2.214}$$

where

$$\alpha' = \frac{A_m L_A I_A \Delta\nu_{DA}}{G_m L_G I_G\, \Delta\nu_{DG}} \exp(x^2_{\text{osc}G} - x^2_{\text{osc}A}) \tag{2.215}$$

and $x_{\text{osc}G}$, $x_{\text{osc}A}$ mean the values of x_G, x_A from Eqs. (2.209) when $\nu = \nu_{\text{osc}}$ (the single-mode oscillator frequency).

The total steady-state power flux out of the single-mode oscillator with the saturable absorber present from both running wave interactions is then

$$P_{\text{out}} = 2P_{\text{total}_G}(\nu_0)\, L_G - 2P_{\text{total}_A}(\nu_0)\, L_A = C_0'(1 - \alpha')\, d_G L_G \tag{2.216}$$

where

$$C_0' = \left(\frac{\hbar^2 c R_1 R_2}{2\pi\langle\mu_0\rangle^2}\right)_G \tag{2.217}$$

from Eqs. (2.157) or (2.136b).

The frequency of the oscillator is determined through solution of a transcendental equation analogous to Eqs. (2.154) and (2.165) in which there are terms from the original gain and absorber profiles and from the gain and absorber holes.

$$\nu_{\text{cav}} = \nu_{\text{osc}} + \frac{\Delta\nu_c}{f}\left[\Delta\phi_G(\nu_{\text{osc}})\, L_G + \Delta\phi_A(\nu_{\text{osc}})\, L_A + \Delta\phi_{HG}(\nu_{\text{osc}})\, L_G + \Delta\phi_{HA}(\nu_{\text{osc}})\, L_A\right] \tag{2.218}$$

in which the first two phase shift terms are given by Eqs. (2.210), f represents the total average fractional energy loss per pass at the oscillator frequency and $\Delta\nu_c$ is the full width of the cavity resonance.

The remaining phase shift terms are due to the "mirror image" holes in the gain and absorber profiles and are given by

$$\begin{aligned} \Delta\phi_{HG}(\nu_{\text{osc}}) &= +d_G\left(\frac{\nu_{mG} - \nu_{\text{osc}}}{\Delta\nu_{LG}}\right)\left[1 + 4\left(\frac{\nu_{mG} - \nu_{\text{osc}}}{\Delta\nu_{LG}}\right)^2\right]^{-1} \\ \Delta\phi_{HA}(\nu_{\text{osc}}) &= -d_A\left(\frac{\nu_{mA} - \nu_{\text{osc}}}{\Delta\nu_{LA}}\right)\left[1 + 4\left(\frac{\nu_{mA} - \nu_{\text{osc}}}{\Delta\nu_{LA}}\right)^2\right]^{-1} \end{aligned} \tag{2.219}$$

The phase shift expressions simplify somewhat (but not much in the general case) when expressed in terms of α and α' defined by Eqs. (2.212) and (2.215).

Finally, it is useful to note that the total small signal gain per pass at frequency ν' in the presence of oscillation at ν_{osc} is

$$\begin{aligned} G_{\text{total}}(\nu') &= G(\nu')\, L_G - A(\nu')\, L_A \\ &- d_G L_G\left\{\left[1 + \left(\frac{\nu_{\text{osc}} - \nu'}{\Delta\nu_{LG}}\right)^2\right]^{-1} + \left[1 + \left(\frac{2\nu_{mG} - \nu_{\text{osc}} - \nu'}{\Delta\nu_{LG}}\right)^2\right]^{-1}\right\} \\ &+ d_A L_A\left\{\left[1 + \left(\frac{\nu_{\text{osc}} - \nu'}{\Delta\nu_{LA}}\right)^2\right]^{-1} + \left[1 + \left(\frac{2\nu_{mA} - \nu_{\text{osc}} - \nu'}{\Delta\nu_{LA}}\right)^2\right]^{-1}\right\} \end{aligned} \tag{2.220}$$

As discussed below, the saturable absorber can lead to mode suppression when $G_{\text{total}}(\nu') < f$ at another resonance frequency ν'.

The equations given above can be solved exactly using numerical techniques similar to those discussed in Sec. 2.22 and 2.23. For a given choice of gain and absorption constants and cavity parameters, one can solve Eq. (2.218) for ν_{cav} using ν_{osc} as the "independent" variable. The unsaturated gain and absorption phase shift terms may be calculated directly from Eqs. (2.208)-(2.210). The value of α for a given choice of ν_{osc} is next determined from Eq. (2.212) and substituted in Eq. (2.213), yielding both $d_L L_G$ and $d_A L_A$. One may then determine the hole phase shift terms from Eqs. (2.219). Substituting those back in Eq. (2.218) completes the solution for ν_{cav} corresponding to a given value of ν_{osc}. Next by substituting the values for ν_{osc} in Eq. (2.215) for α', one can then determine the output power of the oscillator from Eq. (2.216). Finally, one can examine the gain at other frequencies through application of Eq. (2.220). As with the similar solutions discussed in the two-isotope case, "negative" hole depths mean the oscillator is below threshold for the particular choice of parameters. There can be a number of remarkable effects which occur in the solution of these equations which are discussed in more detail in the following sections.

2.31. Frequency Stabilization and Mode Suppression with Saturable Absorbers‡

As noted recently by a number of different people, some strikingly interesting and useful results can be obtained through the addition of intracavity saturable absorbers to inhomogeneously-broadened gas lasers. Some of the earliest work on this problem was done by Lee and Skolnick (1967), who observed an "inverse Lamp dip" in a helium-neon laser containing a neon absorption cell. They pointed out the usefulness of such a phenomenon as an absolute reference for frequency stabilization purposes, but apparently did not appreciate the usefulness of the effect for mode suppression at that time. The basic phenomenon involved is just the same as that which occurs in the normal Lamb dip and may be understood quantitatively from Eq. (2.213). If the absorber Lorentz width is very narrow compared to the Lorentz width in the gain section and α in Eq. (2.213) is appropriately adjusted, the Lorentzian function associated with the absorber will dominate near the center of the absorber resonance. That is, there will be a sharp dip in the absorption and a net increase in over-all gain (and power) at $\nu_{osc} = \nu_{mA}$. The normal tuning dip will thus have a sharp peak in the middle.

‡ Some of the material in this section was reproduced in Bennett (1970b) by permission of the editors.

The form of the power tuning curve simplifies considerably when the absorber and amplifier Doppler widths and line centers are the same. In that case, Eq. (2.216) becomes (Bennett, 1970)

$$P_{\text{out}} = C_0'[G(\nu_{\text{osc}})\, L_G(1-\alpha') - f]\,(1-\alpha')$$
$$\times \left\{1 - \alpha + \left[1 + 4\left(\frac{\nu_{\text{osc}} - \nu_m}{\Delta\nu_{LG}}\right)^2\right]^{-1} - \alpha\left[1 + 4\left(\frac{\nu_{\text{osc}} - \nu_m}{\Delta\nu_{LA}}\right)^2\right]^{-1}\right\}^{-1}. \tag{2.221}$$

where α is no longer frequency-dependent. The form of Eq. (2.221) is illustrated in Fig. 2-32 for several values of α (assumed equal to α'), and relative values of the linewidths similar to those obtainable with the argon ion laser from the current broadening effect.

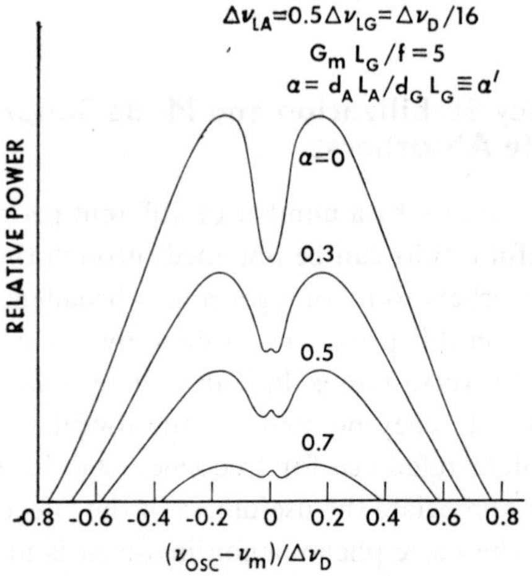

FIGURE 2-32. Form of power tuning curve with a saturable absorber in the cavity for various values of the absorption constant. The absorber and amplifier are assumed to have the same resonance frequencies and Doppler widths.

There are two important virtues of the effect from the frequency stabilization point of view. Obviously the narrower the absorber Lorentz width is, the more sensitive is the error signal which can be used to stabilize the laser. In the case of the molecular absorption bands which fall in coincidence with the 3.39 micron helium-neon laser transition, the absorber Lorentz widths can in principle approach values ≈ 1 kHz. The second virtue is that it is generally much easier to obtain absorption than gain; hence saturable absorbers can usually be operated at much lower pressures and with minimum excitation. Consequently, the pressure- and plasma-induced frequency shifts that plagued the stabilization methods discussed in Sec. 2.27 can largely be avoided. These advantages become spectacularly-important in the recent stabilization experiments using the coincidence between the methane absorption lines and the 3.39 micron helium-neon laser transitions carried out by Barger and Hall (1969).

Much of the early work on saturable absorption in gas lasers was done by Russian scientists. Because a great deal of the early Russian work has regrettably been ignored in the American literature on the subject, it is worth reviewing some of these developments.

In July of 1967 the author visited two different laboratories in the Soviet Union where experiments were in progress on the use of saturable absorption for frequency stabilization purposes in gas lasers. Specifically, V. P. Chebotayev and his colleagues (especially V. N. Lisitzin) were pursuing this techniques using a pure neon discharge tube as an intra-cavity absorber on the 0.63 micron helium-neon laser transition in the Institute for the Physics of Semiconductors at Novosibirsk, and N. G. Basov and his colleagues at the Lebed'ev Institute in Moscow were pursuing a scheme in which saturable absorption in an external beam of formaldehyde molecules was to be used as the frequency reference with one of the strong gas laser transitions in the 3-micron range. Preliminary results of the experiments and the proposals to use saturable absorption for both frequency stabilization and mode selection were presented by Chebotayev and Lisitzin, and by Basov and Letokhov at the Quantum Electronics Conference held in Erevane (EPEBAHE) USSR in 1967.

The work on mode selection and frequency stabilization by Chebotayev and Lisitzin was highly successful and accounts of this work appeared in the Soviet literature early in 1968 (Chebotayev and Lisitzin, 1968). They submitted a report of their work together with I. M. Beterov to the Miami Conference on Quantum Electronics in the Spring of 1968, which was unfortunately included for publication "by title only". The full paper by Chebotayev, Beterov and Lisitzin (1968) was subsequently published in the American literature in the Fall of that year.

The initial experiment by Basov and Letokhov was inherently more complex and was apparently abandoned in its original form (Basov, private communication). However, a detailed review of different possible frequency stabilization methods, including those based on nonlinear absorption, was given by Basov and Letokhov in 1968. (Also, see the more recent review by Letokhov, 1971.)

Subsequently, Lee, Schoefer and Barker (1968) and Feld, Javan and Lee (1968) also noted the usefulness of saturable absorbers for mode suppression.

The basic form of the apparatus used by Chebotayev *et al.* (1968) and Lisitzin (1968) to study the mode isolation effect with saturable absorbers is illustrated in Fig. 2-33. A pure neon discharge cell within the laser cavity was used to provide resonant absorption on the same transitions which provide amplification in the helium-neon mixture. By controlling the pressure, discharge conditions and temperature, significantly smaller Lorentz widths can be obtained in the pure neon absorber than in the helium-neon gain section. (See Fig. 2-24.)

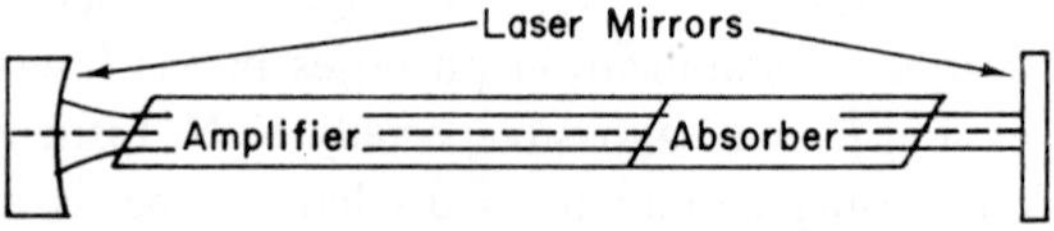

FIGURE 2-33. Apparatus used by Chebotayev *et al.* (Chebotayev and Lisitzin, 1968; Chebotayev, Beterov and Lisitzin, 1968; Lisitzin, 1968) to obtain mode isolation through saturable absorption.

The mode suppression effect comes quantitatively from Eq. (2.220). For example, if the two Doppler widths are the same, and the absorber resonance and gain resonance are the same, when the oscillator is tuned to line center, Eq. (2.220) becomes (Bennett, 1970)

$$G_{\text{total}}(\nu') = G(\nu')\, L_G(1 - \alpha') - \left[\frac{G_m L_G(1 - \alpha') - f}{(1 - \alpha)}\right]\left\{\left[1 + \left(\frac{\nu_m - \nu'}{\Delta\nu_{LG}}\right)^2\right]^{-1} - \alpha\left[1 + \left(\frac{\nu_m - \nu'}{\Delta\nu_{LA}}\right)^2\right]^{-1}\right\}. \tag{2.222}$$

The functional form of Eq. (2.222) is very similar to that of the tuning dip itself and is illustrated in Fig. 2-34 for one choice of linewidth ratios and several values of α (again defined equal to α'). Note that for $\alpha > 0.7$ in Fig. 2-34, the oscillation at line center reduces the gain below threshold for all other frequencies. This discovery of Chebotayev and Lisitzin is of considerable practical importance. For example, with the 6328 Å helium-neon laser transition, Chebotayev and Lisitzin were able to nearly 80 percent of the total available multi-mode power out in a single mode using a pure neon absorber at $\lesssim 0.3$ Torr.

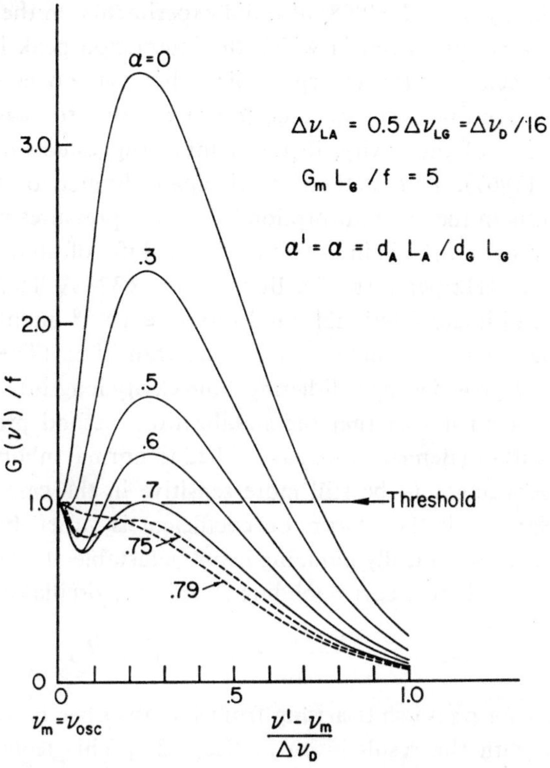

FIGURE 2-34. Gain at frequency ν' when the laser is oscillating at line center with a saturable absorber present in the cavity. (The parameters are the same as those listed in Fig. 2-32.) Note that for values of the absorption parameter greater than 0.7 the gain at all other frequencies is below threshold, whereas without the absorber present ($\alpha = 0$) the gain is above threshold over a wide range of frequency and multi-mode oscillation occurs

The mode suppression effect was originally analyzed in some detail by Kazantsev, Rautian and Surdutovich (1968) and subsequently by Feld, Javan and Lee (1968) using modifications of the Lamb theory. Chebotayev and Lisitzin discussed their experimental results using the theoretical expressions by Kazantsev *et al.* However, the basic cause of the phenomenon is easily understood in terms of the hole-burning effects discussed in Sec. 2.30. By inspection of Eq. (2.220), it is seen that the mode suppression effect arises primarily because for $\Delta\nu_{LA} \ll \Delta\nu_{LG}$ the absorption dip has long been passed in regions of the line where both the unsaturated absorption profile and the gain holes are still subtracting heavily from the total gain.

Chebotayev *et al.* (1968) also did experiments on the use of a low pressure neon absorption cell in which the interaction peak in the output power near the center of the absorption line (due to the overlap of the two absorption holes in Fig. 2-31) was used to provide an error signal for servo-loop stabilization of the cavity, in the manner suggested initially by Lee and Skolnick (1967). This method is principally limited by the variation of pressure shifts in the neon absorption line at the pressures required for a sufficient error signal. [Lisitzin (1968) reported shifts of about 6 MHz/torr of neon and 21 MHz per torr of helium on the 6328 Å line.] In practice it is likely that ultimate fractional stabilization $\approx 10^{-10}$ might be obtained in that way on the 6328 Å helium-neon laser transition. (The error signal in this work was provided by "dithering" the cavity length.)

It is worth noting that the stabilization method previously suggested by the author (Bennett, 1963) as applied to normal inhomogeneously-broadened lasers ought to be still more sensitive in the case of saturable absorption. Namely, if the absorption coefficient is varied by an amount δA_m (for example, by optically pumping neon metastable atoms to the lower laser level), there will be a corresponding shift in the oscillator of amount

$$\delta\nu_{\text{osc}} \approx -\frac{c/2L}{2\pi\Delta\nu_{LA}}(\nu_{\text{osc}} - \nu_m)\,\delta A_m L_A \qquad (2.223)$$

due to the varying phase shift arising from the absorber holes. [This is in exact analogy with the result given in Sec. 2.29.] This technique should again provide a more sensitive indication of line center and again has the advantage that the effect of the dither signal on the oscillator frequency vanishes to first order at the stabilization point. Hence the shift in oscillator frequency with dither signal over the whole center of the line should vary as shown in Fig. 2-35. Note that a reciprocal, mutual self-stabilization

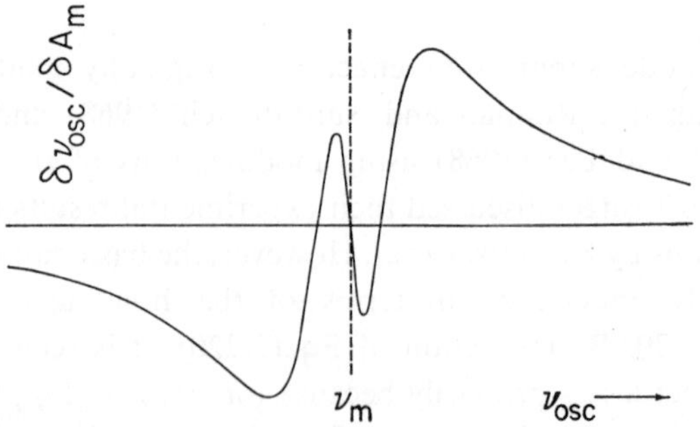

FIGURE 2-35. Error signal that could be obtained by dithering the absorption coefficient in a modified version of the earlier frequency stabilization scheme suggested by the author (Bennett, 1963).

scheme could be used with this approach (see Fig. 2-36) in which laser # 1 is dithered at f_1 and laser # 2 is dithered at f_2 and the reference signals to stabilize the two lasers are obtained from two separate phase-sensitive detectors (or "lock-in" circuits).

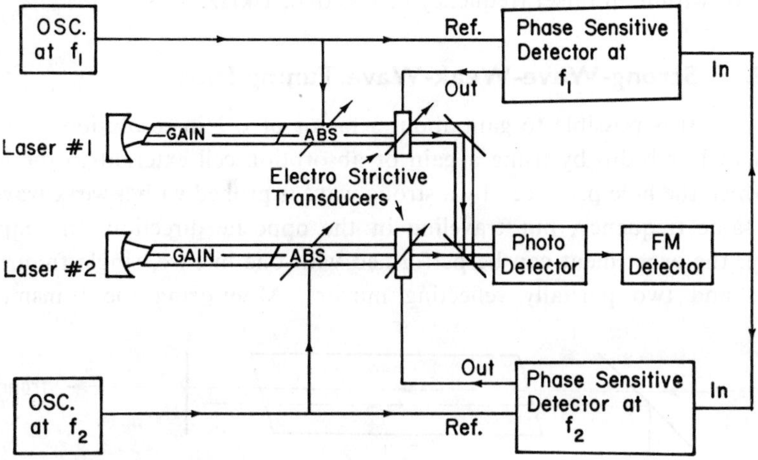

FIGURE 2.36. Reciprocal mutual stabilization scheme proposed by the author and based on simultaneously dithering the absorption in two lasers.

The most sensitive degree of stabilization that has yet been achieved with a gas laser has been carried out recently by R. L. Barger and J. L. Hall (1969). They have made use of the coincidence between the 3.39-micron helium–neon laser transition and the 3.39-micron vibration–rotation absorption line of methane. This absorption line in CH_4 [the *P*(7) line of the ν_3 band] starts from a level which is well populated thermally, even at liquid-nitrogen temperatures, and the upper state has a natural lifetime $\approx$ 10 msec. Consequently the collision-free Lorentz width of the absorber is extremely small. At the same time, the large relative thermal population in the lower level yields extremely large absorption coefficients at low pressures ($\approx$ 0.18 cm^{-1}/Torr). The absorption resonance is higher than the neon transition by 100 MHz. However, Barger and Hall use sufficiently high helium pressure ($\approx$ 12 Torr) to shift the laser transition to the center of the methane line—a practice permitted by the unusually large gain coefficient for this laser transition. This practice also increases the Lorentz width in the laser medium and assures single-mode oscillation. They also note, as shown by Nehara *et al.*, that the methane line is free from appreciable Stark shift. This accumulation of properties makes the system particularly ideal for frequency stabilization.

In practice, Barger and Hall found absorption Lorentz widths ≈ 100 kHz and pressure broadening coefficients for the absorber ≈ 8 kHz/mTorr. Using a methane absorption cell at millitorr pressures for a reference, they have so far stabilized three independent lasers to ≈ 1 part in 10^{11} or within an offset frequency of less than 1 kHz.

2.32. Strong-Wave-Weak-Wave Tuning Dips

It is possible to gain about a factor of $\sqrt{2}$ in resolution over the normal Lamb dip by using a gain or absorption cell external to the laser in which the hole produced by a strong wave is probed with a weak wave of the same frequency, but traveling in the opposite direction. In simplest form, the experiment can be performed with one intense, single frequency laser and two partially reflecting mirrors. Monitoring the transmitted

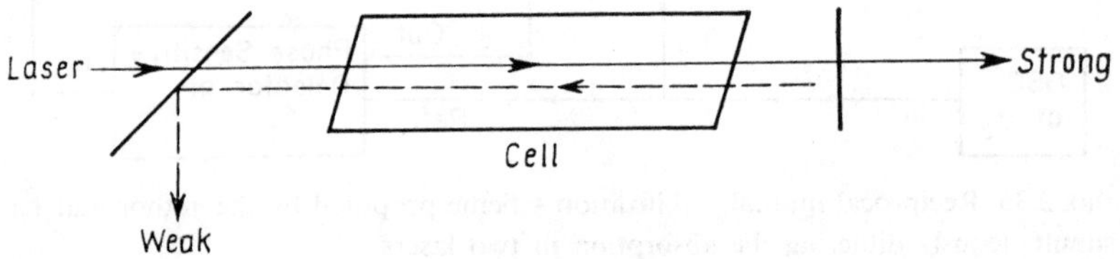

"strong" wave permits measuring the gain or absorption of the "weak" wave on the return path through the cell. [More realistic versions of the experiment would include "chopping" the strong signal and using "lock-in" detection to determine the gain from an unmodulated weak signal, thereby isolating the effects of the hole from the rest of the line profile in the test cell.]

This type of situation in intrinsically different from the normal single-mode Lamb dip in that only one hole is burnt at ν_0 (by the strong wave) in the Doppler profile in the test cell and, more important, the hole depth does not arise from the steady-state gain saturation requirement [Eq. (2.141)] in the oscillator. The small-signal gain (or absorption) in the wing of this hole at $2\nu_m - \nu_0$ is being probed by the weak running wave. (Here ν_m refers to the center frequency of the Doppler profile of the gain or absorption line in the test cell). The gain (or absorption) in the wing of this hole is therefore given by the exact expression in Eq. (2.135) with the substitutions

$$x_2 = \frac{2\sqrt{\ln 2}}{\Delta\nu_D}(2\nu_m - \nu_0 - \nu_m) = \frac{2\sqrt{\ln 2}}{\Delta\nu_D}(\nu_m - \nu_0) = -x_1 = S/2 \tag{2.224}$$

where $\Delta\nu_D$ is the Doppler width for the transition in the test cell. The quantities y_1, y_2 are as defined before and involve the power-broadened Lorentz width- and zero-field Lorentz width-to-Doppler width ratios. The depth of the gain (or absorption) coefficient hole at frequency ν_0 depends on the intensity of the strong wave through the quantity $D_m = D(\nu_m)$ given in Eq. (2.130), in which all quantities now refer to the characteristics of the atoms in the test cell.

In the absence of power broadening in the test cell due to the strong wave, the equations simplify considerably. Letting $y_1 = y_2 \equiv y$, it is seen that the change in the small-signal gain (or absorption) coefficient for the weak wave, induced by the oppositely traveling strong wave, is

$$\Delta G(\nu)\,(\text{or } \Delta A(\nu)) = \frac{\pi\langle\mu_0\rangle^2 D_m}{\hbar\lambda}\left[\frac{U(x,y) + \left(\dfrac{y^3}{x}\right)V(x,y)}{1 + 4\left(\dfrac{\nu_m - \nu_0}{\Delta\nu_{L0}}\right)^2}\right] \tag{2.225}$$

in which

$$x = \frac{2\sqrt{\ln 2}}{\Delta\nu_D}(\nu_m - \nu_0) \quad \text{and} \quad y = \sqrt{\ln 2}\left(\frac{\Delta\nu_{L0}}{\Delta\nu_D}\right). \tag{2.226}$$

and $U(x, y)$ and $V(x, y)$ are the real and imaginary parts of the function $w(z)$ tabulated by Faddeyeva and Terent'ev (1961) [see Sec. 2.7]. By coincidence the numerator in Eq. (2.225) is multiplied by exactly the same Lorentzian factor that enters in the denominator of the normal single-mode tuning dip expression [see Eq. (2.156)]. As discussed in connection with Eq. (2.136), the departures from the Lorentzian form in Eq. (2.225) are of order $(\Delta\nu_{L0}/\Delta\nu_D)^2$. Noting that

$$\lim_{x\to 0}\left(\frac{V(x,y)}{x}\right) = \frac{2}{\sqrt{\pi}} - yU(0,y) \tag{2.227}$$

and applying the definition of $D_m = D(\nu_m)$ from Eq. (2.130), it may be seen that the maximum value of the observed change (for constant intensity flux in the strong wave) occurs at $\nu_0 = \nu_m$ and is given by

$$\Delta G_m(\text{or } \Delta A_m) = \frac{8\pi^2\langle\mu_0\rangle^4}{\hbar^3\lambda c R_1 R_2}[N_2(0) - N_1(0)]\,I_m + \text{Order}\,(y^3) \tag{2.228}$$

where $[N_2(0) - N_1(0)]$ is the effective inversion density in the test cell in the absence of the field per unit frequency interval at the center of the line [in the sense of Eq. (2.40)] and I_m represents the intensity flux in the strong wave at frequency ν_m.

Hence in this type of experiment the change in gain for the weak wave is

$$\Delta G(\nu_0) = \frac{\Delta G_m I(\nu_0)}{I_m}\left[1 + 4\left(\frac{\nu_m - \nu_0}{\Delta\nu_{LO}}\right)^2\right]^{-1} + \text{Order}\,(y^2) \qquad (2.229)$$

where ΔG_m is given by Eq. (2.228). Obviously, whether or not gain or absorption is involved is determined by the effective population inversion density, hence the sign in Eq. (2.228).

The frequency-dependent form of Eq. (2.229) can be obtained directly from the discussion of hole-burning effects in the Doppler limit given by Bennett (1963). One merely has to note that a single Lorentzian hole is burnt in the gain curve seen in the $+z$ direction at frequency ν_0 with full width $2\Delta\nu_{LO}$ in the absence of power broadening [see Eq. (2.139)]. This Lorentzian gain hole is being probed by the oppositely-travelling weak wave at a point in the gain profile seen in the $-z$ direction which appears shifted by amount $2(\nu_m - \nu_0)$ from the center of the hole due to the Doppler effect. Substituting the latter frequency displacement in Eq. (2.139) results immediately in the Lorentzian form in Eq. (2.229).

Although the same *Lorentz widths* are involved in both Eq. (2.229) and in the normal, non-power-broadened single-mode Lamb dip [Eq. (2.142)] the *frequency dependence* of the two expressions upon the (same) Lorentz width is different. For the purposes of comparison, let us assume an infinitely-wide Doppler width, in which case we may neglect the frequency dependence in the numerator of Eq. (2.142). In order to compare the widths of the resonance curves that an experimenter would actually see between the two cases it is necessary to allow for the difference in behavior between the two expressions both at resonance and in the extreme wings of the resonance. Hence, in order to make a fair comparison between the two expressions we have to subtract *1* from the normal Lamb dip expression [apart from the numerator in Eq. (2.142)] and multiply the result by *−2*. When this is done, it is readily seen that the modified Lamp dip expression in the extreme Doppler limit is given by a Lorentzian function whose full width at half-maximum response is $\sqrt{2}\,\Delta\nu_L$. That is, even without the effects of power broadening, the resonance width in the normal single-mode tuning dip is about 40 percent broader than the width obtainable in the external type of tuning dip experiment under present discussion. Although the strong-wave-weak-wave experiment has the virtue of permitting an exact solution, the principle difference in resolution arises from the gain saturation requirement inside the single-mode laser [Eq. (2.141)] rather than the neglect of the hole depth produced by the "weak" wave in the external cell experiment. Of course, if the saturation parameter for the "weak" wave were to approach unity, very considerable departures from the frequency dependence in Eqs. (2.225) and (2.229) would result.

Use of this resonance effect to generate an error signal in laser frequency stabilization experiments from an external cell has been discussed by a number of authors (e.g., Basov and Letokhov, 1968; Chebotayev *et al.*, 1968; Barger and Hall, 1969; Letokhov, 1971). Use of the technique has also been made in tuned-laser absorption spectroscopy (see, for example, Rabinowitz, Keller and LaTourrette, 1969; Shimizu, 1969; Hänsch, Levenson and Schawlow, 1971; the review of Russian work by Letokhov, 1971; and Hänsch, Shahin and Schawlow, 1971).

As discussed by Basov and Letokhov (1968) and Letokhov (1971), the use of external interaction cells in which the two optical paths through the cell are widely-separated in principle permits obtaining a still-further increase in resolution through the optical analogue of the Ramsey (1950) separated-field resonance technique. This type of experiment might be done either with a beam of absorbing molecules traversing the two field regions or by depending on the atoms' thermal motion at sufficiently low pressures. However, Barger and Hall (1969) note that the relative phase of the excitation in the two regions would be difficult to control with enough accuracy in the optical analogue of the Ramsey method to make the technique practical.

2.33. The Self-Stabilization of Laser Frequencies with Saturable Absorbers‡

One of the most interesting and least-exploited methods of laser frequency control consists of the self-stabilization technique proposed independently by Letokhov and Cheboetayev (1969). Although perhaps less suited by itself for the long-term absolute stabilization requirements of wavelength standards, the self-stabilization method has an inherent response time which is much faster than that of any currently-known "active" method and does not require elaborate servo loops or "dithering" techniques. As reviewed by Letokhov (1971), the theory of the self-stabilization effect was first given in the Soviet literature (Letohkov, 1967, 1968; Letokhov and Chebotayev, 1969) using the density matrix formulation and appropriate extensions of the Lamb (1964) theory.

‡ Some of the material from this section was reproduced in Bennett (1972) by permission of the editors.

It is interesting to note that the self-stabilization effect also comes directly out of the simpler hole-burning theory discussed in Sec. 2.30. The effect may be seen most easily by numerical analysis of Eq. (2.218). As previously described this equation may be readily solved for ν_{cav} by using ν_{osc} as the independent variable. One may then may then turn the results around and plot ν_{osc} as a function of ν_{cav}. In this manner, what is otherwise a horrible transcendental equation for the oscillator frequency, may be easily solved. The results obtained have a great deal of similarity to those discussed previously in connection with Fig. 2-16 in the normal single-mode case. In particular if the laser can be operated far above threshold in the region where both the absorption is great and the condition

$$\Delta\nu_{LA} \ll \Delta\nu_{LG} \approx \Delta\nu_{\text{cav}} \ll \Delta\nu_{DA} \text{ or } \Delta\nu_{DG} \qquad (2.230)$$

is satisfied, a frequency tuning characteristic of the type illustrated in Fig. 2-37 is obtained. Near the center of the line, the opposite of the normal (gain) hole repulsion effect is obtained. That is, the "mirror image" absorber hole strongly attracts (through its phase shift term) the oscillator frequency towards the center of the absorber resonance. The result is a flat plateau region near the center of the absorber resonance of the type illustrated in Fig. 2-37 which is, of course, the basis of the self-stabilization effect. One can determine an approximate expression for the slope of the plateau region by differentiating Eq. (2.218) for ν_{cav} in respect to ν_{osc} in the limit given by condition (2.230) and the additional requirements that both $G_m L_G \gg f$ and α be close to unity. In those limits it may be seen that near the absorber resonance,

$$\frac{\Delta\nu_{\text{osc}}}{\Delta\nu_{\text{cav}}} \approx \left(\frac{\Delta\nu_{LA}}{\Delta\nu_c}\right)\left(\frac{f}{G_m L_G}\right)\left(\frac{1-\alpha}{\alpha}\right). \qquad (2.231)$$

This relation implies that far above threshold in the limit (2.230), the slope of the plateau region can be made arbitrarily close to zero. In other words, in that circumstance the oscillator frequency is totally determined by the absorber resonance and independent of the cavity frequency, near the absorber resonance. As indicated in Fig. 2-37, the plateau extends over an interval $\approx \Delta\nu_{LA}$ (the absorber Lorentz width). The data shown in Fig. 2-37 were computed by the author from Eq. (2.218) for the linewidth parameters shown and under the assumption that the absorber and amplifier resonance frequencies and Doppler widths were the same. Although the linewidth data in Fig. 2-37 correspond roughly to values easily obtainable with the 6328 Å helium-neon laser transition, satisfying both the large gain (ten times threshold without the absorber) and large cavity width simultaneously would generally be very difficult to accomplish in practice. The numbers used in

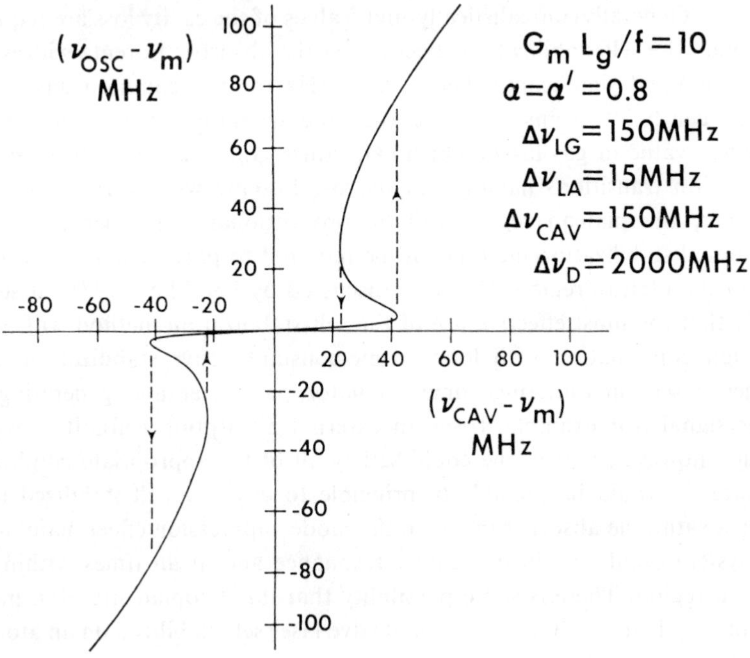

FIGURE 2-37. Illustration of the self-stabilizing effect obtainable from a single-mode gas laser using an internal saturable absorber. The solid curve is based on computer solutions to Eq. (2.218) for the parameters listed in the figure. The dashed lines indicate the hysteresis effects which would occur as the oscillator is tuned through the central region of the line.

Fig. 2-37 for the latter quantities were chosen merely to illustrate the nature of the effect in a hypothetical situation. As indicated by the vertical dashed arrows in Fig. 2-37, significant hysteresis effects would occur in the triple-valued regions of the frequency characteristic. These hysteresis effects are entirely analogous to those discussed in connection with Fig. 2-16 under similar conditions of excitation in the normal single mode case. (In fact Figs. 2-16 and 2-37 have been computed using the same linewidth parameters.) It should be noted that these are really "third-order" hysteresis effects, in contrast to the higher-order hysteresis effects analyzed by Kasantsev *et al.* (1968) and discussed below.

Using a very long and lossy helium-neon laser cavity with a pure neon absorption cell, Chebotayev and Lisitzin (1968) were apparently able to observe some self-stabilization effects within $\approx$ 10 MHz of the neon resonance line. However, this system is complicated by relative frequency shifts between the absorber and amplifier of a similar magnitude at the pressures required to produce a substantial difference between the two Lorentz widths.

Generally, unrealistically-high values of the cavity loss are required to obtain the self-stabilization effect unless the absorber Lorentz widths can be maintained significantly below ≈ 1 MHz for the common gas lasers. Consequently, it seems probable that the technique will mainly be of practical value in gas lasers which can utilize coincidences with molecular absorption transitions having very narrow Lorentz widths. In such cases, however, the plateau region will be proportionately smaller and some additional stabilization method will be required to prevent long-term drifts out of the plateau region. Hence as discussed by Letokhov (1971), it seems likely that the most effective use of the self-stabilization method will occur through combination with longer time-constant active stabilization techniques based on dithering some oscillator parameter and generating an error signal (for example, from an external absorption cell). It is worth noting, however, that if one could satisfy all of the appropriate conditions at once, it would be possible in principle to design a self-stabilized laser using a saturable absorber in which the mode-suppression effect maintained the system oscillating in one cavity resonance and at all times within the plateau region. There is some possibility that this Utopian situation might be obtainable through use of a tunable dye laser self-stabilized on an atomic, gas phase resonant absorber.

2.34. Threshold Hysteresis Effects with Intra-cavity Saturable Absorbers

As discussed by Chebotayev *et al.* (1968), Lisitzin (1968) and Chebotayev and Lisitzin (1969), the introduction of the saturable absorber leads to additional hysteresis effects in the tuning characteristic of the oscillator near threshold. The results in Sec. 2.30 represent the steady-state situation above oscillation threshold. As discussed by Kasantsev *et al.* (1968), fourth-order terms in the field-dependent expressions for the gain and absorption coefficients a result in these additional hysteresis effects.

The basic origin of these hysteresis effects can be observed in the field-dependent expression for the small signal gain coefficient in the absence of Doppler shifts for a running wave given by Eq. (2.42) in Sec. 2.4. The expression for the gain coefficient was of the form

$$G = \frac{1}{I}\frac{\partial I}{\partial x} = \frac{G_0}{1 + b_G + c_G I}, \tag{2.232}$$

where G_0 increases with the inversion in the absence of the field, b_G is frequency-dependent in the Lorentzian width, and c_G is the coefficient of the saturation term and involves the product of the phase interruption rates for the upper and lower levels, i.e.

$$c_G \propto (1/R_1 R_2)_{\text{amplifier}}$$

This last term was neglected in our treatment of oscillators without power broadening, although its first-order effects are included indirectly through the hole-burning process above threshold.

For an absorbing medium, the corresponding expression for the absorption coefficient would be

$$A = \frac{1}{I}\frac{\partial I}{\partial x} = \frac{A_0}{1 + b_A + c_A I}, \tag{2.233}$$

where

$$c_A \propto (1/R_1 R_2)_{\text{absorber}}$$

and may be characterized by different phase interruption rates than those in the amplifier. The net effect of having a gain and absorption tube in series is illustrated by Fig. 2-38 for a case where $c_A > c_G$ and $G_0 > A_0$. That is, the net gain coefficient $(G - A)$ can go through a maximum in its dependence on intensity.

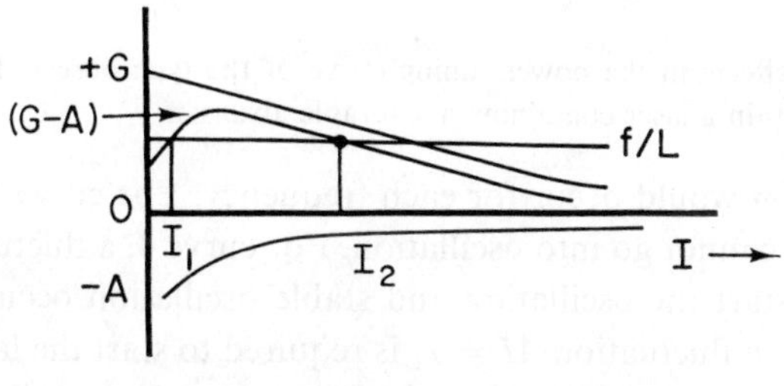

FIGURE 2-38. Qualitative behavior of net gain coefficient in the presence of a saturable absorber [Eqs. (2.232) and (2.233)] (after Kazantsev *et al.*, 1968).

If threshold loss in the laser occurs at the horizontal line f/L, then it is clear that there would be two different stable conditions in the laser: namely, the point where $I = 0$ (no oscillation) and the point where $I = I_2$. Hence if the laser were not oscillating and a sudden fluctuation of intensity of amount $\Delta I = I_1$ arose, the system would immediately go into stable oscillation at intensity I_2. However, if the system were oscillating at intensity I_2, a fluctuation in intensity of amount $\Delta I = -I_1$ could not throw it out of oscillation. The hysteresis effects observed by Chebotayev and Lisitzin may be understood by noting that a whole family of such different charactersitics will occur as a function of frequency as the cavity is tuned in from the edge of the Doppler distribution towards the center of the line, as illustrated at the bottom of Fig. 2-39. The circles indicate points where

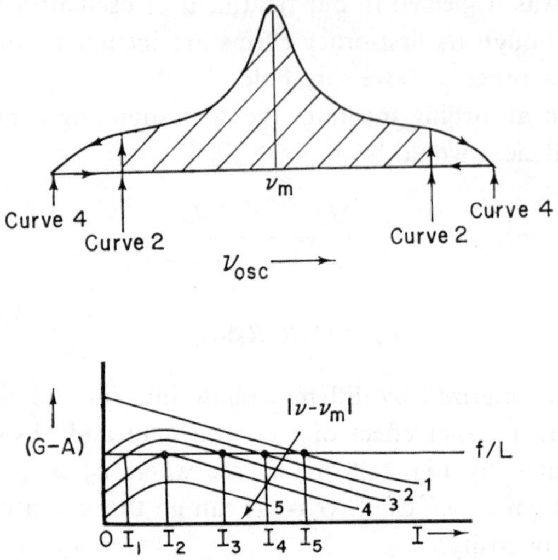

FIGURE 2-39. Bottom: Family of net gain curves in the presence of a saturable absorber which may be used to explain the hysteresis effects illustrated at the top (after Kazantsev *et al.*, 1968).
Top: Hysteresis effects in the power tuning curve of the type observed by Chebotayev and Lisitzin (1969) in a laser containing a saturable absorber.

stable oscillation would occur for each frequency. For curve 5 (ν_{osc} far from ν_m), the system cannot go into oscillation. For curve 4, a fluctuation $\Delta I = I_2$ is required to start the oscillation and stable oscillation occurs at intensity I_2. For curve 3, a fluctuation $\Delta I = I_1$ is required to start the laser and stable oscillation occurs at I_3. For curve 2, the laser goes into oscillation even for $\Delta I = 0$ and stable oscillation occurs at I_4. Curve 1 (stable oscillation at I_5) is nearest to the center of the line and the laser is well above threshold. These hysteresis effects can show up for some values of the gain and absorption coefficient in the single-mode tuning curve as illustrated at the top of Fig. 2-39. However, the shaded region is adequately described by the hole-burning analysis given above in Sec. 2.30.

2.35. Hole-Burning Effects in Three-Level Laser Spectroscopy

The helium–neon laser was the first gas system to oscillate at optical frequencies and has continued to be of substantial value for a wide variety of applications since that time. Comprehensive discussions of this system were given by the author (Bennett, 1962a, 1965a) and a fairly complete tabulation of laser transitions observed in this system is continued in the Appendix to Bennett (1965a).

The system is particularly noteworthy for the large number of stable cw laser transitions which can be made to oscillate with relatively little difficulty in the milliwatt range with little discharge power. The strongest transitions range from the red portion of the visible spectrum to the middle infrared.

One particularly useful aspect of the system arises in the large number of instances in which two laser transitions involve a common excited state. Some of the more important transitions illustrating this aspect of the system are shown in Fig. 2-40 and have been used in a number of recent experiments to determine homogeneous linewidth parameters, in relatively-clean experiments involving interaction with running waves from single-mode sources rather than within the complicated environment of an oscillating laser.

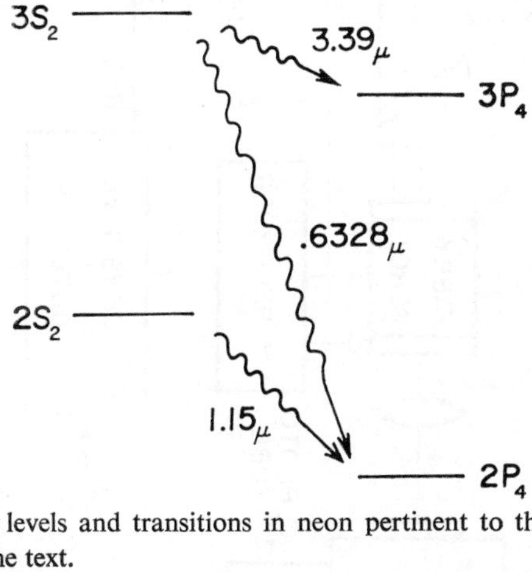

FIGURE 2-40. Energy levels and transitions in neon pertinent to three-level laser experiments discussed in the text.

One of the first of these experiments was conducted by Bennett, Chebotayev and Knutson (1967a) [also see the discussion in Bennett (1968)]. Here the relationship of the high-gain 3.39-micron transition and the red 0.6328-micron line was used in an experiment illustrated schematically in Fig. 2-41. A single-frequency running wave was generated by a helium–neon laser tuned to the center of the Doppler line at 3.39-microns and amplified. The running wave was next used to burn a hole in the velocity distribution of upper-state atoms in the upper laser state ($3s_2$ level in Paschen notation).

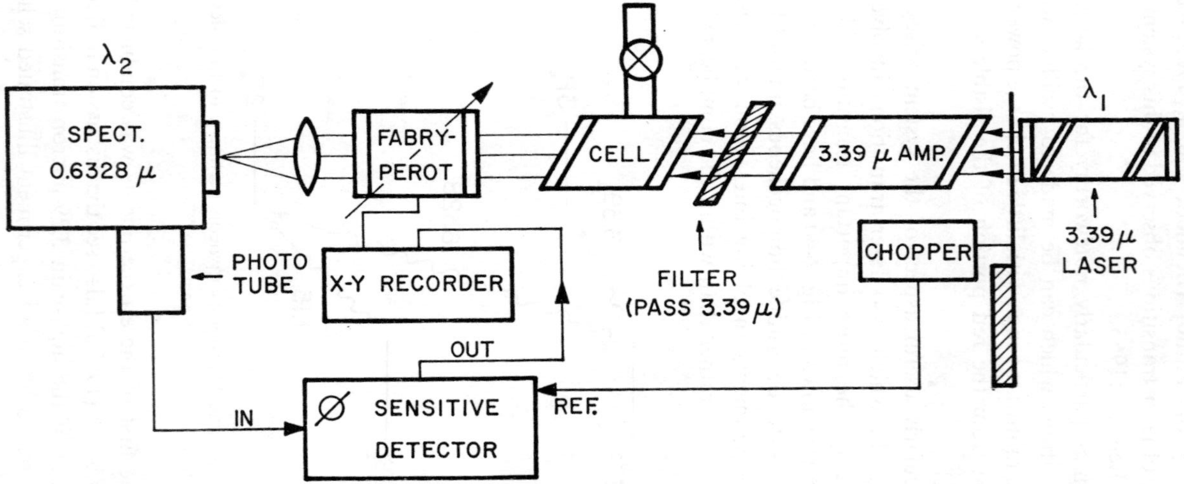

FIGURE 2-41. Schematic diagram of three-level experiment conducted by Bennett, Chebotayev and Knutson, 1967 (also see, Bennett, 1968).

The shape of this hole was then observed in spontaneous emission on the 0.6328-micron line through use of a Fabry–Perot interferometer by looking at the difference in spontaneous emission with and without the laser in the forward direction (achieved by modulating the 3.39-micron laser and using standard phase-sensitive detection methods). Some actual data from this experiment, showing the shape of the hole observed in the line, are given in Fig. 2-42, in which the Fabry–Perot has been swept through a little more than one free spectral range ($c/2L$) for purposes of frequency calibration. The shape of the hole involves the linewidths for both transitions. In the limit that collisions destroy the frequency correlation effects discussed below, the resulting line shape (including the initial Doppler distribution) has been expressed in closed form in terms of tabulated functions (Bennett, 1968). The form of the resultant line shape simplifies considerably, of course, for large Doppler widths and the limit involved here where the two wavelengths are widely different.

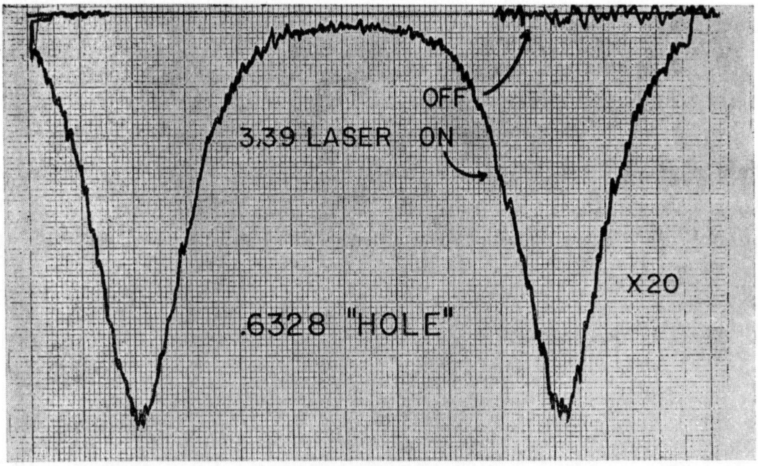

FIGURE 2-42. Hole burned in 6328 Å spontaneous emission profile as observed by Bennett, Chebotayev and Knutson, 1967; (also see Bennett, 1968).

Ignoring correlation effects, the form of the line shape in this experiment may be obtained in the following manner:

Stimulated emission by the running wave with frequency ν_1 at 3.39 microns (λ_1) produces a hole in the upper-state velocity distribution centered at velocity $v_{z1} = \lambda_1(\nu_{m1} - \nu_1)$. The shape of this hole in the velocity distribution is given by,

$$\Delta N(v_z) \propto \left[1 + 4\left(\frac{v_z - v_{z1}}{\lambda_1 \Delta\nu_{L1}}\right)^2\right]^{-1} \exp(-mv_z^2/2kT) \qquad (2.234)$$

where $\Delta\nu_{L1}$ is the full Lorentz width a half-maximum response (including the effects of power broadening) for the transition at λ_1 and is given by Eq. (2.124). Equation (2.234) follows from Eq. (2.129) through application of the Doppler-shift relations. The actual depth of the population hole is not important here, but can be determined from Eq. (2.130). The magnitude of the saturation term in the denominator of Eq. (2.130) does, however, permit a direct experimental determination of the power-broadening correction to the Lorentz width at λ_1 through use of Eq. (2.124).

The apparent frequency distribution of atoms in the velocity hole as observed through the Doppler effect at 0.63 microns (λ_2) is of the form

$$\Delta N(\nu_2') \propto \left[1 + 4\left(\frac{\lambda_2}{\lambda_1}\right)^2\left(\frac{\nu_{20} - \nu_2'}{\Delta\nu_{L1}}\right)^2\right]^{-1} \exp[-m\lambda_2^2(\nu_{m2} - \nu_2')^2/2kT] \tag{2.235}$$

where

$$\nu_{m2} - \nu_{20} \equiv v_{z1}/\lambda_{12} \tag{2.236}$$

is the frequency offset from the center of the profile at λ_2 corresponding to the location of the hole burnt at λ_1.

The hole seen in the spontaneous emission profile at λ_2 is obtained by multiplying Eq. (2.235) by the Lorentzian response function at λ_2 (which involves the non-power-broadened Lorentz width, $\Delta\nu_{L02}$) and integrating over ν_2'. Thus the spontaneous emission hole seen as a function of frequency on the transition at λ_2, due to the running wave interaction at λ_1 is

$$I_2(\nu_2 - \nu_{20}) \propto \int_{-\infty}^{\infty} \frac{d\nu_2' \exp[-m\lambda_2^2(\nu_{m2} - \nu_2')^2/2kT]}{[1 + 4(\nu_2' - \nu_2)^2/(\Delta\nu_{L02})^2]\,[1 + 4(\lambda_2/\lambda_1{}^2)(\nu_{20} - \nu_2')^2/(\Delta\nu_{L1})^2]}. \tag{2.237}$$

This integral is given exactly in terms of the line shape expression contained in the large bracketed term in Eq. (2.135) with the substitutions

$$x_2 = \frac{2\sqrt{\ln 2}}{\Delta\nu_D}(\nu_2 - \nu_{2m}); \quad y_2 = \sqrt{\ln 2}\,\frac{\Delta\nu_{L02}}{\Delta\nu_D};$$

$$x_1 = \frac{2\sqrt{\ln 2}}{\Delta\nu_D}(\nu_{20} - \nu_{2m}); \quad y_1 = \sqrt{\ln 2}\,\frac{\lambda_1\Delta\nu_{L1}}{\lambda_2\Delta\nu_D}; \tag{2.238}$$

$$S = \frac{2\sqrt{\ln 2}}{\Delta\nu_D}(\nu_2 - \nu_{20}) \equiv x_2 - x_1; \text{ and } \Delta\nu_D = 2\lambda_2\sqrt{m\ln 2/2kT}.$$

The exact expression is much too cumbersome to be of any practical value in the experimental determination of linebreadth parameters when the values of the Lorentz widths and Doppler widths are comparable, but nevertheless different. The exact expression simplifies considerably in various limiting cases. For example, if the Doppler width is large compared to both Lorentz widths, if the laser frequency is tuned to the center of the line and if $y_1 \gg y_2$, the exact expression reduces to

$$I(\nu_2 - \nu_{2m}) \propto \left[1 + 4\left(\frac{\lambda_2}{\lambda_1}\right)^2\left(\frac{\nu_2 - \nu_{2m}}{\Delta\nu_{L1}}\right)^2\right]^{-1} \exp\left[-4\ln 2\left(\frac{\nu_2 - \nu_{2m}}{\Delta\nu_D}\right)^2\right]. \tag{2.239}$$

Equation (2.239) can, of course, be seen directly from Eq. (2.237) by noting that the sharper Lorentzian acts like a δ-function. This approximation isn't too bad for the 3.39-micron 0.6328 micron experiment. Using approximation (2.239), Bennett, Chebotayev and Knutson (1967a) obtained values for Lorentz broadening of the 3.39 micron neon line of about 60 MHz/torr of helium.

The mirror image of this experiment was done by Hansch and Toschek (1968) with running waves on the 0.6328- and 1.15-micron transitions. Here a single-frequency running wave at 0.6328 was used to burn a hole in the $3s_2$ population distribution, thereby enhancing the population distribution at the line center on the $2p_4$ state. The shape of the bump induced in this way on the $2p_4$ velocity distribution was determined by measuring the gain in the forward direction on the 1.15-micron transition as a function of frequency, using a second, single-frequency laser source.

Similarly, another experiment was carried out by Shank and Schwarz (1968), using two running-wave laser sources at 0.6328. One of these was used to produce an enhanced population distribution on the $2p_4$ state, and the other was used to measure the gain at 0.6328 as a function of frequency. This approach is a little more limited in that is it impossible to have the two laser sources precisely colinear without strong interference effects and the analysis for the interaction is only valid in the limit of low intensities. However, these considerations should not introduce significant difficulties with the method, and consistent data for line broadening of the 0.6328-micron transition were taken as a function of pressure in this experiment.

Similar studies of induced bumps on the $2p_4$ velocity distribution have also been made in ocillating lasers by a number of investigators (Schweitzer, Birky and White, 1967; Cordover, Bonczyk and Javan, 1967a; 1967b; Holt, 1967; 1968), although the analysis is necessarily more complicated due to the presence of both running waves comprising the standing wave in the laser cavity. Correspondingly, two bumps occur in the lower-state velocity distribution when the mode is detuned from line center. In one of these studies, Holt (1968) has shown that an *asymmetry* in the bump height that exists in the latter case would be implied by correlation effects in the Weisskopf–Wigner and Heitler theories of the spontaneous emission process (Holt, 1967).

The first explanation of the bump height and width asymmetry in three-level experiments of this type was given by Helen Holt (1967) and will be followed here. The meaning of frequency correlation effects in this type of experiment can be understood by considering atoms at rest which emit frequencies ν and ν' successively (see Fig. 2-43). The intensity distri-

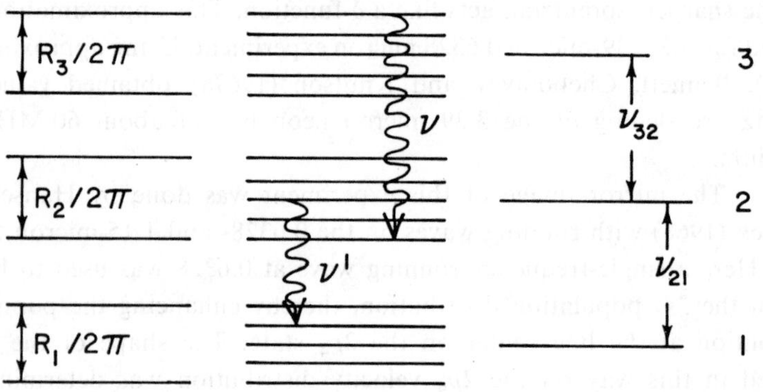

FIGURE 2-43. Schematic energy-level diagram (after Holt, 1967).

bution on the transition 3 → 2 is normally centered at ν_{32} and Lorentzian with full width at half-maximum response

$$\Delta\nu_{32} = \frac{R_2 + R_3}{2\pi}.$$

Similarly, the intensity distribution on the transition 2 → 1 is normally centered at ν_{21} with

$$\Delta\nu_{21} = \frac{R_1 + R_2}{2\pi}.$$

However, there are correlation effects in the cascade process if the same atom emits both photons in the transition 3 → 2 → 1. Namely, if $\nu > \nu_{32}$

the atom will tend to emit a frequency $\nu' < \nu_{21}$ such that

$$\nu + \nu' = \nu_{32} + \nu_{21}$$

within

$$\Delta\nu_{\text{total}} = \frac{R_1 + R_3}{2\pi}$$

and not within $\Delta\nu_{32} + \Delta\nu_{21}$. The width $\Delta\nu_{\text{total}}$ is the same width that would have been obtained if the atom had directly undergone the transition 3 → 1. Hence the frequency emitted in the transition 2 → 1 depends on the frequency emitted in the transition 3 → 2.

Laser experiments can show up such correlation effects rather strongly because it is possible to detect only those atoms which have undergone stimulated emission in the first step of a cascade response and hence which have all been forced to emit their energy at a precisely defined frequency. The initial frequency of stimulated emission is precisely the same in the laboratory frame, but is a function of v_z in the atom's rest frame.

Consider a constant velocity distribution of atoms $N(v_z)$ interacting with a running wave of frequency $\nu > \nu_{32}$ in the z direction as the first step of the cascade sequence in Fig. 2-33. As discussed previously in Sec. 2.17, this interaction leads to a Lorentzian-shaped hole being burned into the velocity distribution of width $\lambda_{32}\,\Delta\nu_{32}$ as shown in Fig. 2-44.

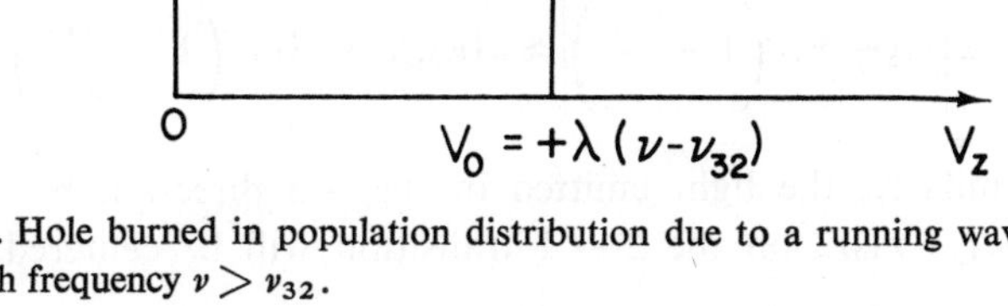

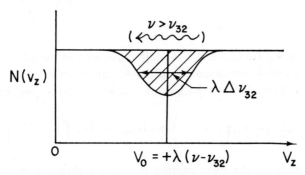

FIGURE 2-44. Hole burned in population distribution due to a running wave in the $-z$ direction with frequency $\nu > \nu_{32}$.

The atoms with v_z precisely equal to $v_0 \equiv \lambda_{32}(\nu - \nu_{32})$ have been forced to undergo stimulated emission at their resonance frequency for the transition 3 → 2. Consequently, they will tend to emit light which is centered about the frequency ν_{21} in their subsequent spontaneous emission on the transition $2 \rightarrow 1$ in their rest frames. Hence light which they emit spontaneously in the $+z$ direction will be centered around the frequency

$$\nu_{+z} \equiv \nu_{21}(1 + v_0/c)$$

within $\Delta\nu_{\text{total}}$ as observed in the laboratory frame. Similarly, light will be emitted with equal intensity by these atoms in the $-z$ direction which will be centered about the frequency

$$\nu_{-z} \equiv \nu_{21}(1 - v_0/c)$$

in the laboratory frame.

Next consider atoms with $v_z = v_0 + \lambda_{32}\,\Delta\nu_{32}/2$. They have been forced to undergo stimulated emission at the frequency $(\nu_{32} - \Delta\nu_{32}/2)$ in their rest frame. Consequently they will tend to emit light spontaneously at a frequency $(\nu_{21} + \delta)$ in their rest frame on the transition $2 \rightarrow 1$, such that

$$(\delta - \Delta\nu_{32}/2) \lesssim \Delta\nu_{\text{total}}/2.$$

The light emitted spontaneously in the $+z$ direction by these atoms with $v_z = v_0 + \lambda_{32}\,\Delta\nu_{32}/2$ will be centered around the frequency

$$\nu'_{+z} = (\nu_{21} + \delta)\left(1 + \frac{v_z}{c}\right) \approx \left(\nu_{21} + \frac{\Delta\nu_{\text{total}} + \Delta\nu_{32}}{2}\right)\left(1 + \frac{v_0}{c} + \frac{\lambda_{32}\Delta\nu_{32}}{2c}\right) \tag{2.240}$$

in the laboratory frame. Hence the full width seen in the laboratory frame of the line emitted spontaneously in the $+z$ direction on the cascade transition $2 \rightarrow 1$ from atoms in the hole at $+v_z$ will be given roughly by

$$\Delta\nu'_{+z} \approx 2\left|\nu'_{+z} - \nu_{21}\left(1 + \frac{v_0}{c}\right)\right| \approx \Delta\nu_{\text{total}} + \Delta\nu_{32}\left(1 + \frac{\lambda_{32}}{\lambda_{21}}\right). \tag{2.241}$$

Similarly, the light emitted in the $-z$ direction by atoms with $v_z = v_0 + \lambda_{32}\,\Delta\nu_{32}/2$ on the $2 \rightarrow 1$ transition will be centered about the frequency

$$\nu'_{-z} = (\nu_{21} + \delta)\left(1 - \frac{v_z}{c}\right) \approx \left(\nu_{21} + \frac{\Delta\nu_{\text{total}} + \Delta\nu_{32}}{2}\right)\left(1 - \frac{v_0}{c} - \frac{\lambda_{32}\Delta\nu_{32}}{2c}\right) \tag{2.242}$$

in the laboratory frame. Therefore, the full width of the line emitted spontaneously in the $-z$ direction on the cascade transit will be roughly

$$\Delta\nu'_{-z} \approx 2\left|\nu'_{-z} - \nu_{21}\left(1 - \frac{v_0}{c}\right)\right| \approx \Delta\nu_{\text{total}} + \Delta\nu_{32}\left(1 - \frac{\lambda_{32}}{\lambda_{21}}\right). \qquad (2.243)$$

Hence the width of the line in the laboratory frame seen on the $2 \to 1$ transition in the $+z$ emission will be broader than the width seen in the $-z$ emission by an amount $\approx 2\Delta\nu_{32}(\lambda_{32}/\lambda_{21})$. Because the total probability of spontaneous emission is the same in both directions on the transitions $2 \to 1$, relative intensity distributions are obtained in the two cases discussed above in which the curve with the narrowest width has the highest peak intensity. Hence there will be an asymmetry in both the width and the peak intensity for the two directions of spontaneous emission relative to the direction of propagation of the running wave which stimulates these atoms to emit on the transition $3 \to 2$.

The above asymmetry is easiest to detect through observation of spontaneous emission in one direction and the use of the standing wave in a laser cavity (detuned from line center for the transition $3 \to 2$) to generate running waves in opposite directions. In this case one gets two spontaneous emission bumps on the cascade transition of the type illustrated in Fig. 2-45. This effect was first observed by Holt (1967) using a scanning

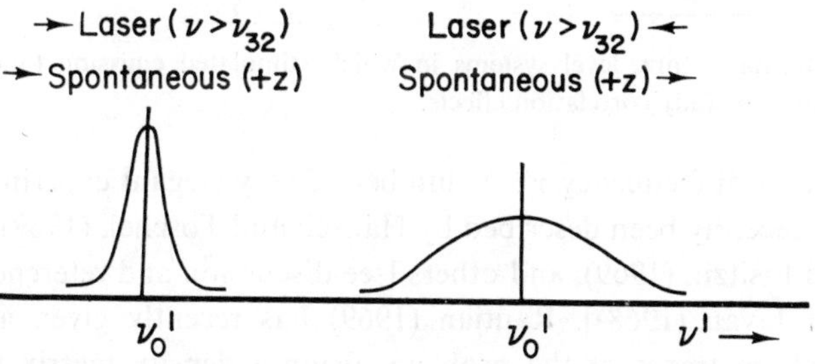

FIGURE 2-45. Asymmetry in the spontaneous emission "bumps" of the type first studied by Holt (1967, 1968).

Fabry–Perot to record the "bumps" on the intensity distributions of the neon $2p_4 \to 1s_4$ spontaneous emission line which were induced through stimulated emission on the $2s_2 \to 2p_4$ laser transition at 1.15 microns when the laser was detuned significantly from the line center. Holt observed a small but definite asymmetry using a lock-in detection system much like that illustrated in Fig. 2-31. The effects of collisions are expected to be quite important under the conditions of this experiment and of course tend to destroy the frequency correlation effects which produce the asymmetry. In the limit of high pressure and $\Delta\nu_D \gg \Delta\nu_L$ the two "bumps" become identical and have full widths which are given by

$$\Delta\nu \approx \frac{\lambda_{32}}{\lambda_{21}}\left(\frac{R_2 + R_2}{2\pi}\right) + \left(\frac{R_2 + R_1}{2\pi}\right),$$

where the relaxation rates include the effects of collisions. As Holt points out, the basic change in the nature of the line shape due to frequency correlation effects at low pressure makes it difficult to conduct a meaningful extrapolation of collision-induced line breadths to low pressure values.

Very similar asymmetry effects can be observed using stimulated emission or absorption as a probe in the standing wave produced in oscillating lasers, with three-level systems such as those illustrated in Fig. 2-46, where oscillation occurs in the cavity at frequency ν and is probed with

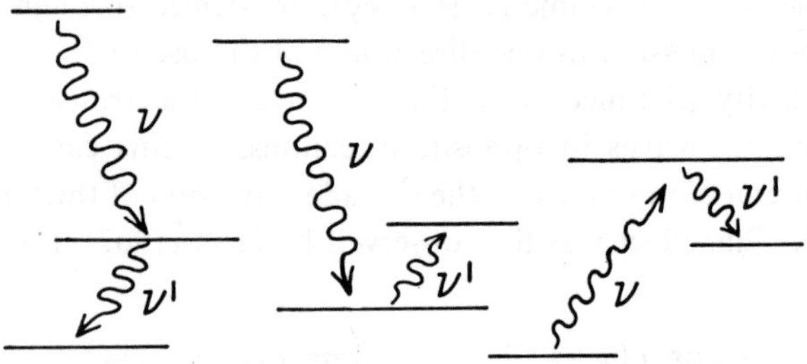

FIGURE 2-46. Basic three-level systems in which stimulated emission (and absorption) has been used to study correlation effects.

another laser at frequency ν'. A number of very elegant experiments of this sort have recently been described by Hänsch and Toschek (1969), by Chebotayev and Lisitzin (1969), and others [see discussion and references given by Feld and Javan (1968)]. Rautian (1969) has recently given an extensive theoretical treatment of the problem, using a density matrix formulation with strong and weak signals at the two frequencies involved. Similar calculations have also been made by a number of other investigators (Feld and Javan, 1968; Feld, 1969; Feldman and Feld, 1970). The technique appears to be a very useful and general one for the investigation of line-breadth parameters in the limit that $\Delta\nu_L \ll \Delta\nu_D$. However, the inclusion of finite Doppler broadening effects leads to rather formidable and cumbersome expressions which will probably limit the practical value of the technique as a method of precision measurement under conditions where the Doppler width is more comparable to the Lorentz width.

ACKNOWLEDGEMENT

The author is indebted to the John S. Guggenheim Foundation for post-doctoral fellowship support during the preparation of the present manuscript.

The author is also indebted to the Air Force Office of Scientific Research and the Army Research Office at Durham for the support of much of the research described in the present manuscript.

The author also wishes to express his thanks to the editors of the journals cited and the authors of the respective articles for permission to reproduce the specific material and figures referenced in the text.

REFERENCES

Abella, I. D., N. A. Kurnit, and S. R. Hartmann (1966), Phys. Rev. **141**, 391.

Arakeljan, V. S., N. V. Karlov, and A. M. Prokhorov, presented at the International Conference on Gas Lasers held at Novosibirsk, U.S.S.R., July 1969 (unpublished).

Arrathoon, R., and D. A. Sealer (1971), Phys. Rev. **4**, 815.

Austen, D. H. (1968), IEEE J. Quantum Electron. **QE-4**, 471.

Ballik, E. A. (1964), PhD dissertation, Oriel College, Oxford, entitled "Experiments with Coherent Radiation".

Ballik, E. A. (1966), Appl. Opt. **5**, 170.

Ballik, E. A. (1971), Canadian J. Phys. **50**, 47.

Barger, R. L., and J. L. Hall (1969), Phys. Rev. Letters **22**, 4.

Basov, N. G., and V. S. Letokhov (1968), Zh. Eksp. Teor. Fiz. **96**, 585.

Beiser, L. (1968), Appl. Phys. Letters **13**, 87.

Beiser, L., CBS Laboratories, private communication.

Bennett, W. R., Jr. (1962a), Appl. Opt. Suppl. **1**, 24.

Bennett, W. R., Jr. (1962b), Phys. Rev. **126**, 580.

Bennett, W. R., Jr. (1963), in *Quantum Electronics III — Paris 1963* edited by P. Grivet and N. Bloembergen (Columbia University Press, New York, 1964), pp. 442–458.

Bennett, W. R., Jr. (1965a), Appl. Opt. Suppl. **2**, 3.

Bennett, W. R., Jr. (1965b), Appl. Opt. Suppl. **2**, 78.

Bennett, W. R., Jr. (1968), reported at the International Conference on Atomic Physics held at New York University in July, 1968; proceedings published in *Atomic Physics*, edited by B. Bederson, V. W. Cohen and F. M. Pichanick (Plenum Press, New York, 1969), pp. 435–473.

Bennett, W. R., Jr. (1970a), Phys. Rev. **A2**, 458.

Bennett, W. R., Jr. (1970b), Comments on Atomic and Molecular Physics **2**, 10.

Bennett, W. R., Jr. (1972), Comments on Atomic and Molecular Physics **3**, 63.

Bennett, W. R., Jr., S. F. Jacobs, J. T. LaTourrette, and P. Rabinowitz (1964), Appl. Phys. Letters **5**, 56.

Bennett, W. R., Jr., P. J. Kindlmann, and G. N. Mercer (1965), Appl. Opt. Suppl. **2**, 34.

BENNETT, W. R., Jr., and P. J. KINDLMANN (1966), Phys. Rev. **149**, 38.

BENNETT, W. R., Jr., E. A. BALLIK, and G. N. MERCER (1966), Phys. Rev. Letters **16**, 603.

BENNETT, W. R., Jr., V. P. CHEBOTAYEV and J. W. KNUTSON, Jr. (1967a), in *Fifth International Conference on the Physics of Electronic and Atomic Collisions*, edited by L. Branscomb (Nauka, Leningrad, U.S.S.R., 1967), p. 521.

BENNETT, W. R., Jr., V. P. CHEBOTAYEV, and J. W. KNUTSON, Jr. (1967b), Phys. Rev. Letters **18**, 688.

BENNETT, W. R., Jr., J. W. KNUTSON, Jr., and R. C. SZE (1969), presented at the International Conference on Gas Lasers held at Novosibirsk, U.S.S.R., July 1969 (unpublished).

BERMAN, P. R., and W. E. LAMB, Jr. (1970), Phys. Rev. **A2**, 2435.

BERMAN, P. R., and W. E. LAMB, Jr. (1971), Phys. Rev. **A4**, 319.

BETEROV, I. M., YU. A. MATYUGIN, S. G. RAUTIAN, and V. P. CHEBOTAYEV (1970), Zh. Eksp. Teor. Fiz. **58**, 1243; Soviet Physics JETP **31**, 668.

BLOCH, F. (1946), Phys. Rev. **70**, 460.

BODE, H. W. (1945), *Network Analysis and Feedback Amplifier Design* (D. Van Nostrand, Co., Inc., Princeton, N.J., 1945).

BORENSTEIN, M., and W. E. LAMB, Jr. (1972), Phys. Rev. **A5**, 1311.

BORN, M. (1923), *Optik* (Julius Springer-Verlag, Berlin, 1933).

BOYD, G. D., and J. P. GORDON (1961), Bell System Tech. J. 40, 489; also see G. D. BOYD in *Advances in Quantum Electronics*, edited by J. Singer (Columbia University Press, New York, 1961), pp. 318–327.

BOYD, G. D., and H. KOGELNIK (1962), Bell System Tech. J. **41**, 1347.

CHEBOTAYEV, V. P., I. M. BETEROV, and V. N. LISITZIN (1968), IEEE. J. Quantum Electron. **QE-4**, 339, abstract 7–13; IEEE J. Quantum Electron. **QE-4**, 788.

CHEBOTAYEV, V. P., and V. N. LISITZIN (1968), Zh. Eksp. Teor. Fiz. **54**, 419.

CHEBOTAYEV, V. P., and V. N. LISITZIN (1969), presented at the International Conference on Gas Lasers held at Novosibirsk, U.S.S.R., July 1969 (unpublished).

CORDOVER, R. H., T. S. JASEJA, and A. JAVAN (1965), Appl. Phys. Letters **7**, 322.

CORDOVER, R. H., P. A. BONCZYK, and A. JAVAN (1967a), Phys. Rev. Letters **18**, 730.

CORDOVER, R. H., P. A. BONCZYK, and A. JAVAN (1967b), Phys. Rev. Letters **18**, 1104.

CORDOVER, R. H., and P. A. BONCZYK (1969), Phys. Rev. **188**, 696.

DEUTSCH, M., and C. S. BROWN (1952), Phys. Rev. **85**, 1047.

DICKE, R. H. (1953), Phys. Rev. **89**, 472.

DICKE, R. H. (1958), U.S. Patent No. 2851652.

DIENES, A. (1968), Phys. Rev. **174**, 400; (1968), Phys. Rev. **174**, 414.

EINSTEIN, A. (1917), Physik Zeit **18**, 121.

FADDEYEVA, V. N., and N. M. TERENT'EV (1961), *Tables of the Function W(z) ... for Complex Argument* (Pergamon Press, New York).

FELD, M. S. (1969), presented at the International Conference on Gas Lasers held at Novosibirsk, U.S.S.R., July 1969 (unpublished).

FELD, M. S., and A. JAVAN (1968), Phys. Rev. Letters **20**, 578.

FELD, M. S., A. JAVAN, and P. H. LEE (1968), Appl. Phys. Letters **13**, 424.

FELDMAN, B. J., and M. S. FELD (1970), Phys. Rev. **1A**, 1376.

FLAMMER, C. (1957), *Spheroidal Wave Functions* (Stanford University Press. Palo Alto, Calif.).

FOLEY, H. M. (1946), Phys. Rev. **69**, 616.

FORK, R. L., L. E. HARGROVE, and M. A. POLLACK (1964), Appl. Phys. Letters **5**, 5.
FORK, R. L., D. R. HERRIOTT, and H. KOGELNIK (1964), Appl. Opt. **3**, 1471.
FORK, R. L., and M. A. POLLACK (1965), Phys. Rev. **139**, A1408.
FORRESTER, A. T., R. A. GUDMUNDSON, and P. O. JOHNSON (1955), Phys. Rev. **99**, 1691.
FOX, A. G., and TINGYE LI (1961), Bell System Tech. J. **40**, 453.
FOX, A. G., and TINGYE LI (1963), Proc. IEEE. **51**, 80.
FRIED, B. D., and S. D. CONTE (1961), *The Plasma Dispersion Function* (Academic Press, Inc., New York, 1961).
GOLDENBERG, H. M., D. KLEPPNER, and N. F. RAMSEY (1961), Phys. Rev. **123**, 530.
GORDON, E. I., and A. D. WHITE (1964), Proc. IEEE **52**, 206.
GORDON, J. P., H. J. ZEIGER, and C. H. TOWNES (1955), Phys. Rev. **99**, 1264.
GOULD, GORDON (1958), unpublished proposal to ARPA.
GRIEM, H. R. (1964), *Plasma Spectroscopy* (McGraw-Hill Book Co., New York, 1964).
GYORFFY, B. L., M. BORENSTEIN, and W. E. LAMB, Jr. (1968), Phys. Rev. **169**, 340.
HAHN, E. L. (1950), Phys. Rev. **77**, 746; **80**, 580.
HÄNSCH, T., and P. TOSCHEK (1968), IEEE J. Quantum Electron, **QE-4**, 467.
HÄNSCH, T. W., and P. E. TOSCHEK (1969), IEEE J. Quantum Electron. **QE-5**, 61.
HÄNSCH, T. W., and P. E. TOSCHEK (1969) presented at the International Conference on Gas Lasers held at Novosibirsk, U.S.S.R., July 1969 (unpublished).
HÄNSCH, T. W., M. D. LEVENSON, and A. L. SCHAWLOW (1971), Phys. Rev. Letters **26**, 946.
HÄNSCH, T. W., I. S. SHAHIN, and A. L. SCHAWLOW (1971), Phys. Rev. Letters **27**, 707.
HERRIOTT, D. R. (1963), Appl. Opt. **2**, 865.
HOLSTEIN, T. (1947), Phys. Rev. **72**, 1212.
HOLSTEIN, T. (1951), Phys. Rev. **83**, 1159.
HOLT, H. K. (1967), Phys. Rev. Letters **19**, 1275.
HOLT, H. K. (1968), Phys. Rev. Letters **20**, 410.
HOLT, H. K. (1970), Phys. Rev. **A2**, 233.
HUGHES, V. W., S. MARDER, and C. S. WU (1957), Phys. Rev. **106**, 934.
JAVAN, A., E. A. BALLIK, and W. L. BOND (1962), J. Opt. Soc. Am. **52**, 96.
JAVAN, A., W. R. BENNETT, Jr., and D. R. HERRIOTT (1961), Phys. Rev. Letters **6**, 106; also see articles by these authors in *Advances in Quantum Electronics*, edited by J. Singer (Columbia University Press, New York, 1961), pp. 18–49.
KAZANTSEV, A. P., S. G. RAUTIAN, and G. I. SURDUTOVICH (1968), Zh. Eksp. Teor. Fiz **54**, 1409 and Soviet Physics JETP **27**, 756.
KLEINMAN, D. A., and P. P. KISLIUK (1962), Bell Syst. Tech. J. **41**, 453.
KNUTSON, J. W., Jr., and W. R. BENNETT, Jr. (1971), Bull. Am. Phys. Soc. **16**, 592.
KOGELNIK, H., and W. W. RIGROD (1962), Proc., IRE **50**, 220.
KOLOMNIKOV, YU. D., YU. V. TROITSKIY, and V. P. CHEBOTAYEV (1965), translated in Radio Eng. and Elect. Phys. **10**, 312.
KRAMERS, H. A. (1927), Atti, Congr. Intern. Fisici, Como **2**, 545, reproduced in *H. A. Kramers-Collected Scientific Papers* (North Holland Publishing Co., Amsterdam, 1956), pp. 333–345.
KRONIG, R. DE L. (1926), Opt. Soc. Am. **12**, 347.
KURNIT, N. A., I. D. ABELLA, and S. R. HARTMANN (1964), Phys. Rev. Letters **13**, 567.
LAMB, W. E., Jr. (1952), Phys. Rev. **85**, 259.
LAMB, W. E., Jr. (1964), Phys. Rev. **134**, A1429.
LAMB, W. E., Jr., and T. M. SANDERS, Jr. (1960), Phys. Rev. **119**, 1901.

LaTourrette, J. T., S. F. Jacobs, and P. Rabinowitz (1964), Appl. Opt. **3**, 981.
Lee, P. H., and M. L. Skolnick (1967), Appl. Phys. Letters **10**, 303.
Lee, P. H., P. B. Schoefer, and W. B. Barker (1968), Appl. Phys. Letters **13**, 373.
Letokhov, V. S. (1967), Zh. Eksp. Pis. Red. **6**, 597 and Soviet Physics JETP Letters **6**, 101.
Letokhov, V. S. (1968), Zh. Eksp. Teor. Fiz. **54**, 1244 and Soviet Physics JETP **7**, 665.
Letokhov, V. S. (1971), Comments on Atomic and Molecular Physics **2**, 181.
Letokhov, V. S., and V. P. Chebotayev (1969), Zh. ETF. Pis Red. **9**, 364 and Soviet Physics JETP Letters **9**, 215.
Lindholm, E. (1945), Ark. Mat. Astron. Fys. **32A**, 17.
Lisitzin, V. N. (1968), Research Thesis, Academ Gorodok, Novosibirsk, U.S.S.R. under the direction of V. P. Chebotayev, entitled "Experimental Investigation of Gas Lasers with Saturable Absorbers as Resonators."
Lisitzin, V. N., and V. P. Chebotayev (1968), Soviet Physics JETP **27**, 227.
McCall, S. L., and E. L. Hahn (1969), Phys. Rev. **183**, 457.
McFarlane, R. A., W. R. Bennett, Jr., and W. E. Lamb, Jr., (1963), Appl. Phys. Letters **2**, 189.
Mikhnenko, G. A., E. D. Protsenko, E. A. Sedoi, and M. P. Sorokin (1970), Optics and Spectroscopy **29**, 65.
Mitchell, A. C. G., and M. W. Zemansky (1934), *Resonance Radiation and Excited Atoms* (Cambridge University Press, Cambridge, England).
Morse, P. H., and H. Feshbach (1953), *Methods of Theoretical Physics* (McGraw-Hill Book Co., New York).
Packard, J. R., W. C. Tait, and G. H. Dierssen (1971), Appl. Phys. Letters **19**, 338.
Pole, R. V. (1965), J. Opt. Soc. Am. **55**, 254.
Pole, R. V., and R. A. Myers (1966), IEEE J. Quantum Electron, **QE-2**, 182.
Pole, R. V., and R. A. Myers (1967), IBM J. Res. Develop. **11**, 502.
Prokhorov, A. M. (1958), Zh. Eksp. Teor. Fiz. **34**, 1658.
Rabi, I. I. (1937), Phys. Rev. **51**, 652.
Rabinowitz, P., R. Keller, and J. T. LaTourrette (1969), Appl. Phys. Letters **14**, 376.
Ramsey, N. F. (1950), Phys. Rev. **78**, 695.
Rautian, S. G. (1966), Soviet Physics JETP **51**, 1176.
Rautian, S. G. (1967), Soviet Physics JETP **24**, 788.
Rautian, S. G. (1969), presented at the International Conference on Gas Lasers held at Novosibirsk, U.S.S.R., July 1969 (unpublished).
Rautian, S. G., and I. I. Sobel'man (1963), Soviet Physics JETP **17**, 635.
Rautian, S. G., and I. I. Sobel'man (1966), J. Quantum. Electron. **QE-2**, 446; also Usp. Fiz. Nauk **90**, 230 and Soviet Physics Usp. **9**, 701 (1967).
Rigrod, W. W. (1969a), J. Appl. Phys. **34**, 2602.
Rigrod, W. W. (1936b), Appl. Phys. Letters **2**, 51.
Rigrod, W. W. (1965), J. Appl. Phys. **36**, 2487.
Rigrod, W. W. (1970), IEEE J. Quant. Electron. **QE-6**, 9.
Rigrod, W. W., H. Kogelnik, D. J. Brangaccio and D. R. Herriott (1962), J. Appl. Phys. **33**, 743.
Rigrod, W. W., and A. M. Johnson (1967), IEEE J. Quant. Electron. **QE-3**, 644.
Schawlow, A. L., and C. H. Townes (1958), Phys. Rev. **112**, 1940.
Schweitzer, W. G., Jr., M. M. Birky, and J. A. White (1967), J. Opt. Soc. Am. **57**, 1226.

Scully, M., and W. E. Lamb, Jr. (1967), Phys. Rev. **159**, 208.

Shank, C. V., and S. E. Schwarz (1968), Appl. Phys. Letters **13**, 113.

Shimizu, F. (1969), Appl. Phys. Letters **14**, 378.

Shimoda, K., and A. Javan (1965), J. Appl. Phys. **36**, 718.

Slepian, D., and H. O. Pollak (1961), Bell System Tech. J. **40**, 43.

Smith, P. W. (1965), IEEE J. Quant. Electron. **QE-1**, 343.

Smith, P. W. (1966), J. Appl. Phys. **37**, 2089.

Smith, P. W. (1968), Appl. Phys. Letters **13**, 235.

Smith, P. W., and T. Hänsch (1971), Phys. Rev. Letters **26**, 740.

Soncini, G., and O. Svelto (1968), IEEE J. Quantum Electron. **QE-4**, 420.

Statz, H., F. A. Horrigan, S. H. Koozekanani, C. L. Tang, and G. F. Koster (1965), J. Appl. Phys. **36**, 2278; errata J. Appl. Phys. **39**, 4045 (1968).

Sze, R. C., Ye. T. Antropov, and W. R. Bennett, Jr., (1972), Appl. Opt. **11**, 197.

Sze, R. C., and W. R. Bennett, Jr. (1972), Phys. Rev. **5A**, 837.

Szöke, A., and A. Javan (1963), Phys. Rev. Letters **10**, 521.

Szöke, A., and A. Javan (1966), Phys. Rev. **145**, 137.

Townes, C. H. (1961), in *Advances in Quantum Electronics*, edited by J. Singer (Columbia University Press, New York, 1961), p. 3.

Van Vleck, J. H. (1948), in *Radiation Laboratory Series* (McGraw-Hill Book Co., New York, Vol. 13, Chapter 8.

Vehara, K., and K. Shimoda (1965), Japan. J. Appl. Phys. **4**, 921.

Voigt, W. (1912), K. Bayer Akad., *Münchner Ber.*, 683 Weisskopf, V. (1932), Z. Phys. **75**, 287.

White, A. D. (1964), Appl. Opt. **3**, 431.

Experimental X-Ray Astronomy

R. NOVICK
Columbia University

Notes by S. Stobbs

Contents

1. INTRODUCTION

1.1. History

X-ray astronomy began in 1948 with the first direct observation of x-ray emission from the Sun (Burnight, 1949). The term x-ray will be used in these lectures to describe photons with energies lying in the approximate range

$$100 \text{ eV} < h\nu < 200 \text{ keV}.$$

The Earth's atmosphere absorbs such photons very strongly (see Fig. 1-1), and it is thus necessary to carry x-ray detectors above an appreciable fraction of the Earth's atmosphere, on high-altitude balloons, rockets, or artificial satellites, before observations can be made. It is for this reason that x-ray astronomy is still in its infancy. Balloons, reaching a height of 30 km above the Earth's surface, can be used for measurements in the high-energy x-ray range (15–200 keV), for which the transmission of the atmosphere at a height of 30 km is sufficiently large. However, for softer x-rays, in the 0.1–15 keV range, rockets and satellites must be used. Rockets can reach heights of about 200 km, but have the disadvantage that the observing time available is extremely short (typically about 4 min above 100 km). It is hoped that the first x-ray satellite will be launched in 1970, and this will greatly increase the observing time in the low-energy range.

1.1.1. *The Discovery of Nonsolar X-Ray Sources*

Observations of solar x-rays continued during the 1950's and produced many interesting results. However, attempts to detect x-rays from nonsolar sources proved unsuccessful. Tentative theoretical calculations were made during this period of the x-ray flux to be expected from such potential x-ray sources as supernovae, the Crab nebula, and magnetic A

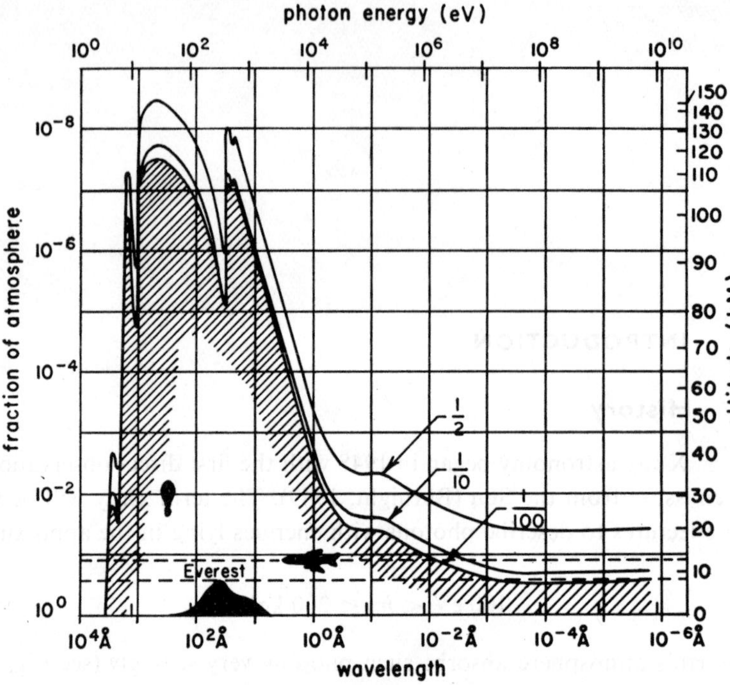

FIGURE 1-1. The transmission of the atmosphere at x-ray energies.

stars. All such calculations predicted that nonsolar sources would be very weak, with strengths typically of the order 10^{-7} times that of the quiet Sun. It was thus clear that much larger, more sensitive detectors were needed.

While such instruments were being developed, it was suggested that it might be possible to detect x-ray emission from the Moon. This could be produced by the high-energy electrons in the solar wind hitting the Moon's surface, or by the excitation of x-ray fluorescence by primary x-rays from the Sun. The first rocket designed to look at the Moon was launched by the American Science and Engineering group (ASE) in October 1961. This first flight was inconclusive, because of instrument failure. A second flight took place in June 1962 and detected a flux of soft x-rays of about 5 photons/cm^2 sec which seemed to be coming preferentially from a limited region of the sky near the position of the center of the Galaxy in the region of the constellation Scorpius. Since this region did not coincide with the Moon's position during the flight, it was concluded that the emission was coming from an object outside the solar system. A constant diffuse background of soft x-rays was also detected. This was the beginning of nonsolar x-ray astronomy.

A third flight, made in October 1962 when the Scorpius region was below the horizon, failed to detect the x-ray source. It was thus confirmed that the source appeared to lie in the general region of Scorpius. It was named Sco X-1 (Giacconi *et al.*, 1962).

1.2. Why Study X-Ray Astronomy?

What information of astrophysical significance can we obtain from a study of x-ray stars? Why do we go to all the trouble to observe these x-rays? There are several reasons:

(1) Because they are there, and unless history fails to repeat itself, we will discover new and unexpected phenomena that will have a profound influence upon our understanding of the Universe.

(2) To study the source mechanisms.

The form of the x-ray spectrum gives important information about the processes which generate the x-rays within the source. Since the nature of these sources is still obscure, a study of the observed x-ray fluxes may help to explain the physical conditions within the source.

(3) To study the interstellar medium.

The interstellar medium absorbs low-energy x-rays. The effective photoelectric cross section, assuming normal cosmic abundances, is shown in Fig. 1-2. As can be seen, this cross section decreases rapidly with increasing energy and shows a series of absorption edges characteristic of individual elements. By observing the absorption of a source as a function of energy, it is possible to measure the density of the interstellar medium. For example, it may be possible, with more refined techniques, to detect the neon edge at 19 Å and so measure the density of interstellar neon directly. X-ray absorption may also help to solve the problem of whether an appreciable concentration of molecular hydrogen (H_2) exists in the interstellar medium. Molecular hydrogen cannot be detected at optical or radio wavelengths but could be detected by its x-ray absorption.

(4) To detect "missing matter".

Various cosmological considerations suggest that the density of the Universe is almost two orders of magnitude greater than the observed smeared-out density from galaxies. It has been suggested that the "missing matter" could be in the form of uncondensed gas (presumably mostly hydrogen) in intergalactic space. If there does exist a large amount of hydro-

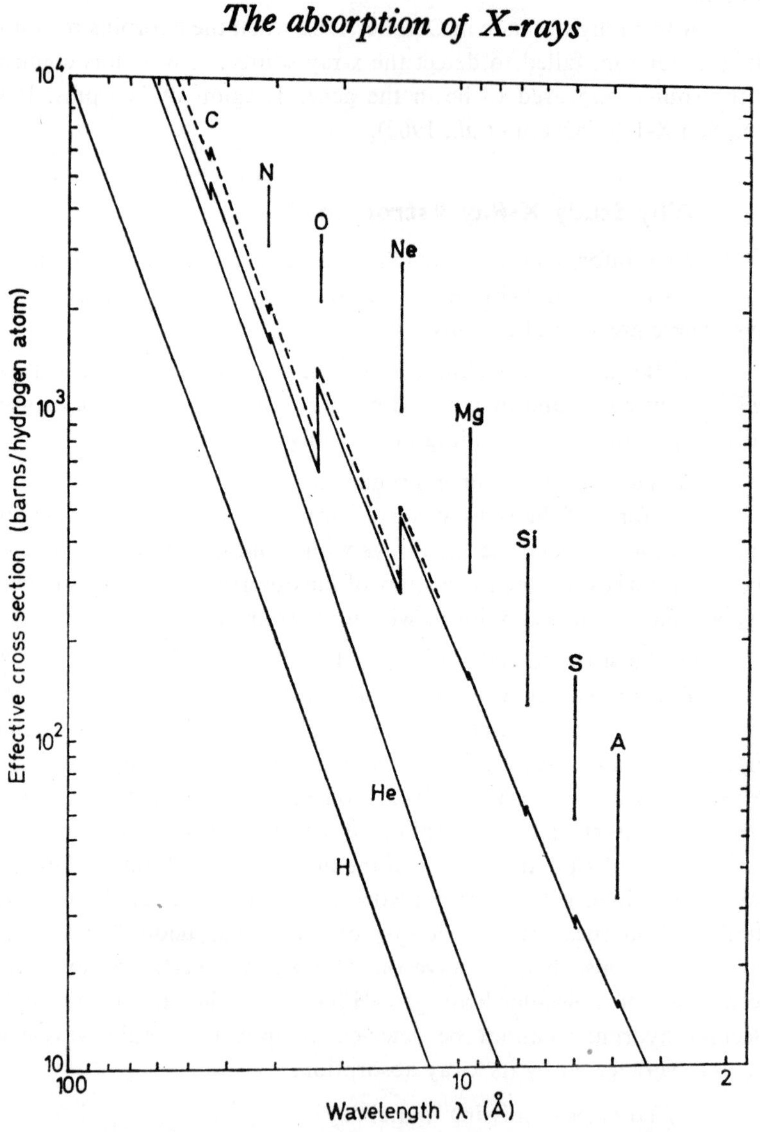

FIGURE 1-2. Photoelectric absorption cross section for x-rays in the interstellar medium, assuming normal cosmic abundances [from K. L. Bell, and A. E. Kingston, Monthly Notices of the Royal Astronomical Society **136**, 241 (1967)].

gen gas between the galaxies, then it has not yet been detected by optical or radio observations. This nondetection severely restricts the possible physical conditions of the gas (see, for example, Gould and Ramsey, 1966). The gas must be a fully ionised plasma with a temperature of between 10^5 °K and 10^7 °K, and a density of not more than 10^{-5} atoms/cm^3. Such a plasma could be detected by its thermal emission in the soft x-ray region. The detection of a soft x-ray flux from the intergalactic gas would be of great cosmological significance.

1.3. Measurements

In order to obtain as much information as possible from the observations of the x-ray flux from a source, it is necessary to determine

(a) The location of the source. (At present there are only three x-ray objects whose locations are known and well defined by optical and radio studies.)

(b) The total intensity of the source in the x-ray region.

(c) Whether the intensity varies with time.

(d) The spectrum of the continuum emission and the structure of any emission lines and absorption edges.

(e) The size, shape, and extent of the source.

(f) The polarization of the radiation.

In order to make these measurements, it is necessary to develop accurate experimental techniques. At present the devices used are very crude compared with the sophisticated instrumentation of optical and radio astronomy.

2. TECHNIQUES

2.1. Rockets and Balloons

Rockets are used for measurements in the energy range

$$100\,\text{eV} < h\nu < 25\,\text{keV}.$$

The high-energy limit is given by the rapid decrease in flux at high energies. The low-energy cutoff is due to interstellar absorption. Figure 2-1 shows how x-rays from the Crab nebula are absorbed by the interstellar medium. High-energy x-rays reach us unattenuated, but wavelengths longer than about 30 Å are not transmitted.

A typical Aerobee-150 rocket carries a payload of about 150 lb to an altitude of 150 miles. It is a two-stage rocket, about 30 ft long and 15 in. in diameter. The first-stage booster engine generates a thrust of 18000 lb and is then ejected. The second stage of the rocket is a liquid-fuel

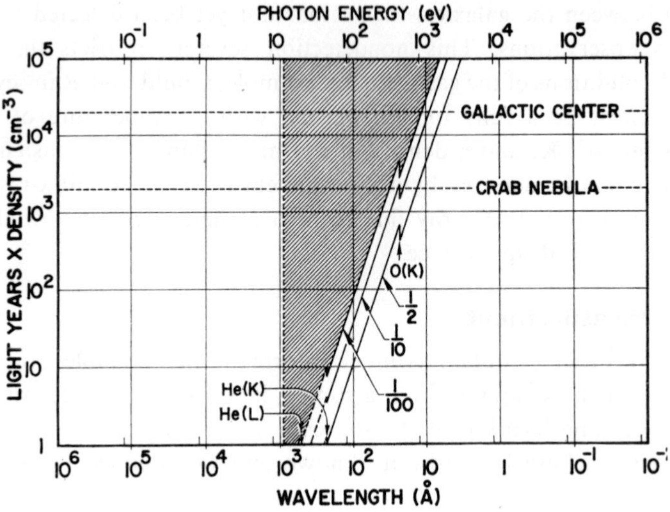

FIGURE 2-1. Interstellar absorption of x-rays from the Crab nebula.

sustainer. Such a rocket is spin stabilized in the early part of the flight and then stabilized with an altitude control system (ACS) or gyrosystem and gas control thrusters. The rocket can be pointed at the source to an accuracy of about 2°, depending on the quality of the gyroscopes. The experimental apparatus is recovered by parachute. The experimental observations are generally sent to the ground by radio, but sometimes specific data are recorded on film. The data transmitted to Earth are recorded on magnetic tape and subjected to computer analysis.

Since the observation time is short (usually about 5 min), the instruments can be powered by batteries. However, since the apparatus generally takes time to stabilize, thermally it is necessary to start the apparatus at least half an hour before the rocket is launched.

2.2. Detectors

There are four types of detectors which have been used for experimental x-ray work:

(a) *Thin-window gas counters*. These are normally used as proportional counters in which the size of the recorded pulse is proportional to the energy of the incident x-ray. These have been widely used for measurements in the energy range

$$0.1 \text{ keV} < h\nu < 20 \text{ keV}.$$

Within this range the efficiency of these counters is very high, depending only on the transmission properties of the thin window and the gas absorption. An efficiency of between 60 percent and 90 percent can be achieved. They have an energy resolution of about 15 percent. This has so far not been sufficient to resolve any emission lines in the x-ray sources.

(b) *Scintillation counters.* Sodium iodide and caesium iodide scintillation detectors have been used for measurements in the high-energy x-ray region (above 20 keV) where proportional counters cease to be efficient. They were first used by Clark in 1964 for his balloon observations of the Crab nebula (Clark, 1965). The efficiency of such devices is very high, but the energy resolution decreases with decreasing photon energy.

(c) *Photoelectric devices.* These are used for low-energy work, where an essentially windowless detector is needed. The device consists of a caesium iodide photocathode (which has an exceedingly large photoelectric yield for soft x-rays) used in conjunction with an electron multiplier. Such detectors have not been widely used.

(d) *Solid-state detectors.* These have been used effectively for nuclear gamma-ray spectroscopy but so far have not been used in x-ray astronomy. Such detectors have a very high energy resolution (about 3 percent) but have a low efficiency in the spectral region of greatest interest; this difficulty is being rapidly overcome, and these detectors will certainly be used extensively in future x-ray studies.

2.3. Proportional Counters

Before we proceed to a discussion of the results obtained from x-ray observations, it is necessary to have a clear understanding of the observational problems involved in x-ray work. It is then possible to make a critical evaluation of the results with a realistic knowledge of the experimental difficulties. In this lecture we will thus discuss in more detail the design and construction of proportional counters, with particular emphasis on the sensitivity of these instruments and how they are used to obtain the spectra and location of sources.

2.3.1. *Construction*

The construction of a proportional counter is shown schematically in Fig. 2-2. The aluminum box is filled with an inert gas (typically argon with a few percent of methane). The window is made of a thin layer of beryllium or Mylar. An x-ray photon of energy $h\nu$ that enters the counter will ionise the gas by ejecting a photoelectron from the most tightly bound level. If I is the ionisation potential for the gas (typically 25 eV), the number n of ion pairs produced by a 1-keV x-ray in the gas is given approximately by

$$n \cong \frac{h\nu}{I} \cong 40.$$

Each ion will emit characteristic line radiation which must be absorbed by a quenching gas; otherwise a continuous discharge will take place. Each electron, accelerating towards the anode, produces a cascade of secondary electrons. The number of electrons continues to increase until they are collected by the anode. Typically each primary electron produces about 10^4 electrons at the anode. Thus each 1-keV x-ray gives a pulse of about 4×10^5 electrons. The delay between the initial ionisation by the x-ray and the production of the amplified signal is of the order of microseconds in typical counters.

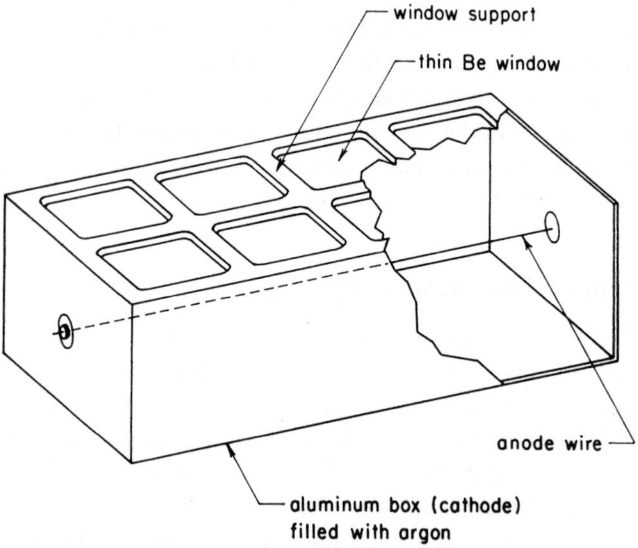

FIGURE 2-2. Schematic construction of a gas proportional counter.

2.3.2. *Energy Resolution*

There is a linear relationship between the x-ray energy and the resultant pulse size since the number of ion pairs is proportional to the photon energy. The energy resolution is limited by the statistical fluctuations of the initial ionisation process. Such fluctuations produce an error ΔV

in the final voltage V of approximately

$$\frac{\Delta V}{V} \cong \frac{n^{1/2}}{n} \cong \left(\frac{I}{h\nu}\right)^{1/2} \cong 15\%.$$

Thus the limit for the energy resolutions of such a counter is typically 15 percent for 1-keV photons.

2.3.3. *Spectral Response*

A highly efficient counter must be constructed so that there is a very high probability that an x-ray will be absorbed by the gas, and a very low probability that it will be absorbed by the counter window. Thus a very thin layer of a low-Z material, such as beryllium or Mylar, is used for the window. These are highly transparent for energies greater than 0.25 keV but appreciably absorb the long-wavelength photons. Window transmission thus limits the spectral sensitivity at low energies.

To obtain maximum absorption within the counter a high-Z gas is chosen (usually argon). A small amount of some polyatomic gas (e.g., methane) is added to absorb the ultraviolet radiation emitted by the noble gas when ionised and to quench the metastable atoms. The photoelectric cross section of argon decreases rapidly with increasing energy. Thus for high energies the gas becomes transparent, and the efficiency of the counter is reduced. This can be partially overcome by using either krypton or xenon and by increasing the pressure.

The quantum efficiency of a counter is not a smooth function of energy, and this must always be remembered when interpreting results. A typical spectral sensitivity curve is shown in Fig. 2-3. The sharp cutoff at high energies is due to the argon becoming transparent, and the slower cutoff at low energies is due to absorption by the counter window. The energy range over which the counter is efficient can be altered by using a different inert gas and by modifying the window material and thickness.

2.3.4. *Sensitivity*

The brightest nonsolar object in the x-ray sky is Sco X-1. This has a flux of 20 photons/cm^2 sec.‡ We wish to estimate how weak a source could be detected above the background by existing gas counters.

As well as detecting x-rays, proportional counters are also sensitive to cosmic rays. Any space rocket is continually bombarded with a flux of

‡ Since Sco X-1 has a flux equivalent to that of the quiet Sun, it has a significant effect on the D layer of the ionosphere. It is thus possible to study Sco X-1 by observing changes occurring in the ionosphere.

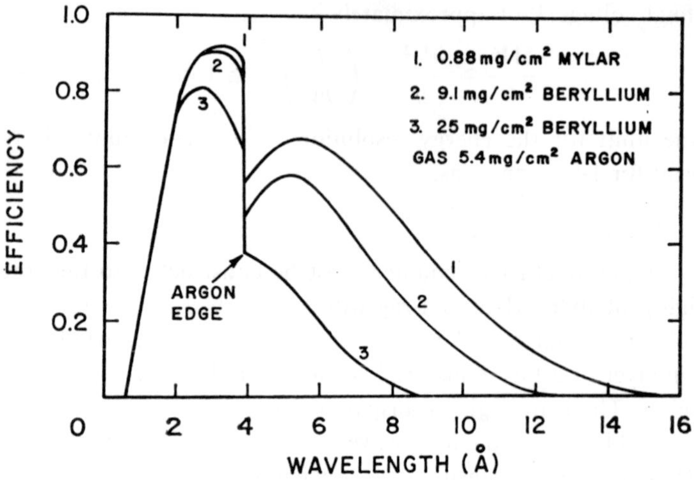

FIGURE 2-3. The efficiency of a gas proportional counter.

energetic protons and γ-rays. The γ-rays are produced both within the Earth's atmosphere and within the rocket itself. These protons and γ-rays ionise the gas within the counter, producing a background.

For example: a 1-BeV proton produces an amount of ionisation equivalent to 2 MeV on going through 1 g of gas. Since the column density of gas within a counter is of the order of a few milligrams, the ionisation energy produced within the counter is $\cong$ 10 keV. Thus a 1-BeV proton gives a pulse equivalent to that of a 10-keV x-ray. There is a substantial flux of cosmic rays, and these give a background count of about 1 count/cm^2 sec, or about 5 percent of the flux from Sco X-1. If this is not significantly reduced, the counter sensitivity is severely limited. Several methods have been used to reduce the background count due to cosmic rays.

(a) *Energy discrimination.* Since the detector signal is proportional to the energy deposited during ionisation, it is possible to eliminate the pulses due to energies well outside the x-ray range. In this way the count from highly energetic cosmic rays can be reduced. This technique is most effective for scintillation counters where the energy deposited by minimum ionising protons is about 1 MeV, well above the x-ray region. With proportional counters the cosmic-ray count can be reduced to about 1/3 count/cm^2 sec with pulse-height analysis.

(b) *Anticoincidence techniques.* If the counter is surrounded by a plastic scintillation counter, energetic protons will be detected both in the gas counter and in the adjacent scintillation counter. Thus signals in the gas counter which are coincident with signals in the surrounding counter can be eliminated electronically. This is a highly efficient method of reducing the background proton count, and most results have been obtained with the use of such devices. By combining a plastic anticoincidence device with pulse-height analysis the background is reduced to about 1/10 count/cm^2 sec.

(c) *Pulse-shape discrimination.* When a keV x-ray enters the counter, it produces a primary electron of energy comparable to that of the x-ray. At a pressure of one atmosphere, the range of such an electron as it produces secondary electrons is of the order of a few microns. Thus the whole event is localised. By contrast, a very energetic cosmic ray produces ion pairs over a track length of a few centimeters. Such a nonlocalised event results in a spread in the times required for the initial electrons to reach the anode, and so the resultant pulse does not show the sudden rise characteristic of a localised initial ionisation (see Fig. 2-4). By observing the rise time of each pulse, the pulses due to cosmic-ray protons can be further reduced, and the net background count can be reduced to 1/100 count/cm^2 sec.

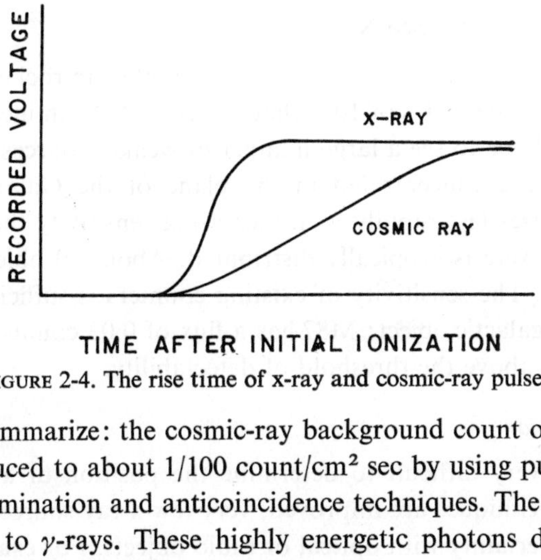

FIGURE 2-4. The rise time of x-ray and cosmic-ray pulses.

To summarize: the cosmic-ray background count of 1 count/cm^2-sec can be reduced to about 1/100 count/cm^2 sec by using pulse-height and rise-time discrimination and anticoincidence techniques. The residual count is mainly due to γ-rays. These highly energetic photons do not interact with the plastic scintillation counter used in anticoincidence devices but may interact with the counter walls producing an energetic electron which enters the gas causing a pulse which is indistinguishable from an x-ray event. NaI and CsI scintillation counters can be used as γ-ray shields, but these are very expensive, and CsI has an especially long decay time. The most efficient γ-ray discriminator yet developed used NaI immersed in CsI but has a very small detector area (Peterson *et al.*, 1966).

(d) *Estimate of sensitivity.* Using the above results we can estimate the minimum flux density that we can detect with available instruments. We shall assume a detector area of 10^3 cm^2. This is not difficult to obtain: the Naval Research Laboratory group (NRL) is developing a detector with an area of 100 sq. ft. A rocket can be readily pointed to observe a source for about 100 sec (about a quarter of the total flight time). During the observation

$$\text{Total background count} \cong \frac{1}{100} \times 10^3 \times 10^2 \text{ counts} \cong 10^3 \text{ counts},$$

$$\text{Fluctuations in background} \cong \sqrt{10^3} \cong 30 \text{ counts}.$$

For a source to be detected, the source count must be greater than three times the standard deviation of the background.

$$\therefore \text{ Detectable signal} \gtrsim 3\sqrt{10^3} \cong 10^2 \text{ counts}.$$

Thus the minimum detectable source flux, ϕ_{source}, is given by

$$\phi_{\text{source}} \times 10^3 \times 10^2 > 10^2 \text{ counts},$$

$$\phi_{\text{source}} > 10^{-3} \text{ photons/cm}^2 \text{ sec},$$

$$\phi_{\text{source}} > 10^{-4} \text{ Sco X-1}.$$

Thus, with the most sensitive proportional counters in rockets we can detect sources with flux densities of 10^{-4} that of Sco X-1. With increasing sensitivity one might hope to see a large number of weaker sources, but since most x-ray sources are concentrated in the plane of the Galaxy, the number detected increases less rapidly with increasing sensitivity than would occur if the sources were isotropically distributed. About 40 objects have so far been observed. The sensitivity of existing counters is sufficient to detect at least one extragalactic object: M87 has a flux of 0.03 counts/cm^2 sec, which is significantly above the threshold of detectability.

2.3.5. *Location*

It is very difficult to determine the position of an x-ray source, and until the techniques are improved, very few x-ray sources can be identified with any certainty with optical or radio objects. For example, the error box on the position of the source Cyg X-2 contains several hundred objects visible on the Palomar Sky Survey 1, making positive optical identification impossible until the uncertainty in position is greatly reduced.

2.3.6. *Two-Dimensional Collimator*

The simplest method of determining position is to use a two-dimensional collimator. This consists of slats of aluminum (which is a good absorber for low-energy x-rays) placed in a two-dimensional grid in front of the counter. The acceptance of such a device is approximately conical giving a field of view of about 1° by 1°. The sensitivity of such a collimator is low, because the effective area is reduced by the presence of the grid, and also because the rocket has a pointing area of the order of 2°, so that the fraction of the time for which it is pointing directly at the source is small.

2.3.7. *One-Dimensional Collimator*

Increased accuracy can be obtained when a scanning rocket with a one-dimensional collimator is used. This gives a rectangular field of view of about 1° by 15°. The rocket scans across the sky through the source and then continues with a scan at right angles to the original direction. The accuracy of such a collimator is not limited by the pointing accuracy of the rocket (about 2°), and the sensitivity is more reasonable, since the effective area is not so greatly reduced. Figure 2-5 shows the scans made across the Cygnus region by the NRL group using a one-dimensional collimator (Byram *et al.*, 1966). The counts recorded during two of the scans are shown in Fig. 2-6. In order to interpret such results, it is necessary to assume a source distribution and to fold this with the instrumental profile to try and fit the observational curves. Such methods can give serious errors and are not good enough for obtaining accurate positions.

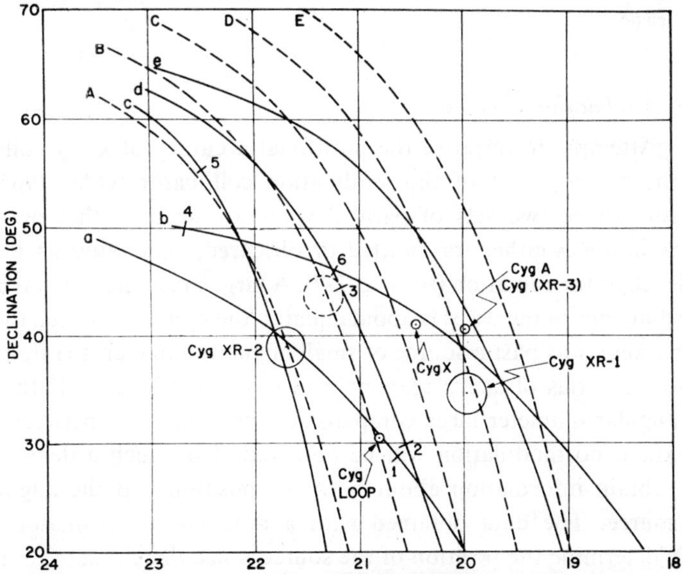

FIGURE 2-5. Scans of the Cygnus region in the NRL flights of June, 1964 (dotted lines) and April, 1965 (solid lines) [from E. T. Byram, T. A. Chubb, and H. Friedman, Science **152**, 67 (1966), © 1966, The American Association for the Advancement of Science].

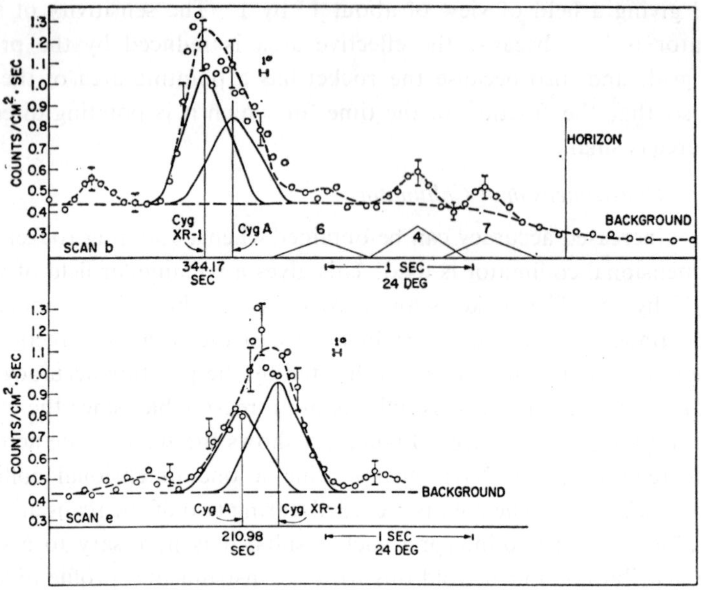

FIGURE 2-6. Counts recorded during two of the scans shown in Fig. 2-5. The solid lines show possible positions for Cyg A and Cyg XR-1 [from E. T. Byram, T. A. Chubb, and H. Friedman, Science **152**, 68 (1966), © 1966, The American Association for the Advancement of Science].

2.3.8. *Modulation Collimators*

Attempts to improve the positional accuracy of x-ray collimators led to the development of the modulation collimator (Oda, 1965). This device consists of two sets of parallel wires arranged so that parallel incident radiation is either transmitted or obscured, depending on the angle of incidence of the radiation (see Fig. 2-7). A large-area detector can be used, and the diameter of the wires is about equal to the spacing between the wires. As the rocket scans past a source of small angular diameter, the transmitted flux shows a series of sharp maxima and minima (Fig. 2-8). If the source has an angular diameter large compared to the separation between successive maxima, no modulation will be observed. Thus such a device can be used to obtain information about both the position and the angular size of the source. The data obtained with a modulation collimator do not uniquely determine the position of the source, since there is some ambiguity as to the center of the observed pattern. Thus it is necessary to use two such detectors of different grid spacings as in a vernier system. By this technique a positional accuracy of 1 arc min is possible.

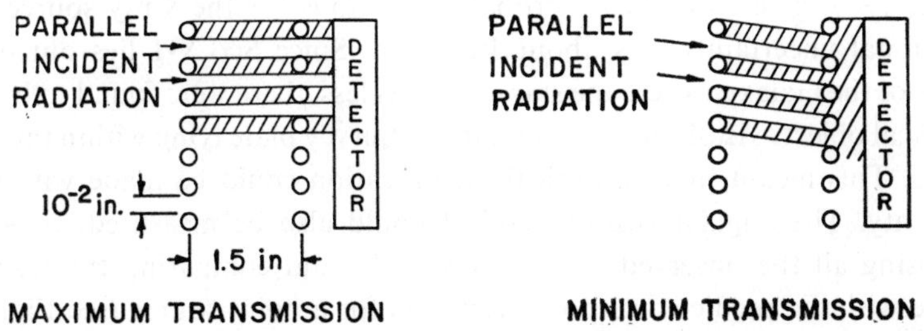

FIGURE 2-7. Schematic diagram of the Oda collimator, showing the conditions for maximum and minimum transmission.

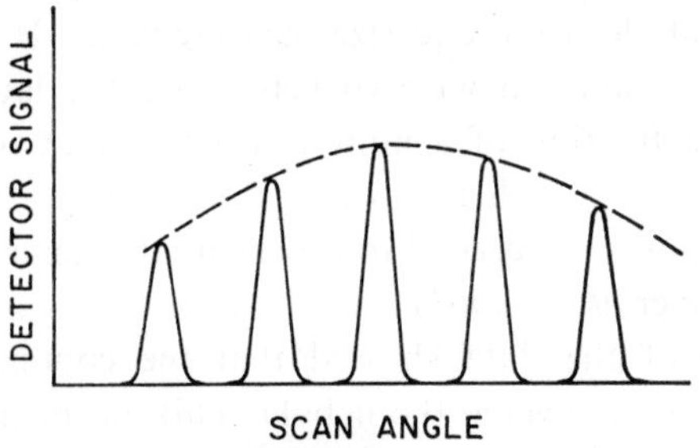

FIGURE 2-8. Oda collimator response for a source of small diameter.

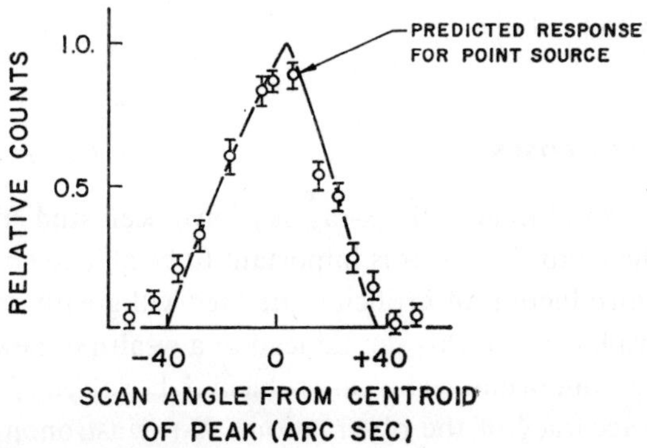

FIGURE 2-9 Superposition of several peaks, showing that the data for Sco X-1 are consistent with a point source.

Measurements on Sco X-1 using a modulation collimator (Gursky *et al.*, 1966) give two equally probable positions for the x-ray source. The positional uncertainty was about 1 arc min. Since Sco X-1 lies out of the plane of the Galaxy, where the star density is less, there were found to be only 3 optical objects visible on the Palomar Sky Survey plate lying within the error boxes. This meant that an optical identification could be made with some certainty. The angular size of Sco X-1 could also be measured, by superimposing all the observed modulation peaks and comparing the resultant shape with the predicted response function for a point source (see Fig. 2-9). This showed that the angular diameter of Sco X-1 could not exceed 20 arc sec.

2.4. Lunar Occultation

Occasionally the Moon will occult an x-ray source, and this can be used for accurate location and size measurements. It is very difficult to observe a lunar occultation with rockets, since the launching has to be timed to within a fraction of a minute in order that the few minutes of observing time coincide with the occultation. This was achieved in 1964 by the NRL group who succeeded in observing the lunar occultation of the Crab nebula (Bowyer *et al.*, 1964).

Analysis of the data showed that the counting rate decreased gradually as the Moon covered the nebula. This meant that the source was an extended object and not a point source, since this would produce an abrupt change in the count. There had been some speculation that the x-rays originated from a neutron star with the Crab, but these results showed conclusively that the x-ray source had dimensions comparable to the optical dimensions of the nebula and also served to unequivocally locate the x-ray source in the Crab nebula.

2.5. X-Ray Lenses

Only two objects in the x-ray sky have been studied in any detail: Sco X-1 and the Crab nebula. It is important to be able to study many more sources. The introduction of artificial satellites will greatly increase the observing time available, and this should lead to a wealth of new observational results. The most important technical problem to be solved is how to increase the positional accuracy of the observations. X-ray astronomy today seems to be in a similar situation to that of radio astronomy before the advent of high angular resolution telescopes which led to such important discoveries.

We thus need to develop some new instrumentation for use in the x-ray region. Below we discuss the prospects for constructing high-resolution x-ray lenses.

There are two properties of x-rays which make the design of an image-forming device to be used at x-ray wavelengths very difficult. The first is that x-rays are readily absorbed by matter; the second is that the index of refraction at x-ray wavelengths is very nearly unity. Thus it is not possible to construct a practical refractive telescope.

A Fresnel zone plate has been used to observe solar x-rays with limited success (Pounds and Russell, 1966). This is not very satisfactory, because the spacing Δ between the zones required for x-ray wavelengths is very small. Typically, $\Delta \approx \sqrt{f\lambda}$ for the first zone, where f is the focal length and λ the wavelength. For $f \approx 10^2$ cm and $\lambda \approx 10^{-8}$ cm, $\Delta \approx 10^{-3}$ cm. Δ decreases rapidly for higher zones. Another problem is that the focal length depends on wavelength. Neither the zone plate nor pin-hole cameras have been widely used.

2.5.1. *Total External Reflection*

Since the index of refraction for x-rays is slightly less than unity, x-rays incident at grazing angles on a surface will be totally reflected. Imaging devices using total external reflection are being developed at ASE (Giacconi *et al.*, 1969). The index of refraction n at x-ray wavelengths can be written

$$n = 1 - \delta - i\beta,$$

where δ and β are small. β is related to the absorption of the material. when absorption is neglected, total external reflection of x-rays will occur for grazing angles less than the critical angle θ_c given by Snell's law:

$$\cos\theta_c = 1 - \delta.$$

Since δ is small, $\theta_c = \sqrt{2\delta}$. For typical materials, the critical grazing angle θ_c lies between 1° and 3°. The reflection efficiency depends both on the angle of incidence and on the absorption parameter β and can be calculated theoretically for ideal materials. Figure 2-10 shows the results of such calculations for different values of β/δ. If appreciable absorption is present, reflection can occur for angles greater than θ_c. For real substances, the reflection curves are more complicated, due to the presence of x-ray absorption edges (see Fig. 2-11).

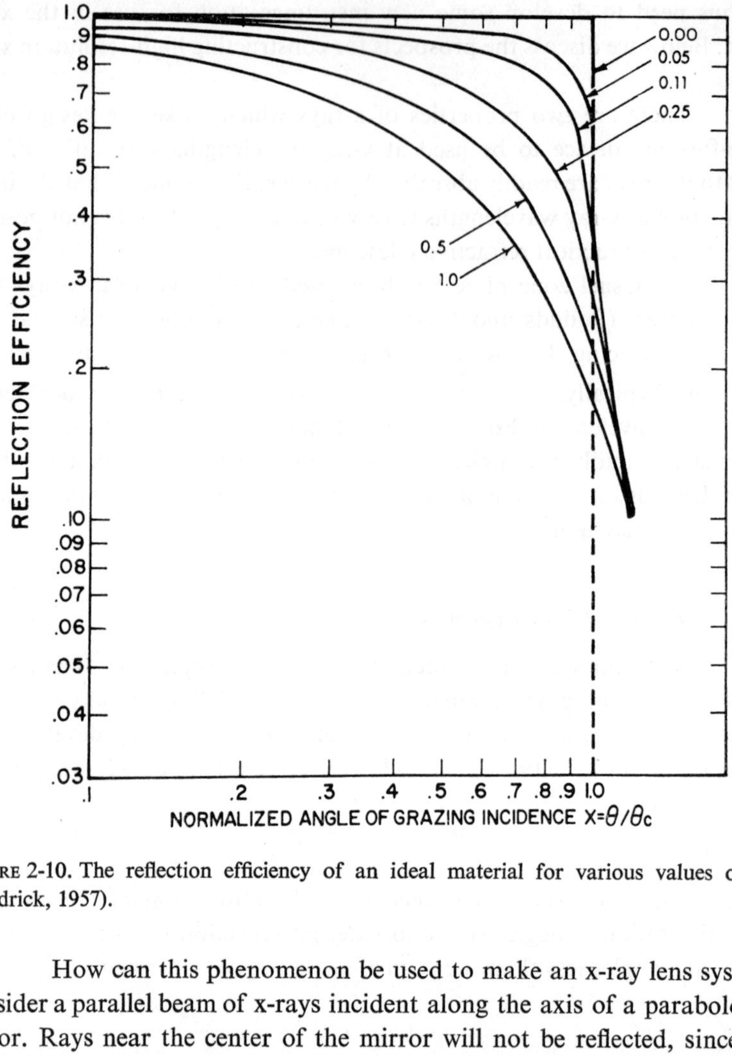

FIGURE 2-10. The reflection efficiency of an ideal material for various values of β/δ (Hendrick, 1957).

How can this phenomenon be used to make an x-ray lens system? Consider a parallel beam of x-rays incident along the axis of a paraboloidal mirror. Rays near the center of the mirror will not be reflected, since the angle of incidence is large. However, rays incident at grazing angles in the far zone will be imaged at the prime focus, and the mirror can act as a focussing device. However, such a system cannot satisfy the Abbé sine condition, and so off-axis rays will be not be focussed at a point, giving severe comatic aberration. Thus such a system will work as a condenser, but not as a focusser. Wolter (1952a, b) showed that this difficulty could be overcome by having two coaxial reflecting surfaces, not simply a single surface. Indeed he showed that one can get lens action with any *even* number of coaxial surfaces but not with an odd number of surfaces. The possible

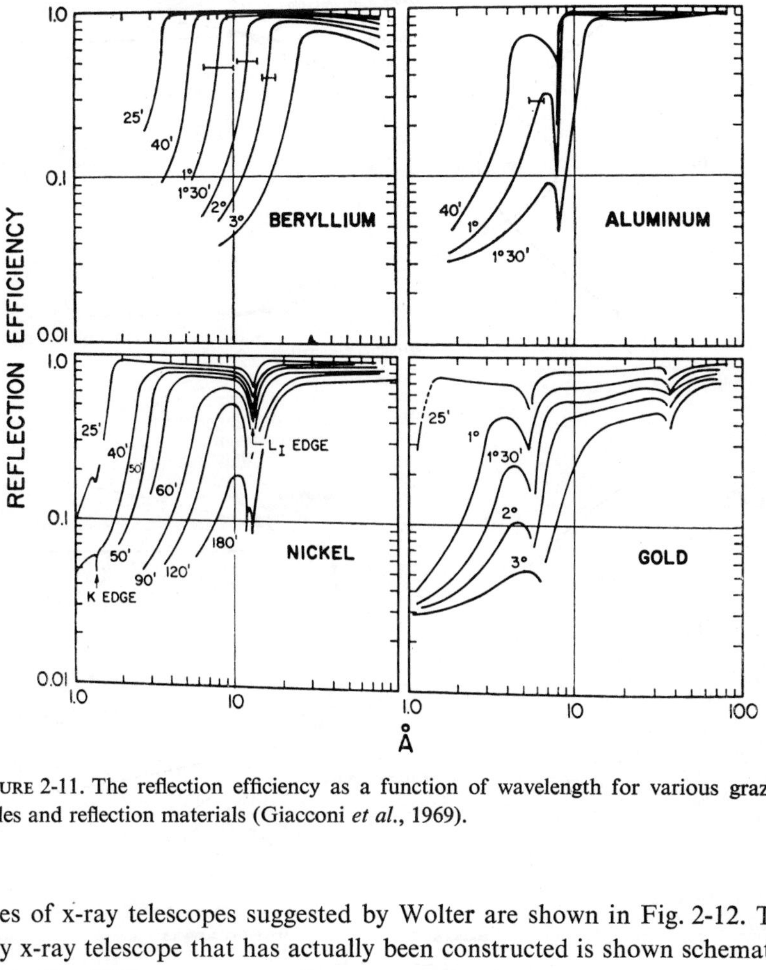

FIGURE 2-11. The reflection efficiency as a function of wavelength for various grazing angles and reflection materials (Giacconi *et al.*, 1969).

types of x-ray telescopes suggested by Wolter are shown in Fig. 2-12. The only x-ray telescope that has actually been constructed is shown schematically in Fig. 2-13.

2.5.2. *Field of View*

As the field of view of such a telescope is increased to include off-axis rays, blurring of the image at the focus increases. This is illustrated in Fig. 2-14. It can be seen that a geometrical resolution of about 1 arc-sec can be achieved with a field of view of about 5 min. Some theoretical calculations of the change in the diameter of the blur circle with increasing angular field of view are shown in Fig. 2-15. Slightly improved resolution can be achieved with a hyperboloidal surface. A geometrical resolution of about 1 arc-sec is a great improvement on the resolution of the gas

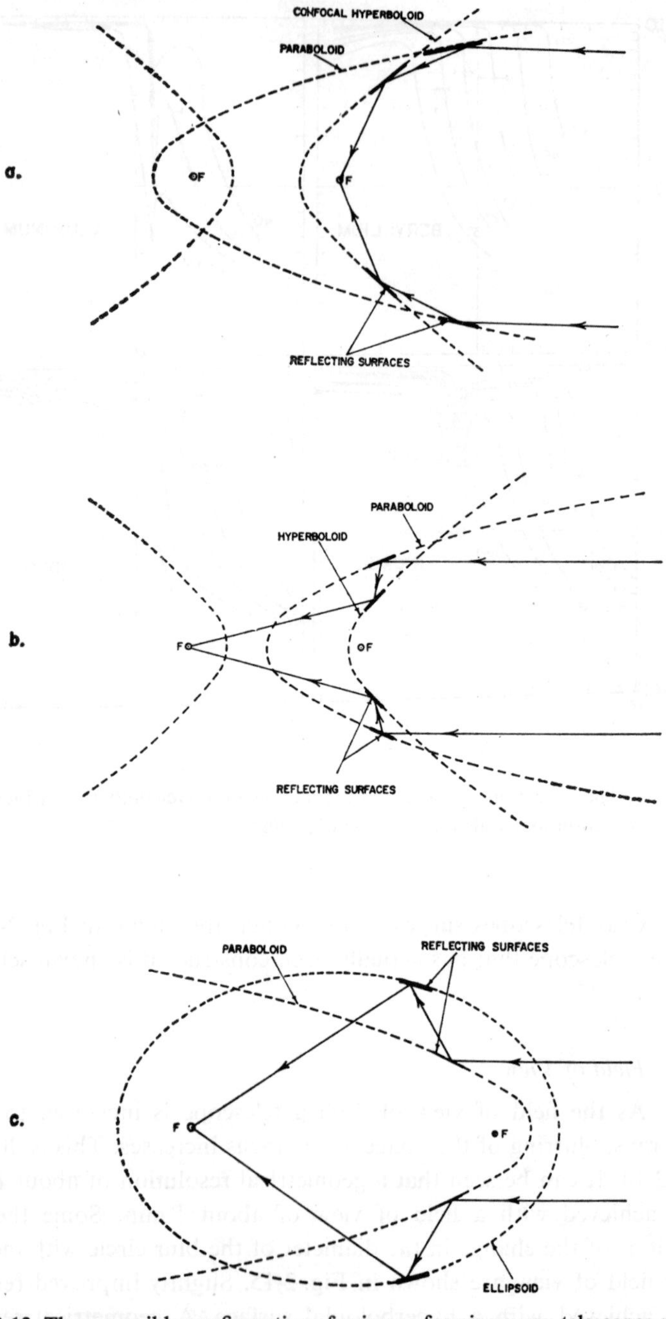

FIGURE 2-12. Three possible configurations for image-forming x-ray telescopes suggested by Wolter (Giacconi *et al.*, 1969).

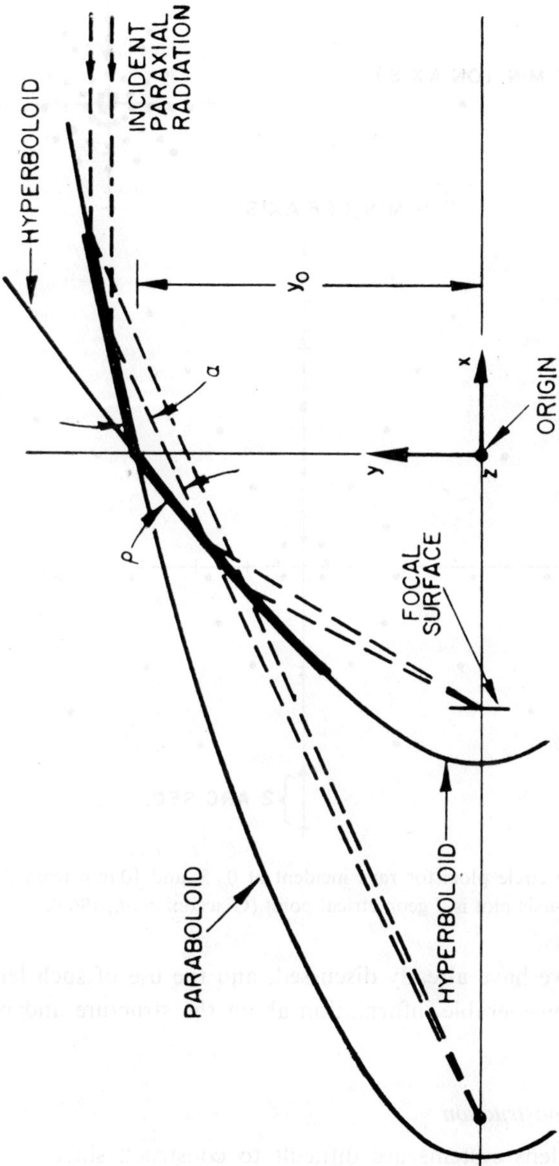

FIGURE 2-13. Schematic cross section of the type of telescope which has been used by the American Science and Engineering group (Giacconi *et al.*, 1969).

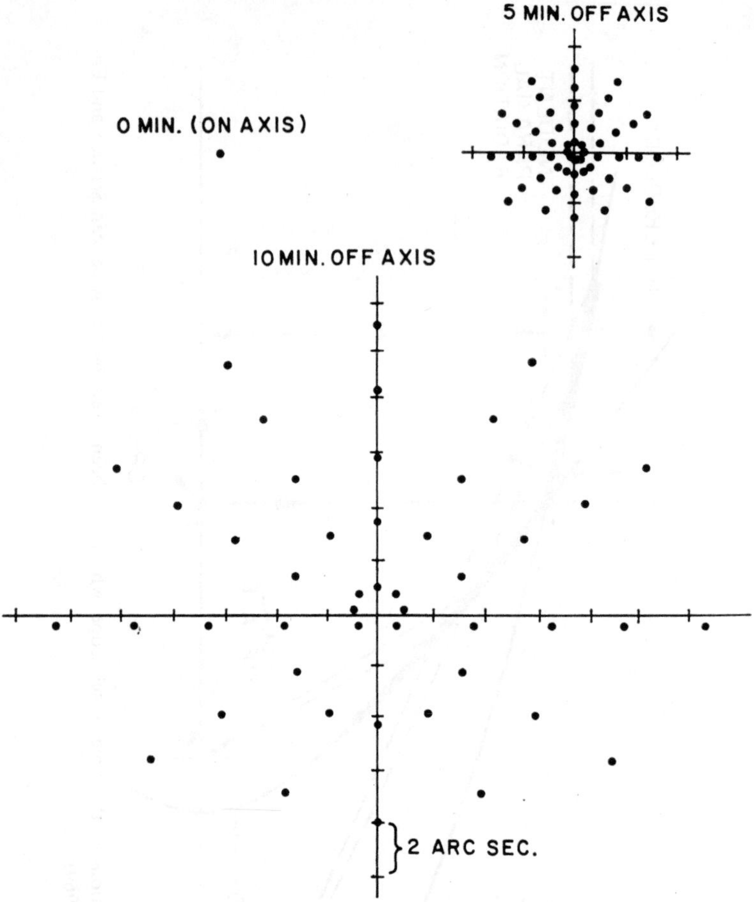

FIGURE 2-14. Blur circle plots for rays incident at 0, 5, and 10 min from the axis of a telescope. The on-axis plot is a geometrical point (Giacconi *et al.*, 1969).

counters that we have already discussed, and the use of such lens systems should yield considerable information about the structure and position of x-ray sources.

2.5.3. *Lens Construction*

Such lens systems are difficult to construct, since the focussing surfaces must be very smooth. One method is to use an optically polished stainless-steel mandrel, which is a negative of the required surface. This is plated with nickel, and the nickel surface is then removed and used in the telescope. Several lenses can be prepared from the same mandrel. The

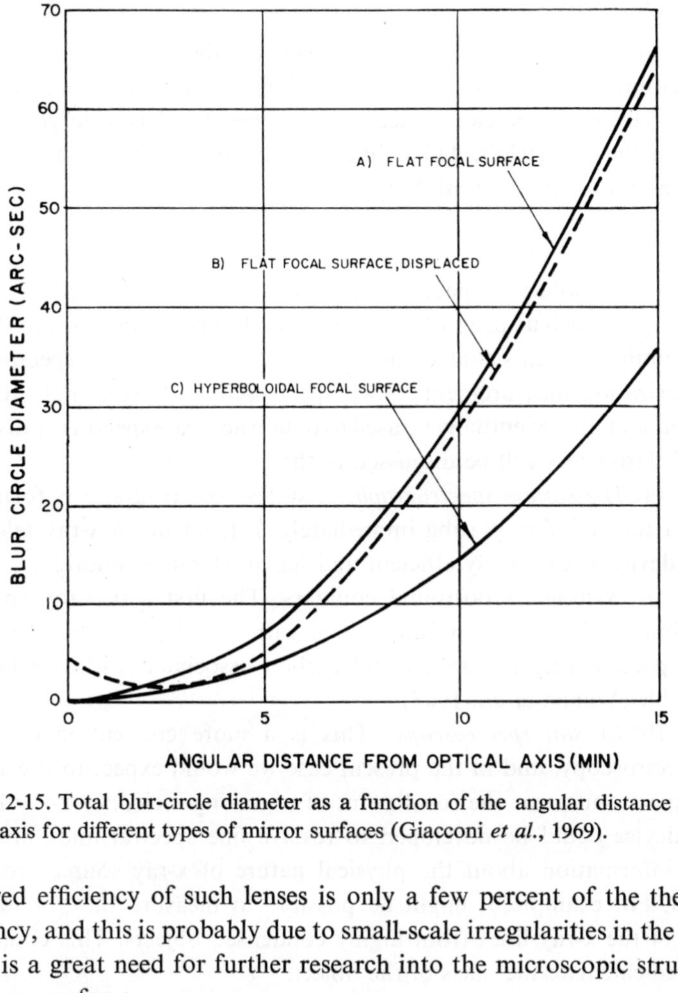

FIGURE 2-15. Total blur-circle diameter as a function of the angular distance from the optical axis for different types of mirror surfaces (Giacconi *et al.*, 1969).

observed efficiency of such lenses is only a few percent of the theoretical efficiency, and this is probably due to small-scale irregularities in the surface. There is a great need for further research into the microscopic structure of these lens surfaces.

Many other problems arise when such lenses are used for astronomical observations. The effective area of the system is small, and so the sensitivity is low. The area can be increased by nesting confocal surfaces, but this leads to mechanical problems. Effective areas of about 10 cm^2 can be achieved (compare gas counters with an area of about 10^3 cm^2). Another difficulty is the need for a detector with very good spatial resolution. Photographic film is good in this respect but has a low efficiency in the x-ray region. It cannot be shielded from cosmic rays, and it integrates over the whole observing time. There is a need for a new kind of detector, but none seem to have the required efficiency.

2.5.4. *Observations*

X-ray telescopes have been flown to observe the Sun (Lindsay, 1965; Paolini *et al.*, 1968). With a resolution of 1 arc-sec certain details of the solar-flare structure can be seen. It is hoped to place a large telescope in orbit in the late 1970's. This will have an effective area of about 10^3 to 10^4 cm^2 and a focal length of 20 ft.

2.5.5. *Auxiliary Devices*

Low-resolution spectral information can be obtained with existing techniques, such as proportional counters, but it is important to develop high-resolution systems that could be used to make precise spectroscopic and polarization measurements. Two spectroscopic devices that are being developed and may eventually be used to obtain accurate spectra are discussed below. Polarimeters will be discussed in the last lecture.

(a) *The slitless spectrograph.* A slitless spectrograph is formed by placing a transmission grating immediately in front of an x-ray telescope. Such a device is extremely efficient and has moderate resolution, certainly better than existing proportional counters. The first soft x-ray spectrum of the Sun was obtained in June, 1968, with a slitless spectrograph, and certain spectral lines of oxygen and carbon associated with a solar flare were visible (Vaiana *et al.*, 1968).

(b) *Crystal spectroscopy.* This is a more conventional form of x-ray spectroscopy, and in the present case we would expect to use a Bragg crystal spectrometer at the focal point of a grazing-incidence x-ray lens. If such a device could be developed to resolve fine spectral lines, much important information about the physical nature of x-ray sources could be obtained. For example, it might be possible to measure the gravitational redshift of the x-ray lines from highly condensed objects. This could then be used to calculate the mass of the object.

Certainly these new techniques will be considerably developed during the next few years.

3. SCORPIUS X-1

3.1. Optical Identification

Sco X-1 has perhaps the unique distinction among astronomical objects of having been observed extensively at x-ray, ultraviolet, visible, infrared, and radio wavelengths. It is thus the x-ray source for which most progress has been made in the understanding of the physical processes which occur within it. We shall thus discuss in some detail the observational and crude theoretical models proposed for Sco X-1, since it may prove to be a prototype for other x-ray sources.

The rocket experiment of the American Science and Engineering group, referred to above, gave the position of Sco X-1 to such precision that only three visible objects were seen within the error boxes. It could be argued that, since Sco X-1 is such a peculiar object, it need not have an optical counterpart. However, simple theoretical arguments can be used to predict the expected optical flux.

The x-ray spectrum of Sco X-1 has an exponential form. This suggests a thermal bremsstrahlung emission mechanism which, for high frequencies, gives a spectrum of the form

$$I(\nu) \propto e^{-h\nu/kT}.$$

Fitting an exponential to the observed spectrum gives an estimate of the temperature T. For Sco X-1 we find

$$kT \simeq 4 \text{ keV}$$

or

$$T \simeq 5 \times 10^7 \text{ °K}.$$

At such a high temperature any gas will be completely ionised. Let us thus assume that the x-ray emission comes from a fully ionised hydrogen plasma. Then the expected visible and infrared spectrum can be predicted from known radiation theory [for details of this theory see, for example, the excellent books by Heitler (1954) and Chandrasekhar (1960)]. If the plasma is opaque to its own radiation, it will have a blackbody spectrum. If it is optically thin, the spectrum will be much flatter (see Fig. 3-1). There will always be a cutoff at very low frequencies due to self-absorption. Thus

$$I(\nu) \propto \nu^2$$

at low frequencies. There will always be a cutoff at high frequencies, due to a limitation in the photon energy available. Thus

$$I(\nu) \propto e^{-h\nu/kT}$$

at high frequencies. Between these two limits the spectrum is approximately flat.

Fitting a curve of this form to the x-ray emission of Sco X-1 gives a flux in the visible region equivalent to that of a star of 13th magnitude. Also, since the spectrum is flat in this region, the flux in the U, V, and B

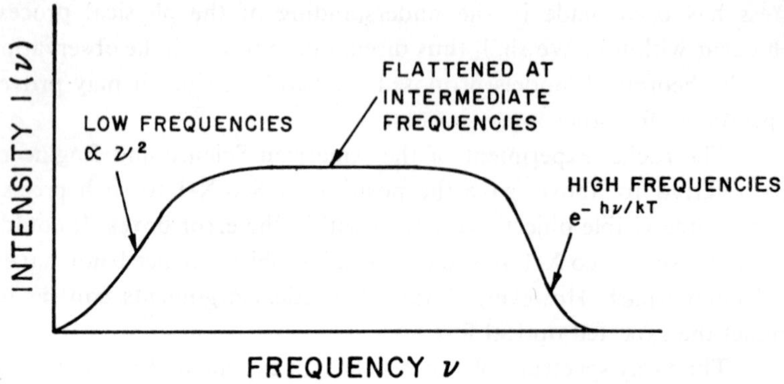

FIGURE 3-1. The spectrum of an optically thin plasma.

filter bands should be the same. Thus the star should look bluer than a normal star. One of the stars within the error boxes on the position of Sco X-1 was of 13th magnitude and was found to have the expected blue excess. It has thus been accepted as the visual counterpart of Sco X-1. Its position is

$$\text{R.A. } 16^{h}17^{m}4^{s}$$

$$\text{Dec } -15^{\circ}\ 31.2'.$$

3.1.1. *Visible Spectrum*

The visible spectrum of Sco X-1 has been studied in great detail (Sandage *et al.*, 1966; Westphal *et al.*, 1968). The optical magnitude changes rapidly and irregularly by ± 0.5 magnitude over periods of a few hours. There is evidence of even more rapid flickering, with a time scale of a few minutes and an amplitude of about 0.02 magnitude. The emission spectrum consists of a few, broad lines showing a large velocity dispersion within the source. There is no observable proper motion. This gives a lower limit to the distance to the source of a few hundred parsecs. Another estimate of distance can be obtained by measuring the equivalent width of the interstellar absorption lines in the spectrum of the source. From the Ca II *K* line, assuming the cosmic abundance of Ca in the interstellar medium,

$$\text{distance} \approx 300 \text{ pc},$$

but this is probably a lower limit only. Other estimates show that the distance d lies within the range

$$300 \text{ pc} \leqq d \leqq 1000 \text{ pc}.$$

3.2. Possible Models

(a) "*Old nova*". The strange optical spectrum of Sco X-1, with its rapid changes in intensity and peculiar spectral color, resembles that of an old nova. These objects have been studied extensively, and it is generally accepted that they are binary systems (see Fig. 3-2). Such a binary system consists of a large red giant star and a dwarf star. As the red star passes through the red-giant phase, it expands considerably, and matter passes from the surface of the star across the saddle point in the gravitational potential and falls towards the smaller star. As it falls, the matter is heated gravitationally. There will be much turbulence within the in-falling gas, and it is not surprising that there are considerable time variations in the intensity of the emitted radiation. The most convincing evidence for such a model for Sco X-1 would be the detection of binary motion: this produces a systematic shift in the frequency of emission lines and a periodic variation in intensity. Such changes have been seen in other old novae but have not yet been detected in Sco X-1.

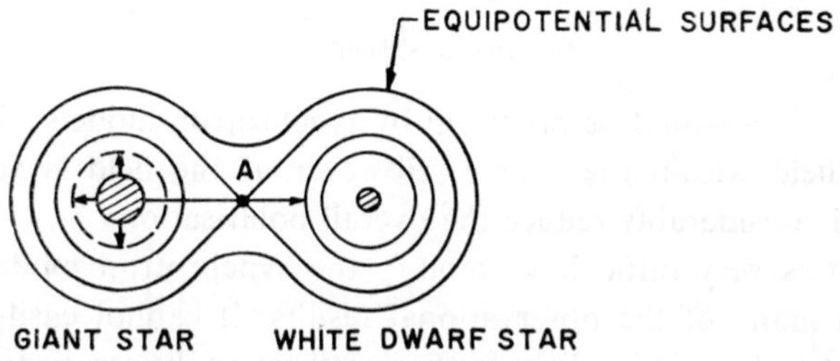

FIGURE 3-2. Schematic diagram of the binary model for an old nova. As the red giant star expands, matter passes across the saddle point A and falls towards the dwarf star.

The x-ray emission from Sco X-1 seems to come from a region with $kT \approx 4$ keV. Can such high temperatures be produced in this model? For a binary system consisting of two ordinary stars, the gravitational heating of the in-falling gas can produce temperatures of only about $kT \approx 10$ eV. However, Shklovsky has shown that if one of the stars is a white dwarf or neutron star, a sufficiently high temperature can be obtained (Shklovsky, 1967, 1968). A proton falling from infinity to the surface of a white dwarf with mass equal to one solar mass gains gravitational energy of about 1 keV.

We thus have a possible model for Sco X-1: a red giant and a white dwarf rotate about their common center of mass. Matter passes from the red giant and falls towards the white dwarf. The in-falling matter gains gravitational energy and produces a turbulent hot plasma in the neighborhood of the white dwarf. This accretion model is very difficult to calculate in any detail. An attempt has been made to solve the equations of motion, including the effects of viscosity and turbulence, but neglecting any radiation effects (Prendergast and Burbidge, 1968). This confirms that the gas does heat up to the required temperature but does not predict the spectrum.

(b) *Synchrotron models*. Synchrotron emission from electrons moving in a magnetic field within a hot plasma could also explain the continuous x-ray emission from Sco X-1. If the electrons within the plasma have a power-law energy distribution, then the resultant synchrotron emission will have a power-law spectrum, which is not observed for Sco X-1. However, by altering the form of the energy distribution of the electrons, it would be possible to reproduce the observed exponential spectrum (Manley, 1966). A crucial result which could be used to distinguish between bremsstrahlung and synchrotron mechanisms would be a measurement of the polarisation of the x-ray or visible emission. Two recent measurements give upper limits to the polarisation of Sco X-1:

$$\text{visible polarisation} < 2\%,$$
$$\text{x-ray polarisation} < 10\%.$$

This is less than would be predicted by synchrotron models with a uniform magnetic field within the source. However, if the field is very irregular, this would considerably reduce the overall polarisation.

It is very difficult to modify the synchrotron model sufficiently to explain many of the observational results. It cannot easily explain the presence of wide emission lines in the spectrum or the erratic time variation. For this reason we shall discuss the old nova model in considerably more detail than this synchrotron model.

3.3. A Simple Theoretical Model for Sco X-1

In this section we will discuss a very simple theoretical model for Sco X-1 which, although very crude, has been remarkably successful in explaining many observational results (see Chodil *et al.*, 1968). Such a model must attempt to explain

(a) The x-ray spectrum and its time variation,

(b) The visible spectrum (continuum and line emission) and its time variation,

(c) The infrared observations,

(d) The observed radio emission,

(e) The ultraviolet spectrum.

3.3.1. *Spectrum of Thermal Bremsstrahlung*

Consider a uniform sphere of pure hydrogen plasma, of radius R (cm) and density ϱ (free electrons/cm^3). We shall assume that the gas is isothermal, with a temperature T ($\sim 5 \times 10^7$ °K). This is, of course, a gross oversimplification and neglects any heating mechanism.

The electrons within the plasma, accelerating in the field of the protons, emit thermal bremsstrahlung radiation (free-free emission). If the density is sufficiently high, the photons will be reabsorbed by the gas (free-free absorption) or elastically scattered by the free electrons (Thomson scattering).

The cross section for bremsstrahlung emission can be calculated using the Born approximation (see Heitler, 1954). The Elwert factor is introduced as a high-energy correction to make the high-frequency limit of the cross section correct. The Gaunt factor is required to correct for the use of plane waves instead of Coulomb wave functions. For a Maxwellian velocity distribution of electrons, when any free-free absorption is neglected, the bremsstrahlung spectrum has the form

$$I(\nu)\,d\nu = 1.68 \times 10^{-15} \frac{Z^2 N_e \varrho}{(kT)^{3/2}}\,d\nu$$

$$\times \int_{h\nu}^{\infty} e^{-E/kT} \left[\ln\left(\frac{\beta + \beta'}{\beta - \beta'}\right)\right] \frac{\beta}{\beta'} \left[\frac{1 - e^{-2\pi Z/137\beta}}{1 - e^{-2\pi Z/137\beta'}}\right] dE, \tag{3.1}$$

where

$$\beta = \left(\frac{2E}{mc^2}\right)^{1/2}, \qquad \beta' = \left(\frac{2(E - h\nu)}{mc^2}\right)^{1/2}.$$

E is the electron energy, and N_e is the total number of electrons. The term kT is in units of keV, and $I(\nu)$ is in keV/keV-sec. At *low frequencies* $h\nu \ll kT$ and so $\beta \approx \beta'$. Thus

$$\text{Elwert factor} = \frac{\beta}{\beta'}\left[\frac{1 - e^{-2\pi Z/137\beta}}{1 - e^{-2\pi Z/137\beta'}}\right] \approx 1.$$

In the low-frequency limit the integral can be solved exactly to give

$$I(\nu)\,d\nu = 1.68 \times 10^{-15} \frac{Z^2 N_e \varrho}{(kT)^{1/2}} \ln\left(\frac{2.25kT}{h\nu}\right) d\nu. \tag{3.2}$$

For low-Z materials and all photon energies, the high-energy Elwert-factor correction is not large. If we put this factor equal to unity, the integral can be directly evaluated in terms of K_0, a modified Bessel function of the second kind:

$$I(\nu)\,d\nu = 1.68 \times 10^{-15} \frac{Z^2 N_e \varrho}{(kT)^{1/2}} \left[e^{-h\nu/kT} K_0\left(\frac{h\nu}{2kT}\right) \right] d\nu. \tag{3.3}$$

This expression is a good approximation for all frequencies and reduces to Eq. (3.2) in the low-frequency limit:

$$h\nu \ll kT, \qquad K_0\left(\frac{h\nu}{2kT}\right) \approx \log\left(\frac{4kT}{h\nu}\right) - 0.557,$$

$$h\nu \gg kT, \qquad K_0\left(\frac{h\nu}{2kT}\right) \approx \left(\frac{\pi}{2}\right)^{1/2} e^{-h\nu/2kT} \left(\frac{2kT}{h\nu}\right)^{1/2}.$$

Thus, at high frequencies, very approximately

$$I(\nu) \propto e^{-h\nu/kT}. \tag{3.4}$$

This exponential approximation was frequently used in early x-ray work. The differences between the three expressions for the spectral intensity $I(\nu)$ [the exact integral, Eq. (3.1), the modified Bessel function, Eq. (3.3), and the exponential form] are illustrated in Table 3-1. These have been computed for a temperature $kT = 5$ keV and $Z = 1$ (Chodil *et al.*, 1968). It can be seen that the differences are not large. However, the temperature derived with the simple exponential form differt significantly, by a factor of about 2, from those derived from the more accurate expressions for the spectral intensity.

TABLE 3-1. Comparison of functions describing bremsstrahlung.[a]

Photon energy (keV)	Exact integral [Eq. (3.1)]	1st approximation [Eq. (3.3)]	Exponential
0.01	6.50	7.00	1.82
0.1	4.35	4.62	1.78
1	2.10	2.17	1.49
3	1.00	1.00	1.00
10	0.157	0.153	0.248
30	1.81×10^{-3}	1.70×10^{-3}	4.51×10^{-3}
100	8.6×10^{-10}	8.0×10^{-10}	3.75×10^{-9}

[a] All intensities are normalised to give $I(\nu) = 1.00$ at $h\nu = 3$ keV.

3.3.2. *Comparison with Observational Data*

Is it possible to derive a unique set of physical parameters for such a model which are consistent with the observational results over the entire spectral range from x-ray wavelengths to radio wavelengths? In order to test the model, it is necessary to compare the intensity of emission in different spectral regions. However, observations at different frequencies must be carried out simultaneously, since the source is highly variable.

The first simultaneous optical and x-ray measurements on Sco X-1 were made on May 18th, 1967 (Chodil *et al.*, 1968; Mark *et al.*, 1969). A rocket was launched from Hawaii to measure the x-ray intensity in the 2–20 keV range. The optical intensity was monitored over the entire period from May 16th to May 20th at the Cerro Tololo Inter-American Observatory in Chile (the variations in optical intensity during this period are shown in Fig. 3-3). The source appears to be highly unstable when at maximum brightness, and more quiescent when the intensity is a minimum. Obviously these very complex time variations cannot be explained by the very simple model we are using.

The results obtained from these simultaneous observations are shown in Fig. 3-4. With the x-ray measurements alone it is possible to fit a thermal bremsstrahlung spectrum of the form given by Eq. (3.3) to the experimental data. To obtain a good fit it is necessary to use an iteration procedure, and much numerical work is involved. The best-fit curve for the x-ray measurements is shown in Fig. 3-4. This gives a plasma temperature of

$$kT = 7 \text{ keV}.$$

Extrapolating the x-ray spectrum to the lower frequencies gives an intensity at optical wavelengths well above the observed intensity. Some of the observed reduction in intensity is probably due to interstellar extinction. This is caused by small dust grains, with diameters of the order of a micron, which scatter and redden the light as it passes through the interstellar medium. It is difficult to estimate the amount of interstellar extinction, ΔV, in the spectrum of Sco X-1, since the distance to the source is very uncertain. However, a study of the stars in the neighborhood of Sco X-1 gives a possible value for ΔV of between 0.75 and 2 magnitudes. This would account for about half the difference between theoretical and observational results. An increase in the interstellar extinction would not be sufficient to reduce the discrepancy, because the spectral shape of the measurements cannot be explained by reddening alone.

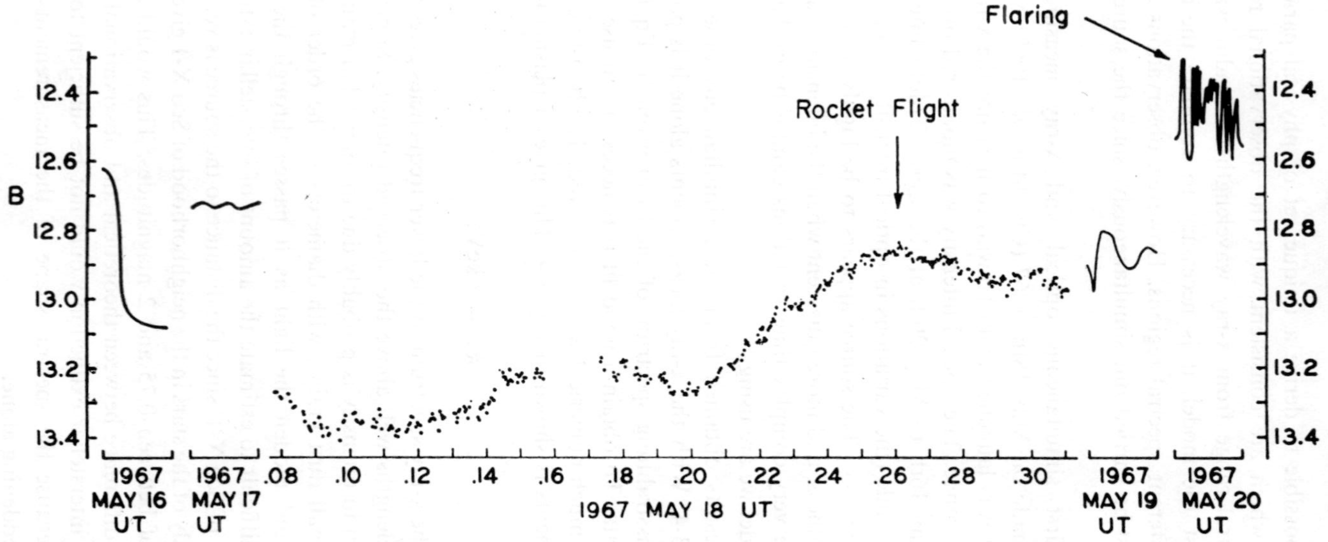

FIGURE 3-3. Optical observations of Sco X-1 during May, 1967 [from Chodil *et al.*, Astrophysical Journal **154**, 645 (1968), © 1968, The University of Chicago Press. All rights reserved].

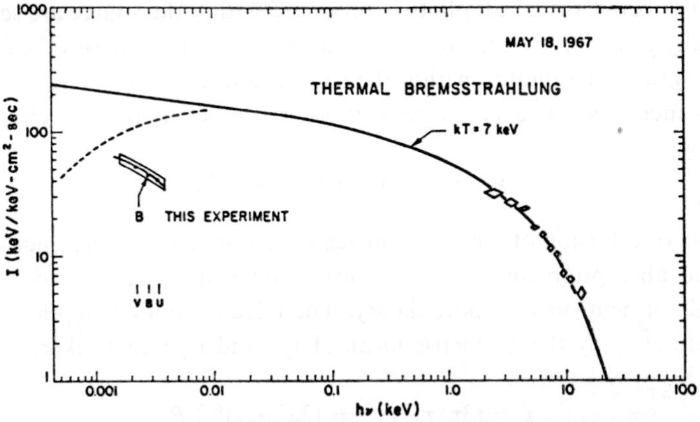

FIGURE 3-4. X-ray and optical spectrum of Sco X-1 on May 18, 1967. The solid line is the thermal bremsstrahlung spectrum with no self-absorption for $kT = 7$ keV. The dashed line is the emission spectrum corrected for free-free absorption. The solid line passing through the optical points is the theoretical intensity after the bremsstrahlung spectrum with self-absorption is corrected for interstellar extinction [from Chodil *et al.*, Astrophysical Journal **154**, 645 (1968), © 1968, The University of Chicago Press. All rights reserved].

3.3.3. *Free-Free Absorption and Thomson Scattering*

It is clearly necessary to modify our simple model of an optically thin plasma in order to explain some of the optical observations of Sco X-1. It appears that the low-frequency radiation is being reabsorbed within the source. The free-free absorption coefficient can be calculated with the Born approximation and is related to the free-free emission cross section (see Heitler). If we neglect elastic Thomson scattering, the absorption coefficient $\varkappa_{ff}$ is given by

$$\varkappa_{ff} = 7.8 \times 10^{-48} \frac{\varrho^2 G}{(kT)^{1/2}\,(h\nu)^3} \frac{h\nu}{kT} \quad \mathrm{cm}^{-1}, \tag{3.5}$$

where G is the Gaunt-factor correction to the plane-wave approximation. For $kT \simeq 7$ keV and $h\nu \simeq 2$ eV, G $\simeq 5$ (G can be calculated from a knowledge of the hydrogen wave functions). From Eq. (3.5) we see that

$$\varkappa_{ff} \propto \frac{1}{\nu^2}.$$

The free-free absorption thus increases at low frequencies. For $kT = 7$ keV and $h\nu = 2.25$ eV (typical of optical radiation within Sco X-1)

$$\varkappa_{ff} = 4.2 \times 10^{-43} \varrho^2 \quad \mathrm{cm}^{-1}. \tag{3.6}$$

As well as being absorbed, photons produced within the source are scattered elastically by the free electrons. This scattering greatly increases the effective path length of the photon within the source, and the absorption coefficient is thus increased. The Thomson scattering coefficient for free electrons is given by

$$\varkappa_{sc} = 6.6 \times 10^{-25} \varrho \quad \text{cm}^{-1}. \tag{3.7}$$

The rate of diffusion of a photon through the source in the presence of both free-free absorption and Thomson scattering can be calculated with the methods of neutron transport theory. The effective optical depth τ is given approximately by the geometric mean of τ_{ff} and τ_{sc} (see Tucker, 1967). In detail

$$\tau = (3\tau_{sc}\tau_{ff})^{1/2} = (3\varkappa_{sc}\varkappa_{ff})^{1/2} R.$$

Using Eqs. (3.6) and (3.7) gives

$$\tau = 9.1 \times 10^{-34} \varrho^{3/2} R \tag{3.8}$$

(for $kT = 7$ keV and $h\nu = 2.25$ eV).

3.4. Observational Results on Sco X-1

(a) *X-ray and optical continuum*. Assuming that the interstellar extinction of Sco X-1 is approximately one magnitude in the visible, we can calculate the optical depth required in the source plasma in order to explain the observed spectral curve. We find that, for the results of the May 18th flight (Fig. 3-4), an optical depth $\tau \simeq 1$ is needed at visible wavelengths. This does, of course, assume that the optical flux and x-ray flux originate within the same region of the source. Such an assumption could be tested by attempting to see whether any correlation exists between the time variations of the optical and x-ray spectra. Experiments are being conducted to test this. The resultant thermal bremsstrahlung spectrum with an optical depth $\tau = 1.0$ in the visible is shown in Fig. 3-4.

If $\tau = 1.0$ at $h\nu = 2.25$ eV, Eq. (3.8) gives us an expression relating the density ϱ and radius R of the source:

$$\varrho^{3/2} R = 1.1 \times 10^{33}. \tag{3.9}$$

Another expression relating ϱ and R can be obtained from the observed total x-ray flux: if the source is at a distance d cm, the measured x-ray flux and Eq. (3.3) require that

$$\varrho^2 R^3 = 1.0 \times 10^{17} d^2. \tag{3.10}$$

If we assume that $d = 300$ pc, Eqs. (3.9) and (3.10) give the following values of ϱ and R:

$$\varrho = 1.2 \times 10^{16} \text{ electrons/cm}^3,$$

$$R = 8.5 \times 10^8 \text{ cm}.$$

The total energy radiated, I, can be obtained by integrating the bremsstrahlung curve (including the effects of self-absorption) over all frequencies. For the model,

$$I = 2 \times 10^{36} \text{ ergs/sec}.$$

If the energy is supplied by the gravitational energy released by material falling towards a highly condensed central star, the mass of the central star required is

$$M_{\text{central}} = 0.1 M_{\odot}$$

(where $M_{\odot}$ is the mass of the Sun), and the rate of accretion of material must be $2.0 \times 10^{-6}\, M_{\odot}$ per year. Such an accretion rate is not unreasonable and suggests a lifetime for the source of about 10^6 years. The mass and radius of the emitting region will change with time, and such changes have been observed. A second rocket was launched from Hawaii on September 29th, 1967, and the observations of Sco X-1 made during the May 18th flight were repeated. During this second flight Sco X-1 was near to maximum optical intensity. The differences between the various parameters of the source model derived from the two sets of observational data are shown in Table 3-2. The temperatures derived from all x-ray observations lie in the range 4–7 keV.

TABLE 3-2.[a]

	September 1967	May 1967
Total I	1.7	1
τ (2.25 eV)	2.5	1
kT (keV)	4	7
ϱ	1.3	1
R	1.1	1
Total mass of plasma	1.7	1

[a] I, τ, ϱ, R, and M have been normalised to unity for May 1967.

(b) *Infrared spectrum.* Observations of Sco X-1 at infrared wavelengths can give more detailed information about the self-absorption occurring within the source. Measurements have been made at Mount

Palomar (Neugebauer *et al.*, 1969) in the spectral range 0.33 to 1.0 μ and 1.65 to 2.2 μ. These were compared with measurements of the optical flux in the range 3325–6050 Å. Substantial time variation was seen, but the spectrum had the general form shown in Fig. 3-5.

In the near-infrared frequency range, when the observations are corrected for interstellar extinction, the spectrum is of the approximate form $I(\nu) \propto \nu^2$. If this is equated to the Rayleigh–Jeans approximation for a blackbody spectrum at temperature T, it is found that the source parameters must satisfy the relation

$$\frac{R^2 T}{d^2} = 10^{-17}\ {}^\circ\text{K}. \tag{3.11}$$

The spectrum becomes approximately flat for higher frequencies (optical wavelengths) where the self-absorption is less.

Combining (3.11) with Eq. (3.10) derived from the total x-ray flux, we can eliminate the distance d, to obtain the relation

$$\frac{\varrho^2 R}{T} = 10^{34}. \tag{3.12}$$

Substituting this in the expression for the free-free absorption coefficient gives an optical depth for free-free absorption (neglecting scattering)

$$\tau_{ff} = \frac{1.8 \times 10^{32} G}{T^{1/2} \nu^2};$$

at $T = 5 \times 10^7$ °K

$$\tau_{ff} = \frac{1.4 \times 10^{29}}{\nu^2}. \tag{3.13}$$

The electron-scattering optical depth is given by

$$\tau_{sc} = 6.6 \times 10^{-25} \varrho R. \qquad \text{[from Eq. (3.7)].}$$

Combining these gives an effective optical depth for different frequencies shown in Table 3-3. The results do not depend critically upon the distance d. The parameters derived from infrared measurements are consistent with those estimated on the basis of the x-ray and optical results alone.

To summarise: the x-ray, optical, and infrared continuum flux can be consistently interpreted in terms of an "old nova" model. Matter passing from a red giant and falling towards a white dwarf star produces a hot plasma with temperature $\simeq 5 \times 10^7$ °K and density $\simeq 10^{16}$ electrons/cm^3,

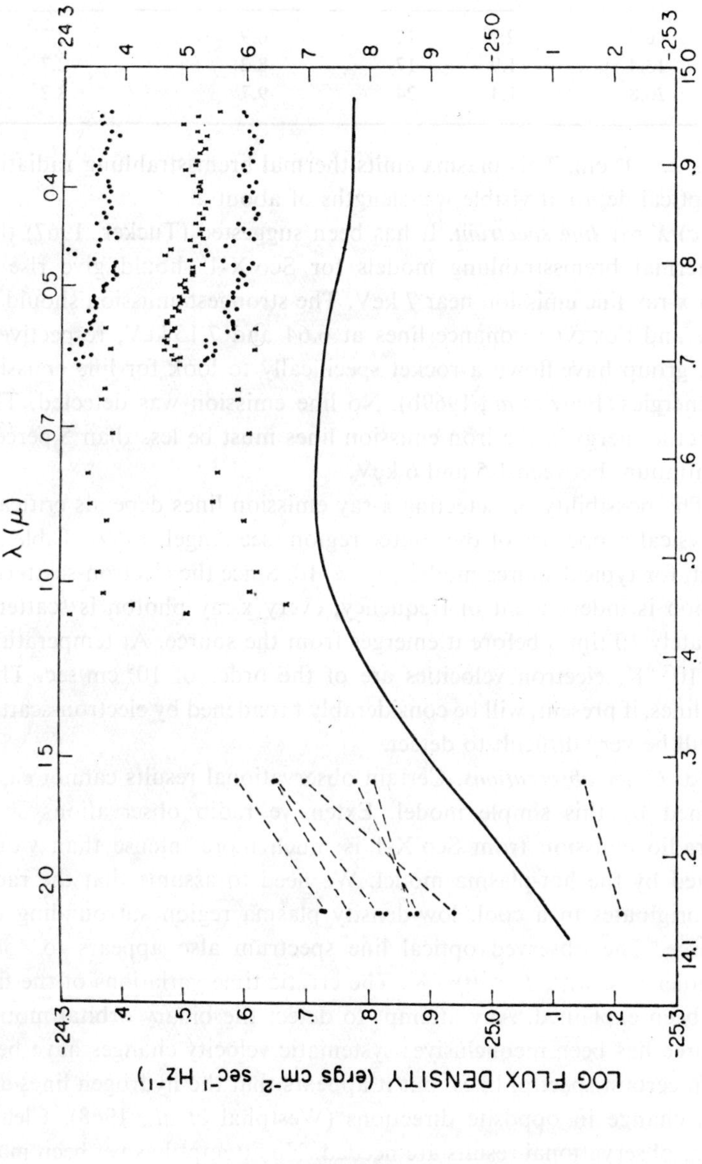

FIGURE 3-5. A sketch of the infrared spectrum of Sco X-1.

TABLE 3-3. Calculated parameters of model.

d (pc)	R (cm) $\times 10^{-8}$	$\varrho \times 10^{-16}$	τ_{sc}	$\tau(\nu = 3.3 \times 10^{14}$ Hz)	$\tau(\nu = 10^{15}$ Hz)
500	6.7	2.7	12	6.9	2.3
1000	13.4	1.9	17	8.2	2.7
2000	26.8	1.4	24	9.7	3.2

and radius $\simeq 10^8$ cm. This plasma emits thermal bremsstrahlung radiation with an optical depth at visible wavelengths of about 2.

(c) *X-ray line spectrum*. It has been suggested (Tucker, 1967) that certain thermal bremsstrahlung models for Sco X-1 should give rise to significant x-ray line emission near 7 keV. The strongest emission should be the FeXXV and FeXXVI resonance lines at 6.64 and 7.15 keV, respectively. The NRL group have flown a rocket specifically to look for line emission at these energies (Fritz *et al.*, 1969b). No line emission was detected. This shows that the energy in the iron emission lines must be less than 5 percent of the continuum between 1.5 and 6 keV.

The possibility of detecting x-ray emission lines depends critically on the physical properties of the source region (see Angel, 1969). Table 3-3 shows that, for typical source models, $\tau_{sc} \simeq 10$. Since the electron-scattering cross section is independent of frequency, every x-ray photon is scattered approximately 10 times before it emerges from the source. At temperatures of about 10^7 °K, electron velocities are of the order of 10^9 cm/sec. Thus any x-ray lines, if present, will be considerably broadened by electron-scattering and will be very difficult to detect.

(d) *Other observations*. Certain observational results cannot easily be explained by this simple model. Extensive radio observations show that the radio emission from Sco X-1 is much more intense than would be predicted by the hot-plasma model. We need to assume that the radio emission originates in a cool, low-density plasma region surrounding the x-ray source. The observed optical line spectrum also appears to come from a cooler gas, with $T < 10^5$ °K. The erratic time variations of the flux have not been explained. Any attempt to detect the binary orbital motion of the source has been inconclusive: systematic velocity changes have been detected in certain spectral lines, but it appears that the hydrogen lines and HeII lines change in opposite directions (Westphal *et al.*, 1968). Clearly many more observational results are needed. No attempt has yet been made to explain the energy injection mechanism. However, even though the observations are very complex, this simple model of a hot plasma has had remarkable success in explaining the basic form of the observed spectrum.

4. THE CRAB NEBULA

4.1. Introduction

The second X-ray source to be located during the early NRL rocket flights was in the Taurus region and was soon identified as the Crab nebula. Its position and size were determined by both lunar occultation and modulation collimator techniques, and it was found to be an extended source of size about equal to that of the optical nebula. The spectrum in the x-ray region is of power-law form, and thus it appears that the Crab may be a prototype for one type of x-ray source while Sco X-1, with its exponential spectrum, is a prototype of another.

4.1.1. *Early Work*

There is a vast amount of observational data on the detailed structure of the Crab nebula at optical and radio frequencies, and we cannot consider the x-ray spectrum without referring to these results. We shall thus begin this lecture with a discussion of the early observations of the Crab nebula and the models which have been proposed to account for the optical and radio spectrum, to see whether these are compatible with the more recent x-ray data. The classic paper on the optical spectrum of the Crab is that of Oort and Walraven (1956), and this gives references to all the earlier work on the subject.

The Crab nebula is the remnant of a supernova observed by Chinese astronomers in 1054 A.D. For a while it was visible during the daytime: it remained visible in the night sky for about 2 years, after which its brightness must have dropped below the 5th magnitude. There are good optical plates of the nebula taken as early as 1899. From a comparison of these early plates with more recent observations, it is obvious that the nebula is changing rapidly. There is an overall expansion of the gas of about 10^3 km/sec, caused by the initial supernova explosion. There are also detailed changes within small regions of the nebula which were first observed by Lampland in 1921 and were more carefully studied by Baade in 1942–3. Baade observed a series of small wisps and knots of gas originating in the region between the central stars and the stationary nebulosity *b* (see Fig. 4-1). These moved rapidly outwards and coalesced with the stationary gas. He estimated that there was about one such ripple every 3 months. He speculated that these ripples could be caused by eruptions within the supernova remnant which he tentatively identified with one of the central stars (the south-preceding star). However, he failed to detect any optical variability

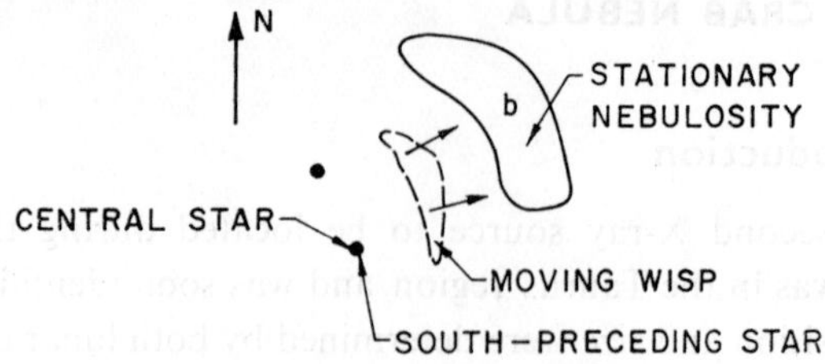

FIGURE 4-1. Moving light ripples near the center of the Crab nebula observed by Baade. The wisps of gas move outwards from the central stars and coalesce with the stationary nebulosity *b*.

of this star and wondered whether the supernova remnant was in fact invisible.

The next important observation was the detection of polarisation in the optical continuum of the Crab. This was first reported by Russian astronomers in 1953–1954 and is discussed in detail by Oort and Walraven, and by Woltjer (1957). The average polarisation across the entire nebula is 9.2 percent. However, within small regions of the source, of size about $\frac{1}{2}$ arc min, the local polarisation is considerably higher, approaching 50 to 80 percent in some regions. The change in the polarisation with increasing aperture size is shown in Table 4-1. Such measurements can be used to obtain the detailed structure of the magnetic field within the source.

TABLE 4-1. The polarisation of the central region of the nebula as a function of aperture size.

Field (min)	Degree of polarisation (%)
1	23
2	20
3	19
4	17.2
5	13.8
6	9.2

4.1.2. *Emission Mechanisms*

These early observations showed that the Crab nebula is a supernova remnant undergoing a large-scale general expansion. There are small regions of great activity within the nebula, perhaps associated with the central south-preceding star. There is strong homogeneous polarisation of the optical emission. What mechanism produces the continuum emission from such a source?

The earliest suggestion was that the continuous radiation was due to free-free transitions (the same as that discussed in the lecture on Sco X-1). However, it was soon realised that there were several difficulties facing such a model:

(i) *Optical polarisation.* Under certain thermal conditions it is possible to get a small amount of polarisation of the emitted free-free radiation, but not the high degree of polarisation observed in the Crab.

(ii) *Emission lines.* There is a noticeable lack of emission lines in the continuum radiation from the amorphous background of the nebula. Prominent lines, especially from hydrogen and helium, would be expected from a thermal source.

(iii) *Radio emission.* The great strength of the emission at radio frequencies cannot be explained.

(iv) *Mass requirements.* Calculations show that a very large mass of gas (about 20 to 30 solar masses) is required to explain the observed flux by means of a thermal model.

(v) *Chinese observations.* The strength of the free-free emission from a source is directly proportional to the density. If we assume that the Crab nebula has expanded uniformly throughout the 900 years since the initial explosion, the density ϱ at a time t years after the outburst is given by

$$\varrho/\varrho_0 = (900/t)^3$$

where ϱ_0 is the present density. Thus the intensity at a time t is proportional to $(900/t)^3$. As the amorphous nebula is now of the 9th magnitude, it must have had a magnitude of -6 for $t = 9$ years. This disagrees with the Chinese observation that the star ceased to be visible after about 2 years.

Because of these difficulties, an alternative model for the Crab had to be sought for. In 1950 Alfvén and Herlofsen suggested that synchrotron emission from electrons accelerating in a magnetic field could account for the radio emission from several galactic radio sources. In 1953, Shklovsky showed that such a mechanism could account for the observed emission from the Crab nebula and did not give rise to the difficulties inherent in the thermal model. We shall thus discuss the synchrotron model of the Crab in more detail. It cannot, of course, explain the fine structure of the wisps and ripples of gas but is remarkably successful in explaining the observed optical, radio, and x-ray emission.

4.2. Synchrotron Emission

The power radiated by an electron moving in a uniform magnetic field can be readily calculated from classical electromagnetic theory [see, for example, Schwinger (1949)]. For a detailed discussion of synchrotron emission mechanisms in galactic radio sources see the excellent review article by Ginzburg and Syrovatskii (1965).

Consider an electron with energy E ergs moving in a magnetic field of strength H gauss. The power radiated by such an electron in the frequency range ν to $\nu + d\nu$ is

$$P(\nu)\,d\nu = \frac{3^{3/2}}{2}\,\frac{e^2}{R}\left(\frac{E}{mc^2}\right)^4 \frac{\nu_0 \nu\, d\nu}{\nu_c^2} \int_{\nu/\nu_c}^{\infty} K_{5/3}(\eta)\, d\eta\,. \tag{4.1}$$

In this expression e is the charge on the electron in esu, m its rest mass, c the velocity of light, and R the radius of curvature of the orbit, all in cgs units. R is given by $R = E/eH_\perp$, where $H_\perp = H\sin\theta$ and θ is the angle between the electron velocity and the magnetic field. ν_0 is the Larmor frequency $= c/2\pi R$ and

$$\nu_c = \frac{3c}{4\pi R}\left(\frac{E}{mc^2}\right)^3. \tag{4.2}$$

Inserting numerical values for the constants and writing the energy of the electron in units of 1 GeV $= 10^9$ eV gives

$$\nu_c = 1.6 \times 10^{13} H_\perp E_{\mathrm{GeV}}^2. \tag{4.3}$$

and

$$\frac{dE_{\mathrm{GeV}}}{dT} = -3.8 \times 10^{-6} H_\perp^2 E_{\mathrm{GeV}}^2. \tag{4.4}$$

From Eq. (4.4) it can be seen that the energy will diminish to half its original value, E_a, in a time

$$T_{1/2} = \frac{8.35 \times 10^{-3}}{H_\perp^2 E_a}\ \text{years}. \tag{4.5}$$

The radiation emitted by a single electron is totally polarised with an electric vector parallel to the radius of curvature of the orbit. The form of the spectral emission $P(\nu)$ is shown in Fig.4-2. $P(\nu)$ has a maximum for $\nu/\nu_c \simeq 0.29$. For low frequencies

$$P(\nu) \propto (\nu/\nu_c)^3.$$

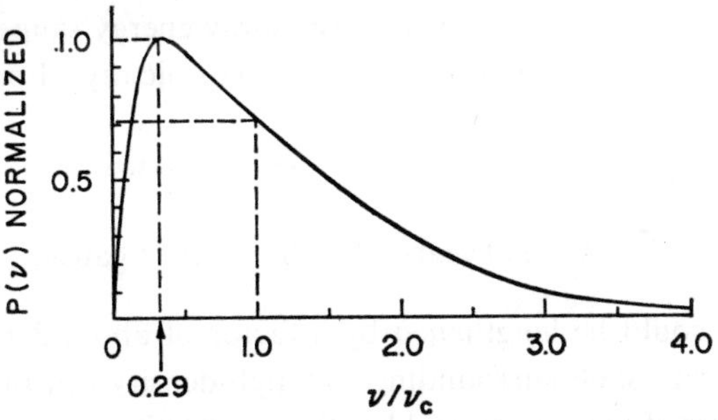

FIGURE 4-2. Normalised synchrotron emission spectrum.

4.2.1. *Comparison with Optical and Radio Observations*

In order that the optical emission from the Crab can be explained by the synchrotron mechanism, the maximum of the spectral emission $P(\nu)$ must lie in the visible region. This requires $\nu_c \simeq 10^{15}$ Hz. (From the recent x-ray and γ-ray observations it is clear that ν_c must be much greater than this. We shall leave the discussion of these results until the next section.) For $\nu_c \simeq 10^{15}$ Hz, Eq. (4.3) gives

$$H_{\perp}E_{\mathrm{GeV}}^2 \simeq 60. \tag{4.6}$$

If we can estimate the magnetic field within the source, Eq. (4.6) can then be used to obtain a value for the electron energies required to produce the optical emission. Various considerations show that the magnetic-field strength should be between 1 and 3×10^{-3} gauss. If we calculate the total energy density of the electrons and protons required to produce the observed optical emission and compare this with the magnetic-energy density $(H^2/8\pi)$ for various magnetic-field strengths, we find that these are equal for a field of about 10^{-3} gauss. If the field is much less than this equipartition value, the electrons will very rapidly escape from the source. If the field is much larger than the 10^{-3} gauss, the half-life of the energetic electrons is considerably reduced [see Eq. (4.5)]. Since the diameter of the Crab nebula is about 6 light years and the velocity of the electrons is about $0.5c$, a lifetime of at least a few decades is required for the electrons to fill the entire volume. This gives an upper limit of 3×10^{-3} gauss for the magnetic-field strength.

For the electrons producing the optical spectrum, Eq. (4.6) with $H = 10^{-3}$ gauss gives

$$E_{\mathrm{GeV}} \simeq 250 \text{ GeV}.$$

The energies required thus lie in the cosmic-ray energy range, and the Crab nebula may prove to be a powerful source of cosmic rays. For such energies, Eq. (4.5) gives

$$T_{1/2} \simeq 30 \text{ years} \quad \text{for } H_{\perp} = 10^{-3} \text{ gauss},$$

$$T_{1/2} \simeq 3000 \text{ years} \quad \text{for } H_{\perp} = 10^{-4} \text{ gauss}.$$

Such lifetimes could be lengthened by a factor of about 3 by the presence of a lower density shell surrounding the high-density central region of the source. However, if the source field has the equipartition value of 10^{-3} gauss, the lifetime of the energetic electrons is considerably less than the historical lifetime of the nebula. Thus some injection mechanism is required to maintain a continuous supply of high-energy electrons: these could not all have been produced during the initial supernova explosion. Until 1968 the origin of this continuous energy supply remained a complete mystery.

4.2.2. *Radio Observations*

The Crab nebula has been studied extensively at radio frequencies [see, for example, Moroz (1964)]. The lunar occultation experiment by Seeger showed that the radio source has dimensions comparable to, but slightly larger than, the optical object. The radio emission is polarised. The observations at optical and radio frequencies can be fitted by a power-law spectrum of the form

$$I(\nu) \propto \nu^{-\alpha},$$

where the spectral index $\alpha \simeq 0.28$ at radio frequencies and $\alpha \simeq 1.5$ at optical frequencies. The total radio luminosity of the Crab is much higher than the total optical luminosity. Defining β as the ratio of the radio intensity at 10^8 Hz to the optical intensity at 7×10^{14} Hz, we find from the observational data

$$\beta_{\text{obs}} \approx 230$$

(correct to about a factor of 2). This is a very difficult measurement to make, because it is necessary to exclude all the radiation from the filaments of gas and the field stars.

4.2.3. *Synchrotron Emission Spectrum*

The spectrum of the radiation produced by synchrotron emission depends critically upon the energy spectrum of the electrons within the source. Since the energy distribution of electrons within the Crab nebula is unknown, it is reasonable to assume that these have a spectrum similar to that of the primary cosmic-ray electrons observed at the Earth. This may, of course, be the wrong assumption to make, since the electron spectrum could be considerably changed by passage through the outer layer of the nebula. However, as an initial hypothesis we shall assume that, within the Crab nebula,

$$n(E)\,dE = kE^{\gamma-1}\,dE,$$

where $n(E)\,dE$ is the number of electrons with energies between E and $E + dE$, and γ and k are constants. The integrated spectrum has the form

$$N(E) = -\frac{k}{\gamma}E^{\gamma},$$

where $N(E)$ is the total number of electrons with energies greater than E. For the observed cosmic-ray spectrum

$$\gamma \simeq -1.8, \qquad 15 < E_{\text{GeV}} < 10^6,$$
$$\gamma \simeq -0.9, \qquad 2 < E_{\text{GeV}} < 15.$$

For lower energies the curve becomes still flatter. In his first attempt to explain the observed optical and radio spectrum of the Crab, Oort used an electron energy distribution very similar to this cosmic-ray spectrum. In fact, he took

$$\begin{aligned} \gamma &= +0.2, \qquad 0.155 < E_{\text{GeV}} < 1.55, \\ \gamma &= -0.8, \qquad 1.55 < E_{\text{GeV}} < 15.5, \qquad (4.7) \\ \gamma &= -1.8, \qquad E_{\text{GeV}} > 15.5. \end{aligned}$$

This energy spectrum must be folded into the expression for the spectral emission for a single electron [Eq. (4.1)]. The result is an approximate power-law spectrum with a spectral index α which increases with increasing frequency. This is of the same form as the observed spectrum. However, this first model predicts too high a flux at radio frequencies. The values of the ratio β obtained from this model are shown in Table 4-2 for different values of the source magnetic field. It can be seen that, for a wide range of field, β is a factor of about 100 too large. In order to reduce the value of β, we must assume that the energy spectrum of the electrons within the nebula is less steep than that of the cosmic rays. It may be that this softer source spectrum is degraded and steepened by passage through the interstellar medium. Oort found that the energy spectrum which gives the best fit to the observations was of the form

$$\begin{aligned} \gamma &= +0.7, & 0.155 < E_{GeV} < 1.55, \\ \gamma &= -0.3, & 1.55 < E_{GeV} < 15.5, \\ \gamma &= -1.3, & E_{GeV} > 15.5. \end{aligned} \tag{4.8}$$

The values of β obtained using this "soft" spectrum are also shown in Table 4-2. These agree with the observed value, $\beta \approx 230$, to within the experimental error.

TABLE 4-2

Magnetic field (gauss)	Computed values of β Cosmic-ray spectrum [Eq. (4.7)]	"Soft" spectrum [Eq. (4.8)]
10^{-4}	14600	509
10^{-3}	7600	434
10^{-2}	11700	881

The total energy of the fast particles within the nebula can be calculated once the energy spectrum is known. Assuming an energy spectrum of the form given by Eq. (4.8), the total energy content of the electrons and protons producing the optical radio emission within the Crab is of the order of 10^{49} ergs. This is considerably greater than the kinetic energy of the expanding gas, which is about 10^{47} ergs. Thus most of the energy within the source goes into accelerating fast particles. The total rate of energy loss is approximately 10^{38} ergs/sec.

Summary

The optical and radio observations of the Crab nebula are consistent with a model of the source in which relativistic electrons, accelerating in a magnetic field, emit synchrotron radiation. For such a model, we require the following parameters:

Magnetic field	$\simeq 10^{-3}$ gauss
Energy of electrons producing optical emission	$\simeq 10^{11}$ eV,
Half-life of these "optical" electrons (thus some injection mechanism is required)	$\simeq$ 30 years
Total energy of charged particles	$\simeq 10^{49}$ ergs,
Total kinetic energy of expansion	$\simeq 10^{47}$ ergs,
Rate of energy loss	$\simeq 10^{38}$ ergs/sec.

The energy spectrum of the electrons within the Crab must be softer than the primary cosmic-ray spectrum in order to explain the observed radio and optical intensities of the source.

4.3. X-Ray Observations

One of the earliest NRL rocket flights detected a strong x-ray source in the Taurus region, which was soon identified as the Crab nebula (Bowyer *et al.*, 1964). The lunar occultation experiment (see Sec. 2.3) established that this is an extended x-ray source, with dimensions comparable to those of the visible object. Can this x-ray emission be explained by our synchrotron model? Certainly the x-ray emission has a power-law spectrum, and this spectrum appears to be a continuation of some reasonable extrapolation of the optical and radio data (see Fig. 4-3). This is exactly what the synchrotron model would predict. The most convincing evidence that the x-rays are produced by synchrotron emission would be the detection of x-ray polarisation. A rocket was flown by the Columbia group on March 6th, 1969, specifically to look for polarisation in the x-ray emission from

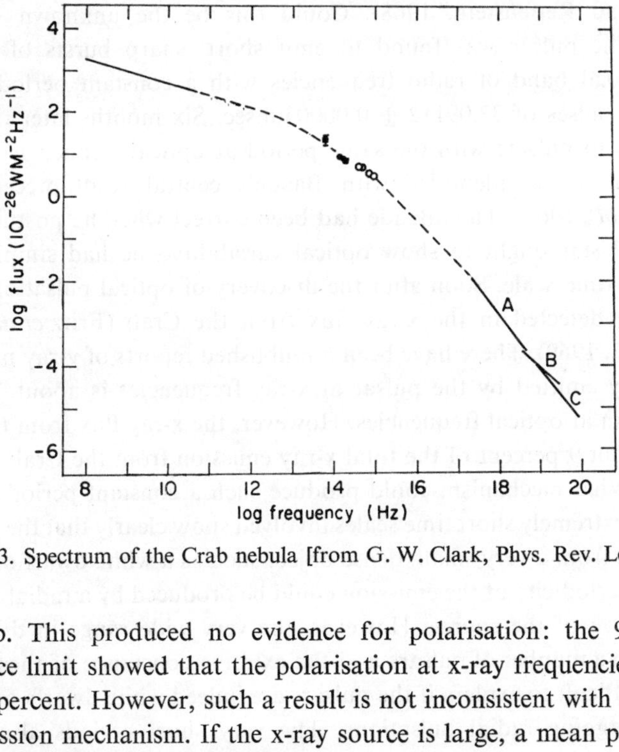

FIGURE 4-3. Spectrum of the Crab nebula [from G. W. Clark, Phys. Rev. Letters **14**, 97 (1963)].

the Crab. This produced no evidence for polarisation: the 99 percent confidence limit showed that the polarisation at x-ray frequencies was less than 27 percent. However, such a result is not inconsistent with a synchrotron emission mechanism. If the x-ray source is large, a mean polarisation of not more than 10–15 percent would be expected. Another flight is scheduled, and this may be decisive.

The spectrum of the Crab has now been extended up to an energy of 100 keV by means of balloon observations (Clark, 1965; Lewin *et al.*,

1968). All such measurements lie reasonably well on the extrapolated spectral curve. Thus synchrotron emission within the Crab seems to produce photons with energies as large as 100 keV or even higher. From Eq. (4.3) we see that in order to produce appreciable emission at x-ray frequencies we require electrons with energies of about 3×10^{13} eV. Such electrons have a half-life of only about 0.3 year within the source, from Eq. (4.5). Thus, if the x-rays are produced by synchrotron emission the problem of an injection mechanism to maintain the flux of high-energy electrons becomes even more severe.

4.4. The Crab Pulsar NP 0532

Renewed speculation about the energy injection mechanism was aroused by the discovery of a pulsating radio source in the Crab nebula (Staelin and Reifenstein, 1968). Could this be the unknown source of energy? The pulsar was found to emit short, sharp bursts of radiation over a broad band of radio frequencies with a constant period between successive pulses of 33.09112 ± 0.00003 msec. Six months later the source was found to pulsate with the same period at optical frequencies, and the optical pulsar was identified with Baade's central south-preceding star (Cocke *et al.*, 1969). Thus Baade had been correct when he postulated that the central star ought to show optical variability—he had simply chosen the wrong time scale. Soon after the discovery of optical pulsations, pulsations were detected in the x-ray flux from the Crab (Fritz *et al.*, 1969a; Bradt *et al.*, 1969). There have been unpublished reports of γ-ray pulsations. The energy emitted by the pulsar at x-ray frequencies is about 100 times greater than at optical frequencies. However, the x-ray flux from the pulsar is only about 9 percent of the total x-ray emission from the Crab nebula.

What mechanism could produce such a constant period of pulsation? The extremely short time scales involved show clearly that the radiation must come from a very small, dense object such as a white dwarf or neutron star. The periodicity of the emission could be produced by a radial pulsation or a rotation of the source. However, the very wide range in the periods observed in a number of pulsars and the extreme constancy of these periods are very difficult to explain if the pulsars are indeed white dwarfs or neutron stars undergoing radial pulsations. The very short periods also exclude a rotating white-dwarf model. Thus we are led to consider whether the pulsars could be small, very dense rotating neutron stars.

Prior to the discovery of pulsars, it had been speculated that neutron stars would be formed during a supernova explosion, when the central core of the star collapsed to form a small object of nuclear densities. During gravitational collapse, the atomic nuclei within the core disintegrate and eventually form a "sea" of free neutrons, protons, and electrons. At nuclear densities ($\simeq 10^{14}$ g/cm^3) the Fermi pressure of the degenerate neutron gas is sufficient to halt the gravitational collapse, and a star of about one solar mass will form a stable object with a radius of about 10 km. The form of the equation of state for matter at such densities is uncertain.

If such an object is formed by adiabatic collapse of a normal main-sequence star, we can make a few very simple calculations. Assuming that angular momentum is conserved, a star of initial radius $\approx 10^{11}$ cm and initial rotation period $\approx 10^6$ sec (comparable to that of the Sun) will have a rotation period of about 0.1–1 msec when collapsed to nuclear densities. This is in reasonable agreement with the observed periods. The period of the Crab pulsar is increasing slowly, with a characteristic time of about 2360 years. This is the same, to within a factor of 2, as the historical lifetime of the nebula. Thus it is possible that a rapidly rotating neutron star was formed during the supernova explosion and that the rotation of this star is gradually slowing down due to interactions with the surrounding plasma.

The magnetic field at the surface of a neutron star may be very large. If the initial main-sequence star has a field of about 100 gauss, this will be "frozen in" during the adiabatic collapse and will give a field of about 10^{12} gauss at the surface of the neutron star. With such a high magnetic field, a rotating neutron star will be a very powerful emitter of low-frequency magnetic-dipole radiation. The characteristic frequency of this emission will be the rotation frequency (about 30 Hz for the Crab pulsar). The axis of the magnetic field of the star need not necessarily coincide with the rotation axis: indeed, the most common configuration seems to be that for which the two axes are perpendicular (see Fig. 4-4). The rate of energy loss by radiation for such a rotating dipole can be calculated classically. This model has been worked out in some detail by Pacini (1968) and by Gunn and Ostriker (1969).

The rate of energy loss from the rotating magnetic dipole is given by

$$\frac{dE}{dt} = -\frac{2\Omega^4}{3c^3}\frac{B_p^2 a^6}{4}, \tag{4.9}$$

where E is the total energy, Ω is the angular rotation velocity, c is the velocity of light, and a is the radius of the star. B_p is the field strength at the

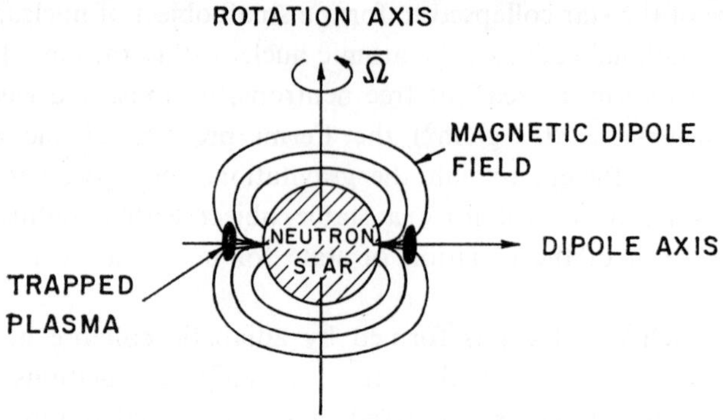

FIGURE 4-4. Rotating-neutron-star model of a pulsar.

pole. The total energy of the star can be estimated from the expression

$$E = \tfrac{1}{2} I \Omega^2, \tag{4.10}$$

where I is the moment of inertia. If we assume that I has the value 1.4×10^{45} g/cm² (a figure obtained from theoretical calculations of a 1.4-solar-mass neutron star), Eq. (4.10) gives a value $E \simeq 3 \times 10^{49}$ ergs for the rotational energy of the Crab pulsar. Furthermore, from Eq. (4.10)

$$\frac{1}{E}\frac{dE}{dt} = 2\frac{1}{\Omega}\frac{d\Omega}{dt}. \tag{4.11}$$

The observed increase in the period of the pulsar, with a characteristic time of about 2360 years, shows that

$$\frac{1}{\Omega}\frac{d\Omega}{dt} = \frac{1}{7} \times 10^{-10}\ \text{sec}^{-1}.$$

Using the value of E obtained above gives a rate of energy loss from the pulsar:

$$\frac{dE}{dt} \simeq 8 \times 10^{38}\ \text{erg/sec}.$$

This value is comparable with Oort's estimate of the amount of energy required to maintain the optical and radio emission from the Crab. We thus see that the pulsar may possibly provide the answer to the as yet unsolved problem of the energy source within the Crab.

How can the low-frequency radiation emitted by the pulsar be used to accelerate electrons to the energies required for the synchrotron emission model of the Crab? In sufficiently strong radiation fields a charged particle is accelerated along the direction of propagation of the radiation, reaching a velocity near to the velocity of light in a small fraction of a wavelength. It then moves with the wave, at constant phase, and continually absorbs energy from the radiation field. In this way very energetic electrons and protons are produced in the plasma surrounding the neutron star. Electron energies in the range 10^{13}–10^{15} eV are readily produced. The maximum electron energy available depends critically on the magnetic-field strength at the surface of the star and on the size of the accelerating region: it may be that all the cosmic-ray particles, with energies up to 10^{20} eV, could be produced in this way. This is certainly a most efficient process for converting mechanical energy into cosmic-ray energy. This fairly crude model of a rotating neutron star can explain, in a simple way, the origin of the energetic particles within the Crab and possibly the origin of cosmic rays in general. It can also explain the wisps and ripples of gas which were seen to emanate from the central south-preceding star, long before it was identified as a pulsar. There are, however, many unexplained phenomena. No explanation has been given for the pulsed emission from the source: this may be synchrotron radiation from electrons trapped within the intense magnetic field surrounding the star (see Fig. 4-4). As the star rotates we see radiation from electrons moving in different orbits. The radiation from each orbit has a characteristic polarisation. Thus we expect to see a pulse of radiation for which the polarisation changes through the pulse. This change in the polarisation has been observed, giving further support to rotating models.

The discovery of pulsars has opened up a whole new area of physics, in which there are many unsolved problems. What is the state of matter within a neutron star? In particular, does it have a superfluid, superconducting core, as suggested by Ginzburg? What is the relativistic magnetohydrodynamic structure of the surrounding plasma? Why do some pulsars suddenly speed up? These, and many other problems, await investigation.

4.5. Gamma Rays from the Crab Nebula

The continuum spectrum from the Crab has been extended to hard x-ray and γ-ray energies, 35–560 keV, using balloon measurements (Haymes *et al.*, 1968). A NaI crystal detector was used. These high-energy measurements appear to match well with a simple extrapolation of the lower energy power-law spectrum.

This experiment was, in fact, the first in which pulsed radiation from the Crab was detected, although the results were not analysed until considerably after the discovery of pulsed radio emission from NP 0532 (Fishman *et al.*, 1969). When the data were reanalysed to look for periodicity, it was found that about 7 percent of the total hard x-ray and γ-ray flux from the nebula was emitted as pulsed radiation, with a double-pulse structure similar to that observed at optical and radio frequencies. The hard x-ray and γ-ray regions contain a major component of the pulsed luminosity of NP 0532.

If electrons with energies as high as 3×10^{13} eV are present in the Crab nebula, producing the x-ray synchrotron emission, these particles may interact with matter in the nebula to produce a flux of γ-rays with energies up to 10^{13} eV. How can such very high energy photons be detected? A γ-ray entering the upper atmosphere interacts to give an electron pair which produces a shower of Čerenkov radiation. Such a shower is seen as a disc of high-intensity radiation spread over an angle of about 1 percent. A typical γ-ray-induced shower lasts for a few nanoseconds. This short time scale allows a γ-ray-induced shower to be distinguished from a high-energy proton-induced shower, which shows different characteristics. In the first experiment designed to look for high-energy γ-rays from the Crab, four mirrors 90 cm in diameter were mounted with their axes parallel and viewed with fast photomultipliers. The photomultiplier signals were amplified and fed into coincidence units, to look for coincidences on a time scale of a few nanoseconds (O'Mongain *et al.*, 1968). The apparatus was designed to detect showers induced by γ-rays with energies greater than 10^{12} eV. Although a positive result was obtained, in that the coincidence rate during observation of the source was three standard deviations above the noise, the random fluctuations present in such an experiment can be so large that the result is not conclusive. The experiment has been repeated, with a positive result which is now about four standard deviations above the noise. This is a somewhat tantalising result, for the reported flux at 10^{13} eV is consistent with the value obtained by an extrapolation of the low-energy spectrum. However, this result is still considered to be preliminary, and the existence of pulsed gamma-ray emission from the Crab pulsar is still in doubt.

The nanosecond light pulses detected by this coincidence technique were interpreted as the Čerenkov radiation shower produced by an energetic γ-ray. However, a possible alternative explanation is that the short light pulses originate within the Crab nebula pulsar itself. The pulsar spectrum is obtained by averaging the emission over many pulsation periods. The radiation might be emitted in a series of sharp, nanosecond bursts: the observed millisecond pulse width would then represent the envelope of the scatter in the arrival time of these sharp pulses. Such pulses would produce counts in the γ-ray Čerenkov detector.

In order to check the alternative hypothesis an experiment was carried out to look directly for nanosecond light pulses from NP 0532 (Hegyi, Novick, and Thaddeus, 1969). A 36-in. telescope with a very small field of view (equivalent to 24 arc sec) was used. With the field of view the background due to atmospheric Čerenkov radiation produced by γ-rays is negligible. The output from two photomultipliers was fed into a coincidence unit to look for correlations on a time scale of 10 nanoseconds. The results show no evidence of any short optical pulses originating in the pulsar.

5. X-RAY POLARIMETRY

5.1. Motivation

The problems of x-ray spectroscopy have not yet been solved. No x-ray lines have yet been observed in the spectra of nonsolar x-ray objects. This is not surprising, since only proportional counters with poor resolution have been used for measurements on such objects. The discovery of the first nonsolar x-ray line will indeed be a great event. The techniques which could be used for such spectroscopic work have already been discussed in Sec. 2. Here we wish to discuss the problems involved in another new aspect of x-ray work: the detection of x-ray polarisation.

The main motivation for this work is undoubtedly a desire to study the x-ray polarisation of the Crab nebula. The synchrotron model for the Crab is most successful in explaining the optical and radio emission—can it also explain the observed x-ray flux? If the x-rays are produced by synchrotron emission, they ought to be polarised to a degree comparable to that of the optical emission. Thus, if the x-rays are produced throughout the entire nebula, we expect an average polarisation of about 9 percent. If, however, they are emitted preferentially from the denser central region, a polarisation as high as 20 percent might result.

Other mechanisms can produce x-ray polarisation, although synchrotron emission is the most obvious choice. Thomson scattering of the x-rays in a nonspherically symmetric thermal plasma can produce a polarisation of about 5 percent (see Angel, 1969). For significant polarisation to occur, a reasonable fraction of the radiation leaving the source must be scattered within the source. This requires that the optical depth for Thomson scattering τ_{sc} must be greater than unity. Maximum polarisation occurs for τ_{sc} in the range

$$1 < \tau_{sc} < 20.$$

This scattering process is thought to be responsible for the optical polarisation of rapidly rotating Be stars: the light is scattered by a disc of ionised hydrogen surrounding the star. In the case of Sco X-1, the observations suggest that $\tau_{sc} \approx 10$. Since it is very probable that the emitting plasma is not spherically symmetric, it is possible that the x-ray flux from certain regions of the source could show a polarisation of about 5 percent. The actual calculation of the expected polarisation is very messy, and Monte Carlo techniques are needed to obtain a solution. However, the possibility that the emission from other types of x-ray sources could be polarised increases the desire for an efficient x-ray polarimeter. Clearly a measurement of the polarisation of the x-ray flux cannot uniquely determine the emission mechanism within the source. All other observational evidence must be considered as well.

5.2. Techniques

The design of an x-ray polarimeter is again limited by the fact that the refractive index for x-rays is so nearly unity. This means that any form of reflection polarimeter such as a Brewster plate is very inefficient. For x-rays incident at grazing angles, the difference between the reflection coefficient for the two different polarisations is so very small that at least 25 reflections are required to obtain a significant separation of the two polarisations. Such a multireflection device has not yet been used and would be very difficult to produce.

Three different physical processes have been used in the design of polarimeters:

(a) *Incoherent scattering*. When x-rays are incident on a gas of free electrons, the radiation is scattered incoherently. The probability of scattering at an angle ψ to the electric vector of the incident radiation is proportional to $\sin^2\psi$. Thus, the distribution of the scattered radiation in the equatorial plane, perpendicular to the direction of the incident radiation, depends on the polarisation of the x-radiation (see Fig. 5-1). The use of this process to detect polarisation is complicated by the fact that in any practical detector the electrons will be bound in atoms and the incident x-radiation can eject a bound electron by photoionisation. This photoelectric absorption reduces the scattering efficiency and makes it difficult to design an efficient practical polarimeter utilising Thomson scattering.

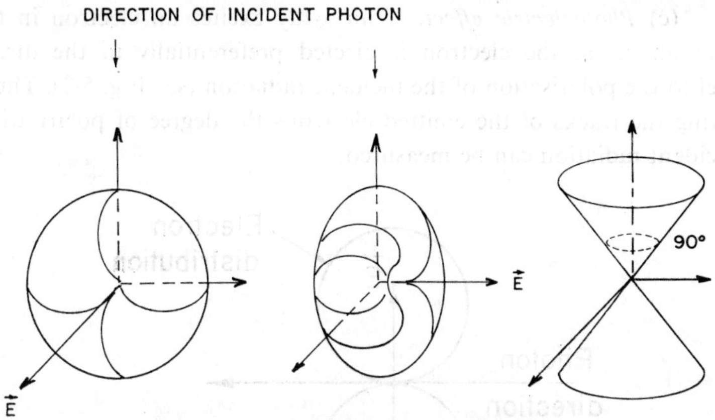

FIGURE 5-1. Sine-squared distribution of scattered photons for two directions of incident polarisation. Those scattered within the 90° cone carry little information about polarisation.

(b) *Coherent scattering*. X-rays are scattered coherently from a crystal lattice. When a parallel beam of x-rays is incident on a plane crystal, radiation with energy E is reflected, where E satisfies the usual Bragg condition

$$E = \frac{hc}{2d \sin\theta}.$$

Here d is the spacing of the crystal planes and θ is the angle of incidence. The reflection coefficient depends upon the polarisation of the incident radiation. If the crystal is rotated about the direction of the incident radiation, any polarisation present results in a modulation of the observed reflected intensity at twice the frequency of rotation.

Borrmann effect: It has been recently observed that certain very perfect crystals show anomalous transmission properties. Such properties are critically dependent on the wavelength and polarisation of the incident radiation, and on the angle of incidence. In principle such a crystal can act as a polaroid sheet, transmitting radiation of a given polarisation and absorbing the oppositely polarised radiation. However, the radiation must be incident very accurately along the Bragg direction. Within a perfect crystal such radiation can be decomposed into two standing wave modes, one of which has a maximum at the position of the atoms, and the other a minimum. If the field at the atoms is small, very little absorption will occur, and thus one of these modes is anomalously transmitted. Although this is a most interesting phenomenon, it probably has very few astrophysical applications, since anomalous transmission will occur only for a very small range of frequencies and angles of incidence.

(c) *Photoelectric effect.* If an x-ray excites an electron in the K shell of an atom, the electron is ejected preferentially in the direction parallel to the polarisation of the incident radiation (see Fig. 5-2). Thus, by observing the tracks of the emitted electrons the degree of polarisation of the incident radiation can be measured.

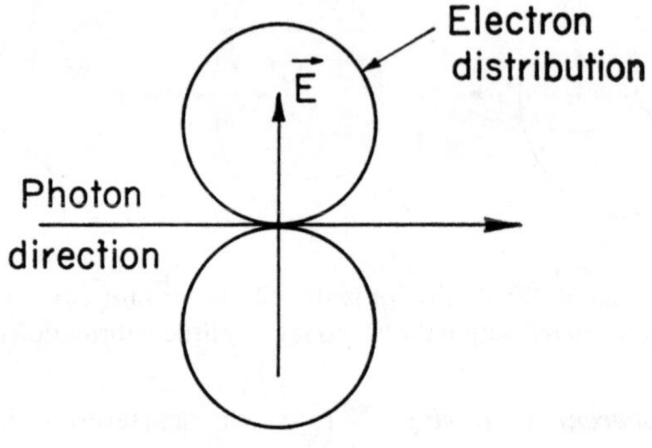

FIGURE 5-2. The angular distribution of the cross section for the photoelectric effect.

5.3. Thomson-Scattering Polarimeters

The first experiment designed to look for polarisation in the x-ray flux from nonsolar sources used a Thomson-scattering polarimeter. An Aerobee-150 sounding rocket was launched on July 27th, 1968, to look for polarisation in the x-ray emission from Sco X-1 (Angel *et al.*, 1969). A second rocket was launched in March 1969 to look specifically at the Crab nebula. In this section we will discuss in more detail the design of the polarimeter used for these observations.

5.3.1. *Detector Material*

If the electrons within the detector material are tightly bound, the incident x-rays can be absorbed as well as scattered, since an x-ray may eject a bound electron. This photoelectric absorption reduces the scattering efficiency of the material. Whether free scattering or photoelectric absorption dominates depends critically on the ratio of the binding energy of the material to the incident photon energy. If the binding energy is much less than the photon energy, essentially free scattering occurs. If, however, the binding energy is much greater than the photon energy, photoelectric absorption dominates. Since the binding energy is proportional to Z^2, the scattering efficiency decreases with increasing nuclear charge Z (see Fig. 5-3). Hydrogen would thus be the most efficient detector material to use. However, a high column density of scattering material is required to ensure that most of the incident photons will be scattered within the detector. For a scattering length of about 10–20 cm, the hydrogen would have to be in the liquid or solid state. Lithium metal is thus most commonly used: this has a scattering length of about 10 cm for 15-keV photons, but the efficiency of this material is low. The efficiency decreases rapidly with increasing wavelength, and a lithium polarimeter can only detect the very high energy tail of the x-ray emission from such sources as Sco X-1 or the Crab nebula (see Fig. 5-4). Lithium hydride can be used to improve the efficiency slightly, but it is a dangerous substance and very hydroscopic.

5.3.2. *Design of a Polarimeter Module*

A lithium block, of length about 10 cm (approximately equal to the scattering length for a 15-keV photon) and width about 5 cm (much less than the scattering length), is surrounded by four proportional counters (see Fig. 5-5). The counters are made as compact as possible; otherwise the effective area of the detector is considerably reduced by the presence of the counters. Thus a high-density counter gas must be used. For a counter thickness of about 2 cm, xenon gas at 3-atm pressure is required to obtain reasonable absorption efficiency. The beryllium windows must be made to withstand this pressure, and thus appreciable absorption will occur within the window. For photon energies of about 5 keV, the window is only about 60 percent efficient.

The difference in the counts recorded by the two X counters and the two Y counters (see Fig. 5-6) gives a measure of the polarisation of the incident radiation. Ideally, if the radiation is 100 percent polarised with the electric vector as shown, the counting rate recorded by the X counters should be zero. However, since the scattering cross section has a typical $\sin^2\psi$ distribution, some photons will be scattered into the X counters. This reduces the sensitivity of the polarimeter. We define the parameter Π for the detector shown in Fig. 5-6 as

$$\Pi = \frac{N_2 - N_1}{N_2 + N_1},$$

where N_1 and N_2 are the counts recorded by the X and Y counters, respectively. Ideally, $\Pi = 1$. For most practical detectors $\Pi \approx 0.3$. Monte Carlo techniques are used to optimise the detector system: the parameters such as the width and length of the lithium block and the size of the proportional counter are varied to obtain the most efficient system. Even so, the efficiency of such a detector is low.

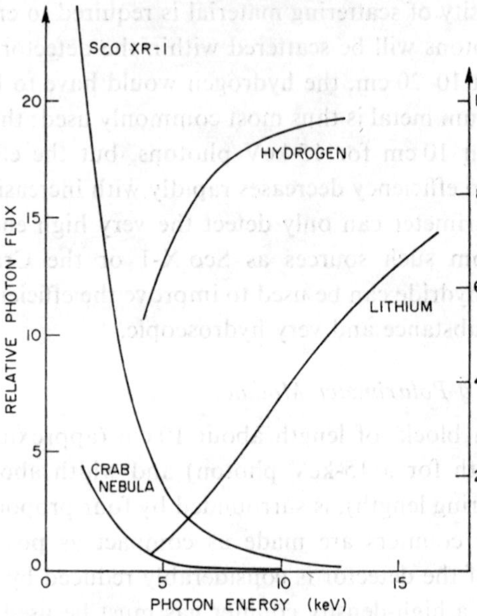

FIGURE 5-3. The scattering efficiency of Li, LiH, and H as a function of wavelength. σ_s is the scattering cross section, and σ_a is the absorption cross section.

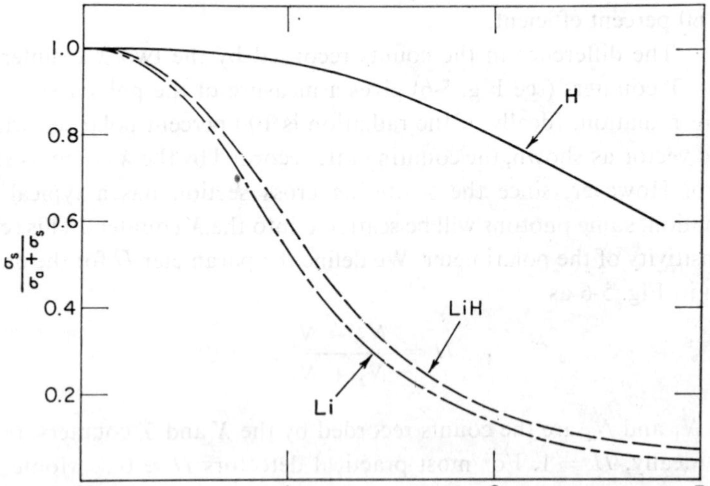

FIGURE 5-4. Li and H efficiency curves compared with the spectra of Sco X-1 and the Crab nebula.

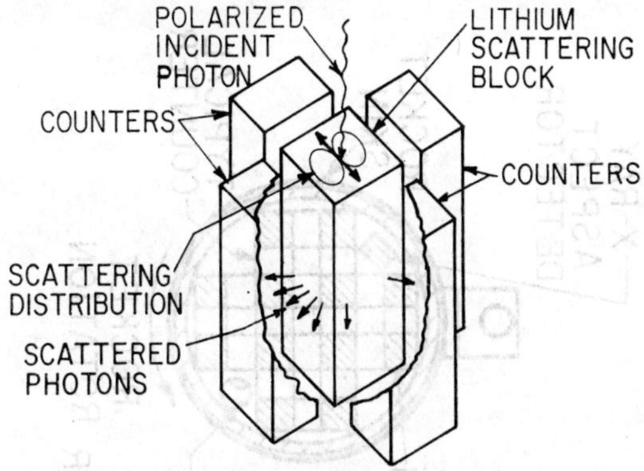

FIGURE 5-5. Schematic representation of the polarimeter (Angel *et al.*, 1969).

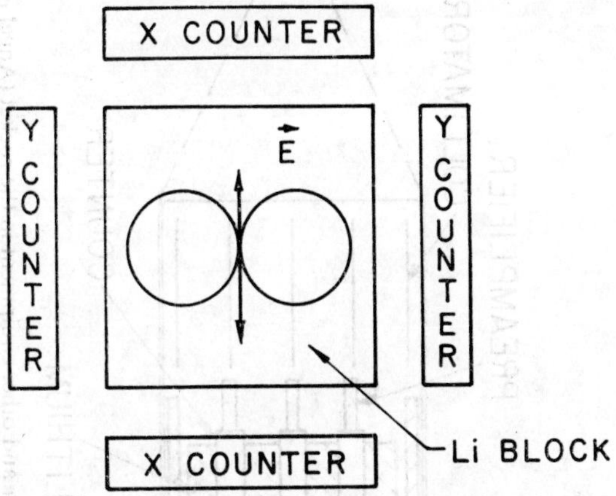

FIGURE 5-6. Diagram to illustrate the definition of the detector efficiency Π. If N_1 is the counting rate in the X counter and N_2 that in the Y counter, then Π is given by $\Pi = (N_2 - N_1)/(N_2 + N_1)$

A series of these lithium-block modules are mounted in a rocket as shown schematically in Fig. 5-7. The entire array is surrounded by a plastic anticoincidence shield to eliminate the counts due to cosmic rays. A pulse-height discriminator is also used. The geometrical area of the total polarimeter is about 900 cm^2. The effective area, however, is considerably less than this and decreases rapidly with decreasing photon energy (see Fig. 5-8).

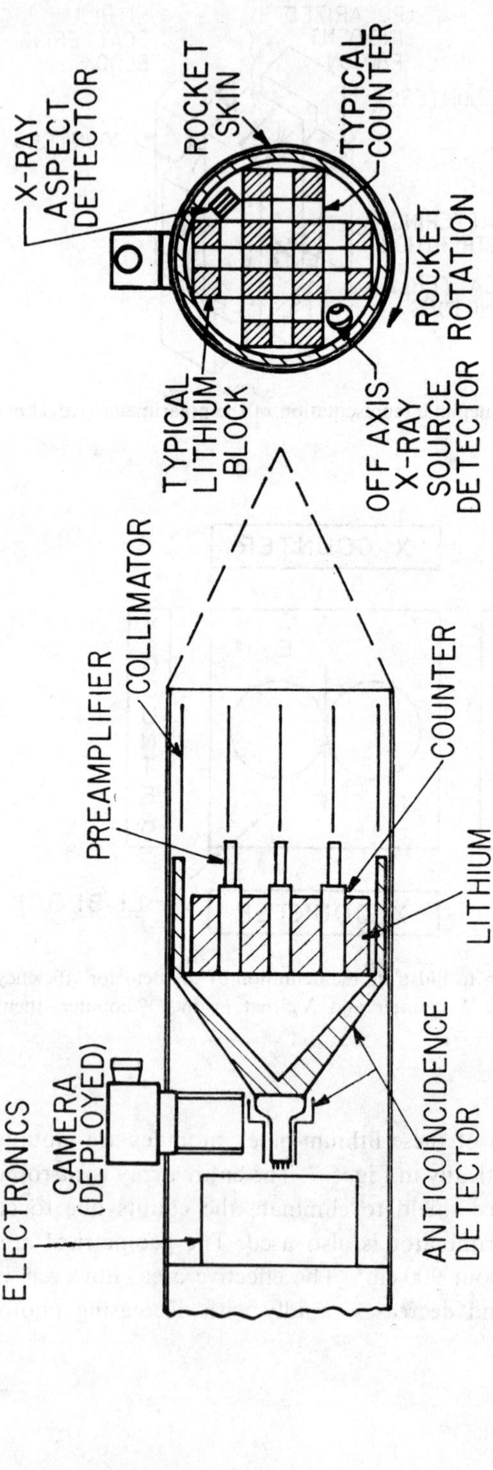

FIGURE 5-7. Mounting of the polarimeter and other equipment in the rocket (Angel *et al.*, 1969).

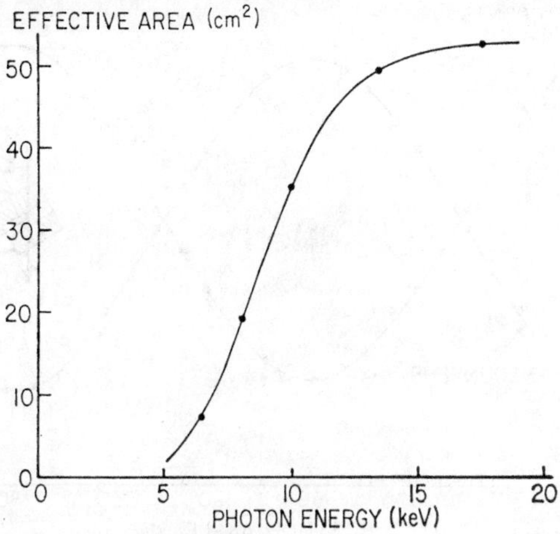

FIGURE 5-8. Variation of the effective area of the polarimeter with photon energy (Angel *et al.*, 1969).

5.3.3. *Detection of Polarisation*

In use, the polarimeter is pointed towards the source and rotated about the line of sight. If the incident radiation is polarised, the counting rate recorded by the counters will be modulated at a frequency equal to twice the rotation frequency of the polarimeter. The modulation observed in the X and Y counters will be in antiphase. This allows the observer to separate true polarisation modulation from time variation of the source. The depth and the phase of the modulation give a measure of the magnitude and position angle of the polarisation vector of the incident radiation. The response of a single polarimeter module to 100 percent linearly polarised bremsstrahlung x-radiation is shown in Fig. 5-9. This was obtained during a laboratory experiment.

Rotating the whole polarimeter about the line of sight removes any false polarisation results due to different sensitivities of the X and Y counters. However, other processes can produce spurious modulation effects. For example, if the x-ray source were slightly off axis, the two counter systems would not be equally illuminated. Laboratory experiments show that the detector must be pointed at the source with an error of not more than 3° to keep these effects small. This accuracy can be achieved with existing rockets. To check the orientation of the rockets, a movie camera is used to take pictures of the star field every 0.25 sec. An x-ray aspect detector with a slat collimator is also used to check the camera results. Spurious results could also be caused by the presence of a nearby weak x-ray source. This might

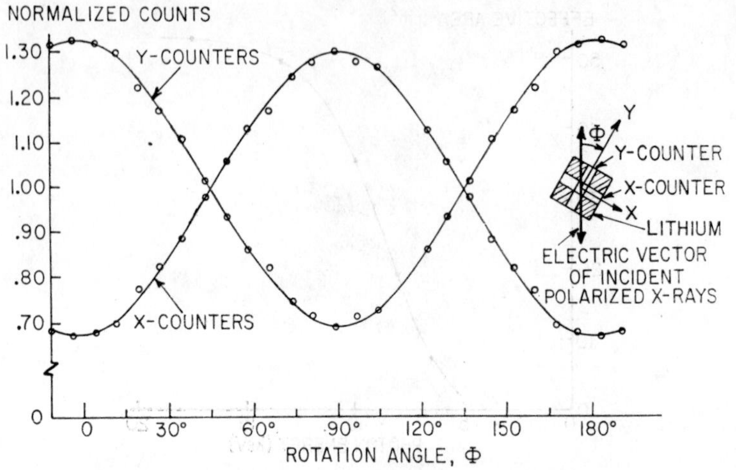

FIGURE 5-9. Variation of the counting rates for the X and Y orthogonal detectors as the polarimeter is rotated with respect to a beam of 100% linearly polarised x-rays (Angel *et al.*, 1969).

fall within the field of view of the polarimeter. Thus an off-axis source detector is also incorporated in the rocket payload. This is a NaI detector, with a field of view extending from 3° to about 12° from the axis of the polarimeter. Another possible source of error could be an anisotropy in the background cosmic-ray counts. This can be checked by balloon measurements.

5.3.4. *Sco X-1 Observations*

Observations of Sco X-1 were made using a lithium-scattering polarimeter in July 1968. A rocket was launched from White Sands Missile Range and was tilted to observe Sco X-1 for about 140 sec. It was then turned to a blank region of the sky, where there were no known x-ray sources, to check for any background modulation. During the observations of ScoX-1 the rocket axis was centered about 2¼° away from the source, but it was still possible to obtain viable results. The optical variations of Sco X-1 were monitored during the flight at both the McDonald and Cerro Tololo Observatories. There was no evidence of any flare activity. The photon counts were analysed into four energy bins, corresponding to the energy ranges of 4.2–8.4 keV, 8.4–12.6 keV, 12.6–16.8 keV, and above 16.8 keV. The crude spectrum obtained by this analysis was consistent with previous measurements. The counts obtained are shown in Fig. 5-10.

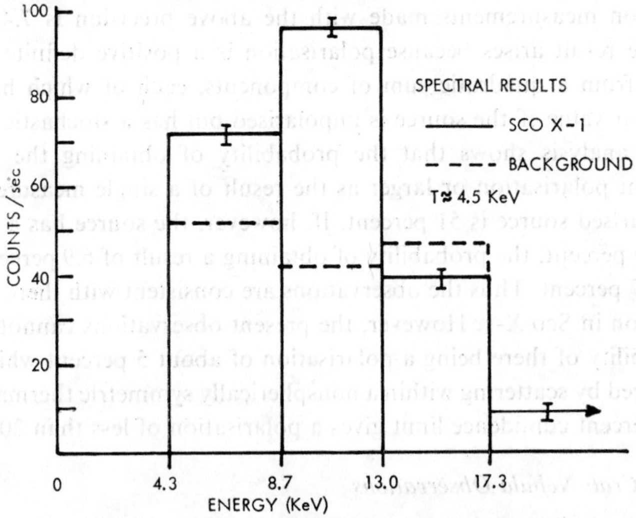

FIGURE 5-10. Spectral results obtained for Sco X-1 and the background count.

The results obtained from observations of Sco X-1 and the background were analyzed for modulation at twice the rotation frequency ω by finding the least-square best fit values of R_0, M_1, and M_2 in the equation

$$R(t) = R_0[1 + M_1 \sin 2\omega(t - t_0) + M_2 \cos 2\omega(t - t_0)].$$

$R(t)$ is the measured counting rate, and t_0 is the time at which the window of the particular counter is parallel to the north-south axis. The observed modulation coefficients M_1 and M_2 are related to the actual modulation by the efficiency parameter Π discussed previously. If N is the total number of counts recorded, the standard deviation δM in the calculated modulation amplitudes is given by

$$\delta M = \left(\frac{2}{N}\right)^{1/2}.$$

In this experiment, $N \approx 10^4$, giving $\delta M \approx 1.4$ percent. Correcting for the factor Π, this gives a statistical limit on the accuracy of each component of the polarisation of about 4 percent. (This is further increased to 6 percent when background corrections are made.)

The results of these observations show that there is certainly no strong evidence for polarisation in the x-ray flux from Sco X-1, and there is no indication of any background polarisation. A straightforward analysis of the data gives a polarisation of magnitude 6.9 percent at a position angle 63° for Sco X-1. However, this result is consistent with an unpolarised source, for which the expected value of the average magnitude of repeated polarisation measurements made with the above precision is 7.4 percent. This finite result arises because polarisation is a positive definite quantity obtained from a quadratic sum of components, each of which has a null expectation value if the source is unpolarised but has a stochastic distribution. An analysis shows that the probability of obtaining the observed 6.9 percent polarisation or larger as the result of a single measurement on an unpolarised source is 51 percent. If, however, the source has a polarisation of 20 percent, the probability of obtaining a result of 6.9 percent or less is only 0.7 percent. Thus the observations are consistent with there being no polarisation in Sco X-1. However, the present observations cannot rule out the possibility of there being a polarisation of about 5 percent which could be produced by scattering within a nonspherically symmetric thermal source. The 99 percent confidence limit gives a polarisation of less than 20 percent.

5.3.5. *Crab Nebula Observations*

A similar rocket was launched in March 1969 to observe the polarisation of the x-ray flux from the Crab nebula. The results obtained are again, inconclusive, since the statistical errors are too large. Applying the analysis discussed above to the combined data from three energy bins, a polarisation of about 10 percent is obtained, which is consistent with the observed optical and radio polarisation measurements (see Fig. 5-11). However, as can be seen from Fig. 5-11 the results obtained from the data from individual energy bins are widely scattered, showing the large errors inherent in such observations. There is certainly no convincing evidence for polarisation in the x-ray flux from the Crab, but the results do not contradict the hypothesis that the x-rays have a similar polarisation to that of the optical and radio emission. The 99 percent confidence limit shows that the x-ray polarisation is less than 27 percent. This is consistent with a synchrotron emission mechanism.

However, until a more accurate measurement is made of the polarisation of the x-radiation from the Crab, it is not possible to prove convincingly that the x-ray flux is a simple extension of the synchrotron emission spectrum observed at optical and radio frequencies. There is a hint, from recent optical data, that the spectrum is decreasing too rapidly with frequency at optical wavelengths to be consistent with the observed x-ray flux. There are also indications that the spectrum cuts off at high γ-ray energies, as would be expected if the x-rays were not produced by synchrotron emission. Thus it is vital to improve existing polarimeters and to develop new techniques, in an attempt to obtain a more accurate measurement of the polarisation of the x-ray flux from the Crab.

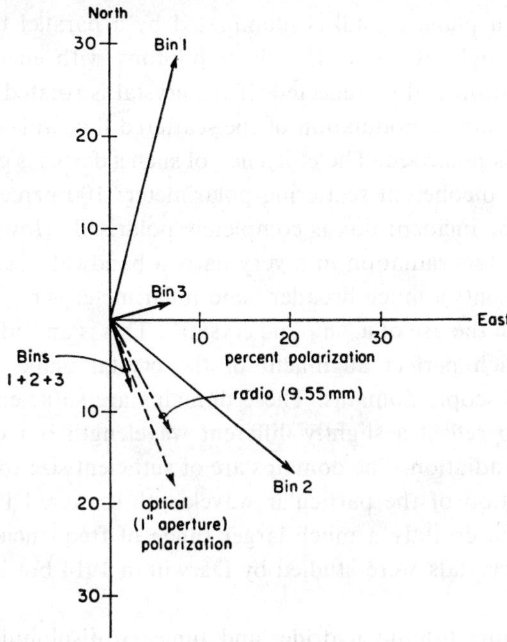

FIGURE 5-11. X-ray polarisation results for the Crab nebula compared with the mean optical and radio polarisation. The length of the vector represents the degree of polarisation, and the angle between N and the direction of the vector is the position angle.

5.4. Recent Polarimetry Developments

Thomson scattering devices are the simplest form of polarimeter but have severe limitations. The accuracy and sensitivity of such devices cannot be greatly improved except by increasing the effective area and the observing time. (The gain increases as the square root of the area and the square root of the time.) The existing Aerobee-150 rockets give an observing time of about 4 min and a total area of 900 cm^2. It is planned to use Aerobee-350 rockets which will give an observing time of about 8 min and an area of 1800 cm^2. This may improve the accuracy sufficiently to obtain a conclusive polarisation measurement for the Crab nebula, but the sensitivity will still not be sufficient to enable polarisation measurements to be made on the weaker x-ray sources. Thus it is necessary to develop new types of polarimeters.

5.4.1. *Bragg Polarimeters*

Polarimeters using coherent scattering from a highly reflecting crystal lattice are being developed for astronomical work (Angel and Weisskopf, 1970). If a plane crystal is illuminated by a parallel beam of x-rays incident at an angle of about 45°, those photons with energies satisfying the Bragg condition will be reflected. If the crystal is rotated about the line of sight to the source, modulation of the scattered flux will occur if the incident radiation is polarised. The efficiency of such a device is greater than the efficiency of an incoherent scattering polarimeter: 100 percent modulation is obtained if the incident flux is completely polarised. However, a perfect crystal only scatters radiation in a very narrow bandwidth, and for astronomical measurements a much broader band polarimeter is required. This can be achieved with the use of a "mosaic crystal". This is an "ideally imperfect crystal" in which perfect alignment of the crystal planes is maintained only over microscopic domains. These domains are sufficiently misaligned for each one to reflect a slightly different wavelength out of the incident broad band of radiation. The domains are of sufficient size to ensure almost complete reflection of the particular wavelength scattered by the domain. Thus, using such crystals, a much larger range of frequencies can be scattered. Mosaic crystals were studied by Darwin in 1914 but have since been neglected.

Graphite, lithium hydride, and tungsten disulphide are suitable, highly reflecting materials which can be made in mosaic form. The photoelectric absorption cross section for such substances is small compared with the coherent scattering cross section. The simplest way to measure the polarisation of an x-ray source would be to place one of these crystals directly behind an aperture stop in the focal plane of an x-ray lens and to reflect the radiation into a small detector. As the whole instrument is rotated, modulation due to polarisation will occur.

Alternatively, a polarisation map of an extended x-ray source such as the Crab nebula could be obtained by placing a mosaic crystal in front of the focus of an x-ray telescope. An image of the source will be formed, and the polarisation of different regions can be measured simultaneously. Such techniques are being actively developed for use in a large satellite. They are probably sufficiently sensitive to measure the polarisation of about fifteen of the brightest x-ray sources, including Sco X-1, the Crab nebula, and M87.

5.4.2. *Photoelectric Polarimeters*

We have seen that photoelectric absorption decreases the efficiency of Thomson-scattering polarimeters. However, the anisotropy of the photoelectric ejection process can itself be used to measure the polarisation of the incident x-radiation (see Fig. 5-2). The electron is ejected preferentially in a direction parallel to the electric vector of the incident radiation. The elastic scattering of the ejected electrons must be reduced to a minimum, otherwise no anisotropic tracks will be observed. Thus a low-Z gas such as hydrogen must be used as the absorbing material. In hydrogen, electrons with energies as low as 2 keV will not be appreciably scattered.

How can these electron tracks be observed? It is not very convenient to use a cloud chamber or bubble chamber in a rocket experiment. One method which has been suggested uses a proportional counter with an anode consisting of a grid of parallel wires (see Fig. 5-12). An electron moving in the direction parallel to the wires will give a single pulse only, whereas one moving in the direction perpendicular to the wires will give multiple pulses. By selecting only single-pulse events, the direction of the electron track can be determined.

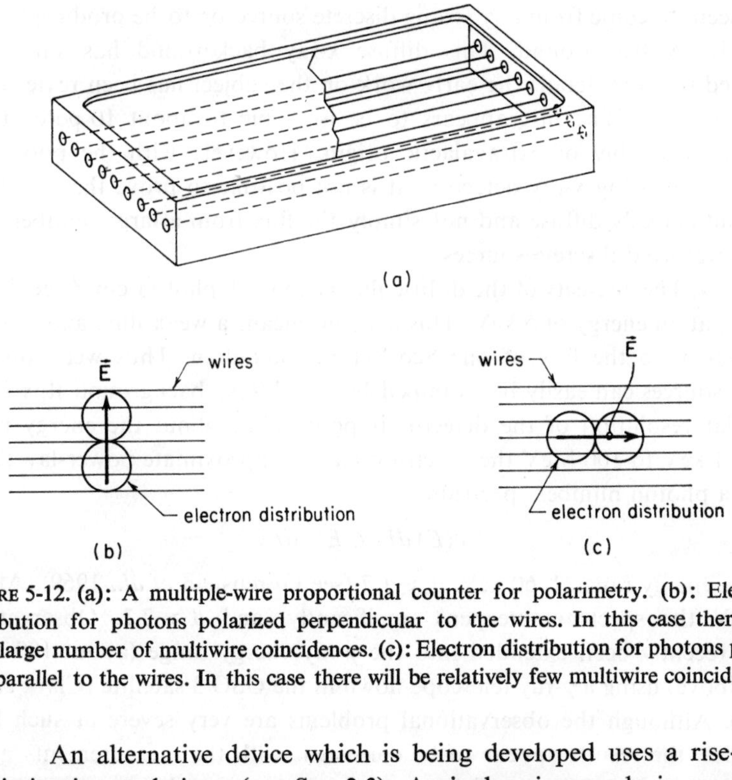

FIGURE 5-12. (a): A multiple-wire proportional counter for polarimetry. (b): Electron distribution for photons polarized perpendicular to the wires. In this case there will be a large number of multiwire coincidences. (c): Electron distribution for photons polarized parallel to the wires. In this case there will be relatively few multiwire coincidences.

An alternative device which is being developed uses a rise-time discrimination technique (see Sec. 2.3). A single wire anode is required. Electrons moving in a direction parallel to the anode will produce a pulse with a very short rise time, whereas electrons moving in a perpendicular direction will give a pulse with a long rise time. By observing the shape of the pulses, the polarisation of the original x-rays can be measured.

These two types of polarimeters which are being developed will, hopefully, give us much more detailed information about the polarisation of some of the brighter x-ray sources. However, there is clearly scope for the application of many other physical ideas in the design of x-ray polarimeters suitable for astronomical work.

6. THE DIFFUSE X-RAY BACKGROUND

6.1. Observations

The very first x-ray rocket, launched ostensibly to observe the Moon, gave indications of the existence of a diffuse x-ray flux which did not seem to come from any strong discrete source or to be produced by the cosmic-ray background. This diffuse x-ray background has since been studied in more detail: the early work on the subject has been reviewed by Gould (1967). The flux appears to be isotropic to about 10 percent and is thus probably of extragalactic origin. However, with the poor resolution of existing x-ray detectors, it is not possible to prove that the background is really diffuse and not simply the flux from a large number of as yet unresolved discrete sources.

The intensity of the diffuse flux is about 1 photon cm^{-2} sec^{-1} sr^{-1} keV^{-1} at an energy of 5 keV. This is by no means a weak flux, and is about equivalent to the flux of one Sco X-1 per steradian. Thus weak discrete x-ray sources can easily be swamped by the diffuse background flux if the angular resolution of the detector is poor. Throughout the energy range from 1 keV to 100 MeV the spectrum has an approximate power-law form, with a photon number spectrum

$$n(E)\,dE \propto E^{-\beta}\,dE.$$

In the energy range 1–60 keV, $\beta \approx 1.7$ (see Gorenstein *et al.*, 1969). Above 40 keV the spectrum steepens significantly, and $\beta \approx 2.2$. Observations have recently been extended into the γ-ray energy range (about 100 MeV and above) using a γ-ray telescope flown in the OSO 3 satellite (Clark *et al.*, 1968). Although the observational problems are very severe at such high energies, the results appear to be consistent with the measurements made at lower energies. Because of the steepness of the spectral curve, the intensity at such high energies is very small (in fact, the cosmic-ray proton intensity is 10^4 times larger than the γ-ray intensity). The measurements placed an upper limit of 1.1×10^{-4} photon cm^{-2} sec^{-1} sr^{-1} on the flux for $E > 100$ MeV. These satellite observations also found evidence for an enhancement of the γ-ray background in the direction of the galactic center.

6.1.1. *Soft X-Ray Flux*

Measurements of the background flux have also been made in the soft x-ray energy range, 1/4 keV to 1 keV (Bowyer *et al.*, 1968; Henry *et al.*, 1968). At the present time, the results for these low energies are very confused and contradictory. There seems to be some evidence that the soft x-ray flux is greater than would be expected from a simple extrapolation of the energy spectrum above 1 keV. If this low-energy excess does exist, it is of considerable astrophysical and cosmological significance. However, accurate measurements at these low energies are hampered by the low efficiency and poor energy resolution of existing x-ray detectors. At energies of less than 1 keV, a thin-window proportional counter acts essentially as a Geiger counter—it has practically no energy resolution. The efficiency of a soft x-ray detector is shown in Fig. 6-1. The detector has a maximum efficiency for energies greater than 1 keV. There is a small window in the low-energy range, below the carbon K edge at 44 Å. The counts recorded in the low-energy region are distinguished from the higher energy counts by a pulse-height discriminator, but this is a difficult technique and can lead to incorrect results.

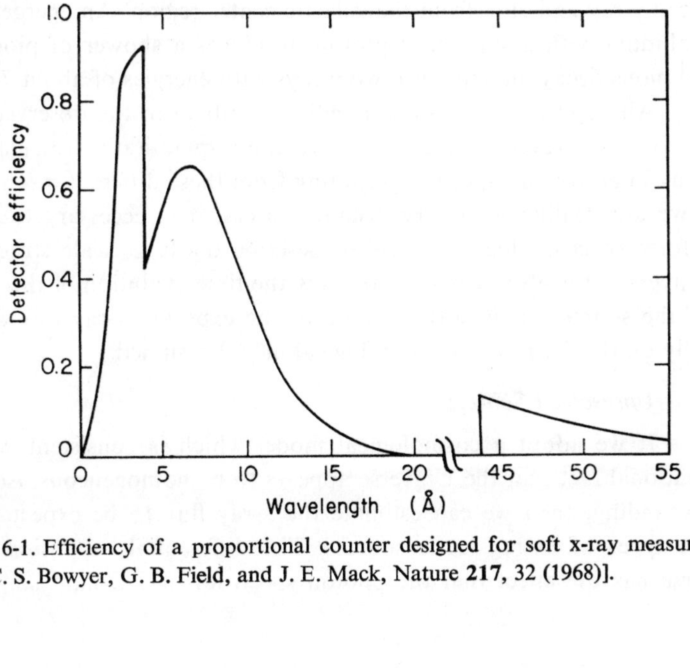

FIGURE 6-1. Efficiency of a proportional counter designed for soft x-ray measurements [from C. S. Bowyer, G. B. Field, and J. E. Mack, Nature **217**, 32 (1968)].

There are other problems: in order to reduce the absorption of the low-energy photons in the counter window, a very thin Mylar window is used. This may allow charged particles from the inner fringes of the radiation belts to enter the detector and give spurious counts. There may also be a background count due to ultraviolet photons, especially L α photons from the Sun. Thus the low-energy excess may be an instrumental effect: the situation is unclear.

6.2. Interpretation of the X-Ray Background above 1 keV

Any theory of the origin of the diffuse x-ray flux must attempt to explain both the intensity and the observed power-law spectrum. There have been a number of proposed interpretations:

(a) *Discrete sources.* The background may be the integrated flux from a large number of unresolved, discrete, extragalactic x-ray sources (Gould and Burbidge, 1963).

(b) *Inverse Compton scattering.* Compton scattering of starlight or the cosmic blackbody radiation off intergalactic relativistic electrons produces photons with x-ray energies (Felten and Morrison, 1966).

(c) *High-energy processes.* The interaction of high-energy cosmic-ray protons with the interstellar or intergalactic medium will result in a flux of high-energy electrons and photons. Such processes are especially important for the production of γ-rays and may well account for the enhanced γ-ray emission from the galactic center region. An energetic proton, colliding with a stationary proton, produces a shower of pions. The neutral pions decay directly into two γ-rays with energies of about 70 MeV.

All these processes undoubtedly contribute to the observed background flux. However, it is difficult to make a quantitative calculation of the x-ray intensity and spectrum resulting from these different mechanisms. Since we are dealing with very distant sources, it is necessary to assume some form of cosmological model to describe the large-scale structure of the Universe. It is also necessary to know the time evolution of the properties of the sources. Thus any estimate of the expected x-ray flux depends critically on the form of the cosmological model assumed.

6.2.1. *Unresolved Sources*

If we adopt a cosmological model which is consistent with the observational fact that the Universe appears to be homogeneous, isotropic, and expanding, then we can estimate the x-ray flux to be expected from various types of extragalactic sources. The differential expansion of the Universe has the effect that the photon spectrum of a distant source observed at the Earth is different from the spectrum emitted at the source. However, in any conventional cosmological model, if a distant source has a power-law emission spectrum, the observed photon spectrum will also be of power-law form, with the same exponent. The observed intensity will depend on the number density of the sources and the photon emission spectrum of each source.

We can first calculate the contribution to the x-ray flux from all normal galaxies like our own. The number density of galaxies is 1×10^{-75} cm^{-3}. If we assume that the emission spectrum is not time dependent, the estimated flux is too small by a factor of 100. This estimate could be increased by assuming that the average x-ray power of a normal galaxy was much greater in the past than it is now (Silk, 1968). This hypothesis is not as arbitrary as it sounds: there is some evidence from the radio source counts that strong radio sources were more powerful emitters at an earlier epoch.

However, there may be a large number of galaxies whose present x-ray emission is much greater than that of our own galaxy. What is needed is a determination of the average x-ray power of a large number of galaxies. Observations of the x-ray flux from the Coma cluster (which contains about 10^3 galaxies) give an average flux much greater than that of our own galaxy (Felten *et al.*, 1966). Indeed, using this value, the estimated background x-ray flux is much greater (by a factor of about 10) than the observed flux. It is thus clear that normal galaxies do contribute to the x-ray background flux, but more information about the intensity of these sources and the cosmological model to be adopted must be available before an accurate estimate of the importance of this contribution can be made.

More exotic x-ray sources may also be responsible for the x-ray background. We might assume that objects such as M87 produce both the x-ray and radio backgrounds. There is some evidence to support this idea: the spectrum of the universal background is similar to the observed spectrum of M87. If $N_r(E)$ is the photon spectrum at radio energies, and $N_x(E)$ the specrum at x-ray energies, the ratio $N_r(E)/N_x(E)_{\text{background}}$ equals the ratio $N_r(E)/N_x(E)_{M87}$ to within a factor of 3. Thus there may be a special class of galaxies, of which M87 is a typical member, which produce most of the observed background.

6.2.2. *Inverse Compton Scattering*

If a low-energy photon is in collision with a high-energy electron, the photon energy is increased. It is attractive to attempt to use this process to account for the high-energy background flux because of the high density of low-energy photons present in the 3 °K blackbody radiation. If this radiation is universal, the number density of low-energy photons throughout the Universe is very large (about 10^3 photons/cm^3). Thus, wherever there is a significant flux of relativistic electrons, the rate of production of x-ray photons by Compton scattering will be large. Radio galaxies are a good source of high-energy electrons. However, the conversion of 3 °K blackbody photons to x-ray photons by Compton scattering is a slow process and, within the high magnetic fields of radio galaxies, most of the relativistic electron energy will be converted into synchrotron emission. For appreciable Compton scattering to occur the high-energy electrons must escape into the intergalactic medium, where the magnetic field is less than 10^{-7} gauss and the synchrotron loss is small.

Thus we are led to consider the Compton scattering of blackbody photons within the intergalactic medium. The form of the x-ray spectrum produced by the process depends critically on the injected electron spectrum. If the electron energy spectrum is of power-law form

$$N_e(E)\, de \propto E^{-m}\, dE,$$

then Compton scattering with blackbody photons will produce a power-law x-ray photon number spectrum

$$n(E)\, dE \propto E^{-\beta}\, dE$$

with $m = 2\beta - 1$. From the high-energy x-ray data $\beta \approx 2.2$, implying $m \approx 3.4$ in the intergalactic medium. This is a much steeper electron spectrum than is observed in most sources where, typically, $m \approx 2.4$. However, there is no independent evidence as to the form of the intergalactic electron spectrum. Electrons escaping from radio sources and entering the intergalactic regions probably suffer considerable energy losses, and these losses will steepen the spectrum. For steady-state conditions, Compton scattering and synchrotron losses lead to an increase in the power-law exponent of unity. Thus if $n \approx 2.4$ within a radio source, we expect $n \approx 3.4$ in the intergalactic medium. This is exactly the exponent required to produce the observed x-ray spectrum.

Can this mechanism produce the observed intensity of x-radiation? If we assume that the number density and intensity of radio galaxies is the same throughout all space, then the estimated flux is about 100 times too small to account for the observed background. We thus need to postulate that the radio galaxies in remote space are about 100 times more powerful or more numerous than in nearby regions of the Universe. One advantage of this assumption is that now most of the x-ray flux originates in very distant sources where the blackbody radiation energy density is much higher. (This is simply a result of the cosmological expansion of the Universe.) Thus the rate of conversion of energy by Compton scattering is increased, and the condition on the allowed magnetic field can be relaxed. Thus it may be that the x-ray flux is produced within the very distant radio galaxies themselves (see Felten and Rees, 1969). The observed break in the x-ray spectrum at about 60 keV, where the slope of the spectrum changes, may then be due to some other energy loss mechanism which competes with the Compton scattering process within the source.

Since there are some difficulties inherent in this electron injection mechanism, are there any other sources for high-energy electrons in the intergalactic medium? If the observed cosmic-ray flux is universal, then a collision of a high-energy cosmic-ray proton with a stationary proton of the intergalactic gas (if it exists) will produce a shower of pions. The charged pions ultimately decay into electrons. The spectrum of these secondary electrons will reflect the primary cosmic-ray spectrum. Since the form of the primary cosmic-ray spectrum in the intergalactic medium is unknown, we can only assume that it is the same as the observed galactic spectrum. If we make this assumption and take an intergalactic gas density of 10^{-5} cm^{-3}, the estimated x-ray flux arising from the secondary electrons agrees well with the observed flux. However, this good agreement is probably somewhat fortnitous.

We see that there are several plausible explanations for the origin of the diffuse background. Each depends critically on somewhat arbitrary assumptions concerning many important questions: (a) the large-scale structure of the Universe; (b) the existence of the cosmic 3 °K blackbody background; (c) the existence of intergalactic matter; and (d) the universality of the cosmic rays. Thus a detailed study of the diffuse background may prove to be most rewarding. It may help to solve some of these fundamental problems.

6.3. The Soft X-Ray Background

We have already discussed the problems inherent in making observations in the low-energy x-ray region ($E < 1$ keV). However, it is interesting to consider some of the implications of such observations, while bearing in mind the uncertainties in the existing measurements.

The high-energy x-ray flux is highly isotropic. However, the photoelectric cross section increases rapidly for low energies, and extragalactic soft x-ray photons are appreciably absorbed by the interstellar gas. Thus the soft x-ray flux appears to be anisotropic, with a maximum intensity in the direction of the galactic pole and a minimum intensity in the plane of the Galaxy. This dependence on galactic latitude can be used to determine the density of the interstellar gas (see Bowyer *et al.*, 1968). The photoelectric cross section σ in the 0.25 to 1 keV energy range is given by

$$\sigma = \sigma_0 \left(\frac{E}{E_0} \right)^{-3.2},$$

where $E_0 = 0.28$ keV and $\sigma_0 = (0.84 + 21.4\zeta) \times 10^{-21}$ cm^2 (Bell and Kingston, 1967). Here ζ is the helium abundance in the interstellar gas, $\zeta = n(\mathrm{He})/n(\mathrm{H})$. Assuming a normal cosmical helium abundance, $\zeta \approx 0.1$, the observed absorption can be used to calculate the column density of interstellar neutral hydrogen. Alternatively, assuming a hydrogen column density determined from 21-cm measurements, the helium abundance can be obtained. The first soft x-ray measurements indicated that the galactic x-ray absorption was considerably less (by a factor of about 3) than would be expected for 21-cm measurements and a normal helium abundance. Either the helium abundance was abnormally low, or the hydrogen column density was less than had previously been assumed. If this discrepancy is real, it may indicate that the interstellar medium is very irregular and clumpy. There are other indications that the 21-cm column density measurements may be in error: rocket observations show that the interstellar L α absorption lines in the spectra of many stars are much weaker than would be expected. The situation is confused, and more soft x-ray measurements are needed.

Since the low-energy flux has a maximum intensity towards the galactic pole, most of the flux must be of extragalactic origin. A galactic soft x-ray source has been discovered in the Cygnus–Vulpecula region (Henry *et al.*, 1968), but the galactic contribution to the observed background is probably small. If there is a significant excess in the background flux at low energies, the most likely origin of this component is the free-free emission from a hot, dense intergalactic gas.

Many people have speculated about the existence of an intergalactic gas of density $\approx 10^{-5}$ cm^{-3} (see Sec. 1.4). If it exists, the lack of intergalactic 21-cm emission or absorption and the lack of intergalactic Lyman-α absorption in the spectra of quasars imply that the gas must be fully ionised. However, if the temperature is much greater than 10^6 °K, the high-energy x-ray emission would be incompatible with the observed flux measurements above 1 keV. Thus, if the gas exists, it must have a temperature of between 10^5 and 10^6 °K. Such a gas would be a strong emitter of soft x-rays (see Field and Henry, 1964; Weymann, 1966, 1967). Tentative calculations on the observed soft x-ray flux show that it is consistent with the free-free emission from a gas of density $\sim 10^{-5}$ cm^{-3} and temperature between 3×10^5 °K and 8×10^5 °K. However, future measurements may give a very different result. Certainly observations of the soft x-ray background will probably settle the question of the existence of a dense intergalactic gas.

REFERENCES

Adams, D. J., B. A. Cooke, K. Evans, and K. A. Pounds (1969), Nature **222**, 757.

Anathakrishnan, S., and K. R. Ramanathan (1969), Nature **223**, 488.

Angel, J. R. P. (1969), Astrophys. J. **158**, 219 (1969).

Angel, J. R. P., R. Novick, P. Vanden Bout, and R. Wolff (1969), Phys. Rev. Letters **22**, 861.

Angel, J. R. P., and M. Weisskopf (1970), Astron. J. **75**, 231.

Bell, K. L., and A. E. Kingston (1967), Monthly Notices Roy. Astron. Soc. **136**, 241.

Bowyer, C. S., E. T. Byram, T. A. Chubb, and H. Friedman (1964), Science **146**, 912.

Bowyer, C. S., G. B. Field, and J. E. Mack (1968), Nature **217**, 32.

Bradt, H., S. Rappaport, W. Mayer, R. E. Nather, B. Warner, M. Macfarlane, and J. Kristian (1969), Nature **222**, 728.

Burnight, T. R. (1949), Phys. Rev. **76**, 165.

Byram, E. T., T. A. Chubb, and H. Friedman (1966), Science **152**, 66.

Chandrasehkar, S. (1960), *Radiative Transfer* (Dover Publications, New York).

Chodil, G., H. Mark, R. Rodrigues, F. D. Seward, C. D. Swift, I. Turiel, W. A. Hiltner, G. Wallenstein, and E. J. Mannery (1968), Astrophys. J. **154**, 645.

Clark, G. (1965), Phys. Rev. Letters **14**, 91.

Clark, G. W., G. P. Garmire, and W. L. Kraushaar (1968), Astrophys. J. Letters **153**, L203.

Cocke, W. J., M. J. Disney, and D. J. Taylor (1969), Nature **221**, 525.

Darwin, C. G. (1914), Phil. Mag. **27**, 315 and 675.

Felten, J. E., R. J. Gould, W. A. Stein, and N. J. Woolf (1966), Astrophys. J. **146**, 955.

Felten, J. E., and P. Morrison (1966), Astrophys. J. **146**, 686.

Felten, J. E., and M. J. Rees (1969), Nature **221**, 924.

Field, G. B., and R. C. Henry (1964), Astrophys. J. **140**, 1002.

Fishman, G. J., F. R. Harnden, and R. C. Haymes (1969), Astrophys. J. Letters **156**, L107.

Fritz, G., R. C. Henry, J. F. Meekins, T. A. Chubb, and H. Friedman (1969a), Science **164**, 709.

Fritz, G., J. F. Meekins, R. C. Henry, and H. Friedman (1969b), Astrophys. J. Letters **156**, L33.

Giacconi, R., H. Gursky, F. Paolini, and B. Rossi (1962), Phys. Rev. Letters **9**, 439.

GIACCONI, R., H. GURSKY, and L. P. VAN SPEYBROECK (1968), Ann. Rev. Astron. Astrophys. **6**, 373.

GIACCONI, R., W. P. REIDY, G. S. VAIANA, L. P. VAN SPEYBROECK, and T. F. ZEHNPFENNIG (1969), Space Sci. Rev. **9**, 3.

GINZBURG, V. L., and S. I. SYROVATSKII (1965), Ann. Rev. Astron. Astrophys. **3**, 297.

GORENSTEIN, P., E. M. KELLOGG, and H. GURSKY (1969), Astrophys. J. **156**, 315.

GOULD, R. J. (1967), Am. J. Phys. **35**, 375.

GOULD, R. J., and G. R. BURBIDGE (1963), Astrophys. J. **138**, 969.

GOULD, R. J., and W. RAMSEY (1966), Astrophys. J. **144**, 587.

GUNN, J. E., and J. P. OSTRIKER (1969), Nature **221**, 454.

GURSKY, H., R. GIACCONI, P. GORENSTEIN, J. R. WATERS, M. ODA, H. BRADT, G. GARMIRE, and B. V. SREEKANTAN (1966), Astrophys. J. **144**, 1249.

HAYMES, R. C., D. V. ELLIS, G. J. FISHMAN, J. D. KURFESS, and W. H. TUCKER (1968), Astrophys. J. Letters **151**, L9.

HEGYI, D., R. NOVICK, and P. THADDEUS (1969), Astrophys. J. Letters **158**, L77.

HEITLER, W. (1954), *The Quantum Theory of Radiation* (Clarendon Press, Oxford), 3rd edition.

HENDRICK, R. W. (1957), J. Opt. Soc. Am. **47**, 165.

HENRY, R. C., G. FRITZ, J. F. MEEKINS, H. FRIEDMAN, and E. T. BYRAM (1968), Astrophys. J. Letters **153**, L11.

HILTNER, W. A., D. E. MOOK, D. J. LUDDEN, and D. GRAHAM (1967), Astrophys. J. Letters **148**, L47.

LINDSAY, J. C. (1965), Ann. Astrophys. **28**, 586.

LEWIN, W. H. G., G. W. CLARK, and W. B. SMITH (1968), Can. J. Phys. **46**, S409.

MANLEY, O. P. (1966), Astrophys. J. **144**, 1253.

MARK, H., R. E. PRICE, R. RODRIGUES, F. D. SEWARD, C. D. SWIFT, and W. A. HILTNER (1969), Astrophys. J. Letters **156**, L67.

MOROZ, V. I. (1964), Astron. J. **70**, 755.

NEUGEBAUER, G., J. B. OKE, E. BECKLIN, and G. GARMIRE (1969), Astrophys. J. **155**, 1.

ODA, M. (1965), Appl. Opt. **4**, 143.

O'MONGAIN, E. P., N. A. PORTER, J. WHITE, D. J. FEGAN, D. M. JENNINGS, and B. G. LAWLESS (1968), Nature **219**, 1348.

OORT, J. H., and TH. WALRAVEN (1956), Bull. Astron. Inst. Neth. **12**, 285.

PACINI, F. (1968), Nature **219**, 145.

PAOLINI, F. R., R. GIACCONI, O. MANLEY, W. P. REIDY, G. S. VAIANA, and T. ZEHNPFENNIG (1968), Astron. J. **73**, S73.

PETERSON, L. E., A. S. JACOBSON, and R. M. PELLINS (1966), Phys. Rev. Letters **16**, 142.

POUNDS, K. A., and P. C. RUSSELL (1966), Space Res. **6**, 34.

PRENDERGAST, K. H., and G. R. BURBIDGE (1968), Astrophys. J. Letters **151**, L83.

SANDAGE, A., P. OSMER, R. GIACCONI, P. GORENSTEIN, H. GURSKY, J. WATERS, H. BRADT, G. GARMIRE, B. V. SREEKANTAN, M. ODA, K. OSAWA, and J. JUGAKU (1966), Astrophys. J. **146**, 316.

SCHWINGER, J. (1949), Phys. Rev. **75**, 1912.

SHKLOVSKY, I. S. (1967), Astrophys. J. Letters **148**, L1.

SHKLOVSKY, I. S. (1968), Sov. Astron.—AJ **11**, 749.

SILK, J. (1968), Astrophys. J. Letters **151**, L19.

STAELIN, D. H., and E. C. REIFENSTEIN (1968), Science **162**, 1481.

TUCKER, W. (1967), Astrophys. J. **148**, 745.

Vaiana, G. S., W. P. Reidy, T. Zehnpfennig, L. van Speybroeck, and R. Giacconi (1968), Science **161**, 564.
Westphal, J. A., A. Sandage, and J. Kristian (1968), Astrophys. J. **154**, 139.
Weymann, R. (1966), Astrophys. J. **145**, 560.
Weymann, R. (1967), Astrophys. J. **147**, 887.
Wolter, H. (1952a), Ann. Physik **10**, 94.
Wolter, H. (1952b), Ann. Physik **10**, 286.
Woltjer, L. (1957), Bull. Astron. Inst. Neth. **478**, 301.

Atomic Processes in Astrophysics

A. DALGARNO
Harvard College Observatory and
Smithsonian Astrophysical Observatory, Cambridge

Contents

In this brief course, I had the choice of presenting a broad but superficial description of the wide variety of atomic processes that must be considered in the interpretation of astrophysical phenomena or of making a small selection of such atomic processes and studying them in greater depth. I decided to follow the latter prescription and to discuss atomic processes involving one- and two-electron atomic or molecular systems. The first of my two main topics is radiative transitions and the second is collision processes.

1. RADIATIVE TRANSITIONS

I shall begin with a brief summary of the quantum theory of radiation. The interaction of photons with an atom is described in standard texts on quantum electrodynamics (cf. Akhiezer and Berestetskii, 1965). The absorption and emission of a single photon in a radiative transition between two stationary states of an atomic system may be derived from the first-order scattering matrix. The absorption and emission of two photons involve the second-order scattering matrix.

Let $\mathbf{r}$ represent the position vectors of the N electrons of an atomic system referred to the nucleus as origin and let $\psi_i(\mathbf{r})$ and $\psi_f(\mathbf{r})$ be the wave functions of the initial and final states with associated eigenvalues E_i and E_f. The first-order scattering-matrix element appropriate to the emission of a photon of energy ω, momentum $\mathbf{k}$, and polarization $\mathbf{e}$, is

$$S_{if}^{(1)} = -2\pi i U_{if}^{(1)}\, \delta(E_f - E_i - \omega), \tag{1.1}$$

where

$$U_{if}^{(1)} = \frac{-e}{(2\omega)^{1/2}} \int \psi_f^*(r)\, \boldsymbol{\alpha}\mathbf{e}\, e^{-i\mathbf{k}\cdot\mathbf{r}} \psi_i(r)\, d\mathbf{r} \tag{1.2}$$

in which $\boldsymbol{\alpha}$ is the Dirac matrix. The transition probability per unit time is given by

$$w_{if} = 2\pi\,|U_{if}|^2\,\delta(E_f - E_i - \omega)\,\varrho_f, \tag{1.3}$$

where ϱ_f is the density of final states.

Expand the plane wave in (2),

$$e^{i\mathbf{k}\cdot\mathbf{r}} = \sum_{LM\lambda} \mathbf{e}\cdot Y_{LM}^{(\lambda)}(\mathbf{k}/k)\,g_L(kr)\,Y_{LM}^{(\lambda)*}(\mathbf{r}/r), \tag{1.4}$$

where $Y_{LM}^{(\lambda)}$ is a vector spherical harmonic and

$$g_L(kr) = (2\pi)^{3/2}\,(-i)^L\,(kr)^{-1/2}\,J_{L+1/2}(kr). \tag{1.5}$$

If the wavelength of the emitted photon is large compared to the dimensions a of the atom, the main contribution to (1.2) comes from r such that $kr \ll 1$, in which case

$$g_L(kr) \sim 4\pi(-ikr)^L/(2L+1)!!. \tag{1.6}$$

Then the probability of emission of a photon of momentum $\mathbf{k}$ and polarization $\mathbf{e}$ into the solid angle $d\Omega$ may be written in the form

$$w_{if}\,d\Omega = \sum_{LM\lambda} \mathbf{e}\,|Y_{LM}^{(\lambda)}(\mathbf{k}/k)|^2\,w_{LM}^{(\lambda)}\,d\Omega, \tag{1.7}$$

where $w_{LM}^{(\lambda)}$ is the probability of emission of a photon in a state of angular momentum L, projection M, and parity $(-)^{L+\lambda+1}$.

Since the wavelength is much larger than a and the electron velocity v is of the order $\omega a/c$, $v/c \ll 1$ and ψ_i and ψ_f can be replaced by non-relativistic Pauli wave functions. (We shall examine this aspect more closely later.) It can then be shown that

$$w_{LM}^{(\lambda)} = \frac{2(L+1)}{L(2L+1)\,[(2L-1)!!]^2}\,\omega^{2L+1}|\langle f|\,Q_{LM}^{(\lambda)}\,|i\rangle|^2, \tag{1.8}$$

where $Q_{LM}^{(1)}$ and $Q_{LM}^{(0)}$ are, respectively, the electric and magnetic multipoles, defined by

$$Q_{LM}^{(1)} = e\left(\frac{1}{2L+1}\right)^{1/2}\sum_{i=1}^{N} r_i^L\,Y_{LM}^{*}(\mathbf{r}_i/r_i) \tag{1.9}$$

and

$$Q_{LM}^{(0)} = \left(\frac{1}{2L+1}\right)^{1/2}\sum_{i=1}^{N} \nabla_i[r_i^L\,Y_{LM}(\mathbf{r}_i/r_i)] \times \left\{\frac{e}{m(L+1)}\right\}\mathbf{l}_i + \boldsymbol{\mu}_i; \tag{1.10}$$

in (1.10), $\mathbf{l}_i = \mathbf{r}_i \times \mathbf{p}_i$ is the orbital angular momentum operator of the ith electron and $\boldsymbol{\mu}_i$ is the spin angular momentum operator $e\boldsymbol{\sigma}_i/2m$.

When the wavelength is large compared to a, only one term of (1.7) contributes significantly to the emission probability. If it is integrated over the solid angle $d\Omega$ and summed over the two independent photon polarizations $\mathbf{e}$ the resulting total emission probability is simply $w_{LM}^{(\lambda)}$.

The leading term of (1.7) is determined by selection rules. If J_i and J_f are the initial and final angular-momentum quantum numbers of the atomic system, M_i and M_f are the initial and final projection quantum numbers, and π_i and π_f are the initial and final parities, then the selection rules are

$$\begin{aligned} M_i - M_f &= M, \\ |J_i - J_f| \leqq L &\leqq J_i + J_f, \\ \pi_i &= \pi_f(-1)^{L+1+\lambda}. \end{aligned} \tag{1.11}$$

Because $\omega^L|\langle f|\, Q_{LM}^{(\lambda)}\, |i\rangle|$ contains $(\omega r)^L$ in the integral, the leading term of (1.7) corresponds to the smallest allowed value of L,

$$L = |J_i - J_f|. \tag{1.12}$$

Allowed transitions are electric-dipole transitions for which $L = \lambda = 1$. The selection rules are

$$J_f = J_i, J_i \pm 1; \qquad \pi_i = -\pi_f, \tag{1.13}$$

but $J_i = 0 = J_f$ is forbidden. For magnetic-dipole transitions, $L = 1$ and $\lambda = 0$. The selection rules are

$$J_f = J_i,\ J_i \pm 1; \qquad \pi_i = \pi_f, \tag{1.14}$$

but $J_i = 0 = J_f$ is forbidden. Indeed $J_i = 0 = J_f$ is forbidden for all multipole transitions, electric or magnetic.

The metastable states of helium and of the ions of the helium isoelectronic sequence are of considerable astrophysical significance. The $2\,^1S_0$ state of helium has zero angular momentum as does the ground $1\,^1S_0$ state. Thus the $2\,^1S_0$ state cannot decay to the ground state by the emission of a single photon. However, it can decay by the simultaneous emission of two electric-dipole photons according to

$$\text{He}\,(2\,^1S_0) \to \text{He}\,(1\,^1S_0) + h\nu_1 + h\nu_2. \tag{1.15}$$

Two-photon transitions become possible when the interaction of the atom and the radiation field is carried to second order. It is clear that the resulting formula for the probability will contain a summation over intermediate states and involve frequency-dependent denominators. The explicit expression for the probability per second for the simultaneous emission of two (electric-dipole) photons with one photon in the frequency range $d\nu_1$ is

$$w\,d\nu_1 = \frac{1024\pi^5 e^4 \nu_1{}^3 \nu_2{}^3}{c^6} |\mathbf{M}|^2_{\text{av}}\, d\nu_1, \tag{1.16}$$

where

$$\mathbf{M} = \mathop{\mathbf{S}}_s \frac{\langle f|\,\mathbf{re}_1\,|s\rangle\,\langle s|\,\mathbf{re}_2\,|i\rangle}{E_i - E_s - h\nu_2} + \mathop{\mathbf{S}}_s \frac{\langle f|\,\mathbf{re}_2\,|s\rangle\,\langle s|\,\mathbf{re}_1\,|i\rangle}{E_i - E_s - h\nu_1}. \tag{1.17}$$

Expression (1.16) has to be averaged over the directions of propagation and the directions of polarization independently for the two photons. Because the photons transport no angular momentum in the $2\,^1S_0$ decay process, a transition $\Delta M_J = 0$ involving no change in projection quantum number is followed by another $\Delta M_J = 0$ transition and $\Delta M_J = \pm 1$ is followed by $\Delta M_J = \mp 1$. Accordingly, a factor $\mathbf{e}_1 \cdot \mathbf{e}_2$ emerges from (1.17). Using the average $(\mathbf{e}_1 \cdot \mathbf{e}_2)^2_{\text{av}} = 1/3$, we can simplify (1.16) to

$$w\,d\nu_1 = \frac{1024\pi^6 e^4 \nu_1^3 \nu_2^3}{3c^6}$$

$$\left| \mathop{\mathbf{S}}_s \langle f|\,p_z\,|s\rangle\,\langle s|\,p_z\,|i\rangle \left\{ \frac{1}{E_i - E_s - h\nu_2} + \frac{1}{E_i - E_s - h\nu_1} \right\} \right|^2 d\nu_1. \tag{1.18}$$

The two-photon decay process gives rise to a broad continuum of photon energies since the only restriction on the photon frequencies ν_1 and ν_2 is the energy conservation requirement that

$$E_i - E_f + h\nu_1 + h\nu_2 = h\nu_T, \text{ say.}$$

The continuum extends from $\nu = 0$ (infinite wavelength) to $\nu = \nu_T$ and it is symmetric about the central frequency $\nu_T/2$, where it reaches its maximum intensity.

The probability of induced emission of two photons in a radiation field of density $\varrho(\nu_1)$ is related to the formula (1.18) for spontaneous emission by a factor $c^3\varrho(\nu_1)/8\pi h\nu_1^3$ and the probability of the absorption of two photons is (Levinson and Nikitin, 1965)

$$w = \frac{16\pi^4}{9h^2} \int \varrho(\nu_T - \nu_1)\,\varrho(\nu_1)\,d\nu_1 |\mathbf{M}|^2. \tag{1.19}$$

The form of the expression for $\mathbf{M}$ is not attractive for computational purposes, especially since it includes an integration over continuum states. However, sufficient data are available for neutral helium to permit approximate estimates. The individual matrix elements appearing in $\mathbf{M}$ are those appropriate to all single-photon electric-dipole transitions for both the initial $2\,^1S$ state and the final $1\,^1S$ state. The signs of the matrix elements derived from experimental or theoretical calculations of single-photon transition probabilities or oscillator strengths are unknown and some arbitrariness is introduced in evaluating (1.18).

1.1. Variational Procedures

It is possible to construct a variational procedure for the evaluation of summations such as those occurring in (1.18). Consider the summation

$$I(\omega) = \mathop{\mathbf{S}}_{s} \frac{\langle\psi_i|L_1|\psi_s\rangle\,\langle\psi_s|L_2|\psi_f\rangle}{E_i - E_s + \omega}, \tag{1.20}$$

where L_1 and L_2 are any two operators. Then clearly

$$I(\omega) = \langle\psi_i L_1|\,(E_i - H + \omega)^{-1}\,|L_2\psi_f\rangle \tag{1.21}$$

is a formal identity where H is the system Hamiltonian and

$$H\psi_s = E_s\psi_s. \tag{1.22}$$

Write

$$(E_i - H + \omega)^{-1}\,L_2\psi_f = \chi_f, \tag{1.23}$$

so that

$$I(\omega) = \langle\psi_i|\,L_1\,|\chi_f\rangle. \tag{1.24}$$

But (1.23) is the inhomogeneous differential equation

$$(H - E_i - \omega)\,\chi_f + L_2\psi_f = 0, \tag{1.25}$$

which we can replace by the variational statement

$$\delta J_2(\omega) = 0, \tag{1.26}$$

where $J_2(\omega)$ is the functional

$$J_2(\omega) = \langle\chi_f|\,H - E_i - \omega\,|\chi_f\rangle + 2\langle\chi_f|\,L_2\,|\psi_f\rangle. \tag{1.27}$$

Let us use as a trial form of χ_f in (1.27) the linear combination

$$\chi_f(\omega) = \sum_{j=1}^{m} a_j(\omega)\,\phi_j, \tag{1.28}$$

where ϕ_j comprise some basis set of the appropriate symmetry. By applying a unitary transformation to $\boldsymbol{\phi}$, we may arrange that the transformed functions (which we continue to call $\boldsymbol{\phi}$) satisfy

$$\begin{aligned} \langle\phi_j|\,H\,|\phi_k\rangle &= \mathscr{E}_j\,\delta_{jk}, \\ \langle\phi_j|\phi_k\rangle &= \delta_{jk}. \end{aligned} \tag{1.29}$$

Then carrying out the variation (1.26), we obtain

$$a_j(\omega) = -\frac{\langle\psi_f|\, L_2\, |\phi_j\rangle}{\mathscr{E}_j - E_i - \omega}, \tag{1.30}$$

$$\chi_f(\omega) = -\sum_{j=1}^{m} \frac{\langle\psi_f|\, L_2\, |\phi_j\rangle \langle\phi_j|}{\mathscr{E}_j - E_i - \omega}, \tag{1.31}$$

and

$$I(\omega) = -\sum_{j=1}^{m} \frac{\langle\psi_f|\, L_2\, |\phi_j\rangle \langle\phi_j|\, L_1\, |\psi_i\rangle}{\mathscr{E}_j - E_i - \omega}. \tag{1.32}$$

The dynamic polarizability of ψ_i is given by a special form of (1.20). The dipole polarizability (or refractive index) contains the summation

$$\alpha_i(\omega) = 2 \mathop{\mathbf{S}}_{s} \frac{(E_s - E_i)|\langle\psi_i|\, \mathbf{r}\, |\psi_s\rangle|^2}{(E_s - E_i)^2 - \omega^2}$$

$$\sim 2 \sum_{j=1}^{m} \frac{(\mathscr{E}_j - E_i)\, |\langle\psi_i|\, \mathbf{r}\, |\phi_j\rangle|^2}{(\mathscr{E}_j - E_i)^2 - \omega^2}. \tag{1.33}$$

The cross section for Rayleigh scattering is determined by the square of the dipole polarizability.

A comparison (Chan and Dalgarno, 1965) of (1.33) with experimental data on the refractive index of helium, reproduced in Fig. 1-1, shows excellent agreement for m as small as 4.

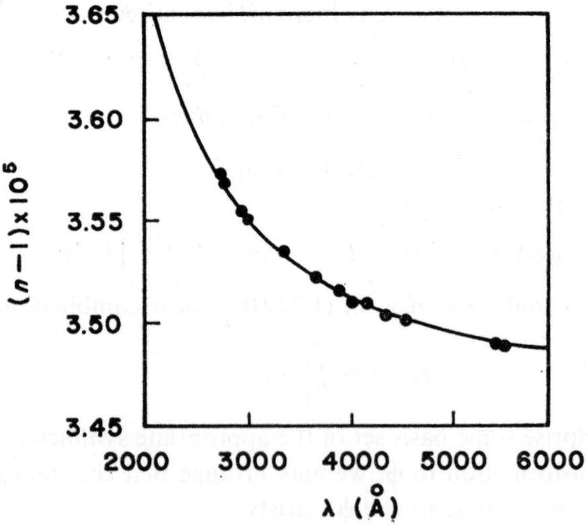

FIGURE 1-1. Values of $n - 1$ for He as a function of wavelength. The full line is theoretical and the circles are the experimental values. From Chan and Dalgarno (1965) by courtesy of the Institute of Physics and the Physical Society.

The variational approximation (1.33) can be written down immediately from the observation that by choosing (1.28), we are in effect approximating the Green's function of the system

$$\mathop{\mathrm{S}}_{s} |\psi_s\rangle \frac{1}{z - H} \langle\psi_s| \sim \sum_{i=1}^{m} |\phi_j\rangle \frac{1}{z - H} \langle\phi_j|. \tag{1.34}$$

Alternatively, we are solving in the subspace spanned by $\boldsymbol{\phi}$, with a Hamiltonian H_{pp} where

$$H_{PP} = PHP, \qquad P = \sum_{j=1}^{m} |\phi_j\rangle \langle\phi_j|. \tag{1.35}$$

A measure of the completeness of the set $\boldsymbol{\phi}$ is provided by evaluating moments of the matrix elements (Sitz and Yaris, 1968, Dalgarno and Epstein, 1969):

$$S_k = \mathop{\mathrm{S}}_{s} (E_s - E_i)^k |\langle\psi_i| L |\psi_s\rangle|^2. \tag{1.36}$$

It is interesting to note that the long-range van der Waals interaction between two atoms a and b can be calculated from (1.33). At large separations R of a and b, the interaction contains a term $-C_{ab}/R^6$ and the coefficient C_{ab} can be written as an external factor times the integral of the product of the dynamic dipole polarizabilities at imaginary frequencies (Casimir and Polder, 1948)

$$C_{ab} = \frac{3}{\pi} \int_0^\infty \alpha_a(i\omega)\, \alpha_b(i\omega)\, d\omega \tag{1.37}$$

But

$$\alpha(i\omega) \sim 2 \sum_{j=1}^{m} \frac{(\mathscr{E}_j - E_i) |\langle\psi_i| \mathbf{r} |\phi_j\rangle|^2}{(\mathscr{E}_j - E_i)^2 + \omega^2}. \tag{1.38}$$

1.2. Two-Photon Decay Calculations

1.2.1. *2^1S States of the Helium Sequence*

Using these variational procedures, accurate calculations have been performed of the two-photon decay rates of the 2^1S states of the helium sequence from HeI to NeIX (Drake, Victor, and Dalgarno, 1969).

The spectrum for HeI, which is typical, is shown in Fig. 1-2 as a function of frequency and in Fig. 1-3 as a function of wavelength. Table 1-1 lists the integrated decay probabilities

$$A = \frac{1}{2} \int_0^{\nu_T} w(\nu_1)\, d\nu_1 \tag{1.39}$$

and the wavelength at which the intensity reaches a maximum. The probabilities increase rapidly with nuclear charge Z. From inspection of the form of (1.18) and recalling that ν increases as Z^2, it is clear that $A \sim Z^6$. In fact, for large Z

$$A \sim 16.4(Z-1)^6 \text{ sec}^{-1}. \tag{1.40}$$

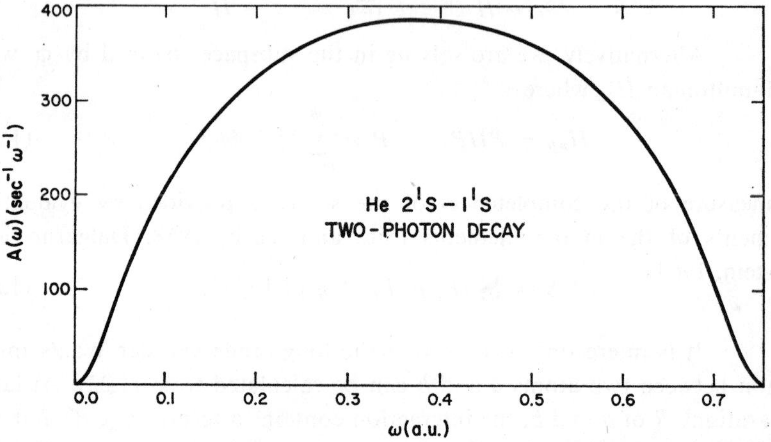

FIGURE 1-2. Photon energy distribution for the $2\,^1S$–$1\,^1S$ two-photon decay fo the HeI; ω is the photon energy in atomic units and $A = 51.3$ sec^{+1}. From Drake, Victor, and Dalgarno (1969) by courtesy of the Physical Review.

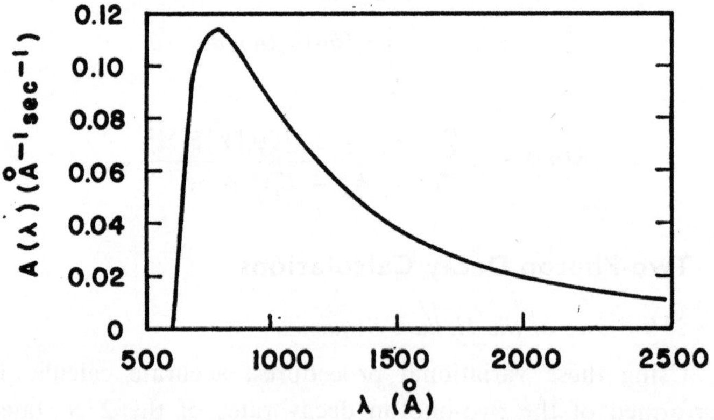

FIGURE 1-3. Photon energy distribution for the $2\,^2S$–$1\,^1S$ two-photon decay of HeI as a function of wavelength. From Drake, Victor, and Dalgarno (1969) by courtesy of the Physical Review.

The two-photon decay of the $2\,^1S$ state of NeIX may have been observed in a laboratory plasma. Figure 1-4, which reproduces the spectrum (Elton, Palumbo, and Griem, 1968), shows a coincidence between the observed spectrum shape and the predicted two-photon emission continuum.

TABLE 1-1. Two-photon decay probabilities of the metastable $2\,^1S$ states of the helium sequence.

System	A (sec^{-1})	λ_{max}
HeI	5.13×10^{1} [a]	771
LiII	1.95×10^{3}	258
BeIII	1.81×10^{4}	128
BIV	9.26×10^{4}	76.4
CV	3.31×10^{5}	50.9
NVI	9.43×10^{5}	36.1
OVII	2.31×10^{6}	27.1
FVIII	5.05×10^{6}	21.0
NeIX	1.00×10^{7}	16.7

[a] Pearl (1970) has recently measured the lifetime of He($2\,^1S$) to be 38 ± 8 msec, a factor of two *longer* than that predicted for two-photon decay.

The ions CV, OVII, and NeIX are important constituents of the solar corona. They have been identified by several allowed and spin-forbidden single-photon transitions (Evans and Pounds, 1968; Widing and Sandlin, 1968).

We assume an electron temperature T_e of 1.5×10^6 °K and an electron density of 10^9 cm^{-3}. The rate of collision-induced transitions

$$X(2\,^1S) + e \rightarrow X(1\,^1S) + e \tag{1.41}$$

is about 10^2 sec^{-1} in the solar corona, so that the two-photon decay is the usual mode of depopulating the metastable ions beyond LiII. The rate of population of the $2\,^1S$ levels may approach the rate of population of the $2\,^1P$ levels in which case the integrated two-photon intensity is comparable to that of the strong-resonance $2\,^1P$–$1\,^1S$ single-photon transition. Figure 1-5 is a reproduction of a measured solar spectrum (Evans and Pounds, 1968) and superimposed upon it is the predicted two-photon profile, modified by instrumental sensitivity (Dalgarno and Drake, 1969). Two-photon emission is unlikely to be a substantial source of soft x-rays in the coronal spectrum, but it provides nevertheless the most efficient mechanism for depopulating the singlet metastable states.

1.2.2. *$2\,^2S$ States of the Hydrogen Sequence*

The earliest calculations of a two-photon process referred to the decay of the metastable H($2\,^2S_{1\,2}$) state (Breit and Teller, 1940). The formula is essentially identical to (1.18) and the spectral shape is a scaled version of Fig. 1-2. It has long been suggested that the two-photon decay of H($2\,^2S_{1/2}$)

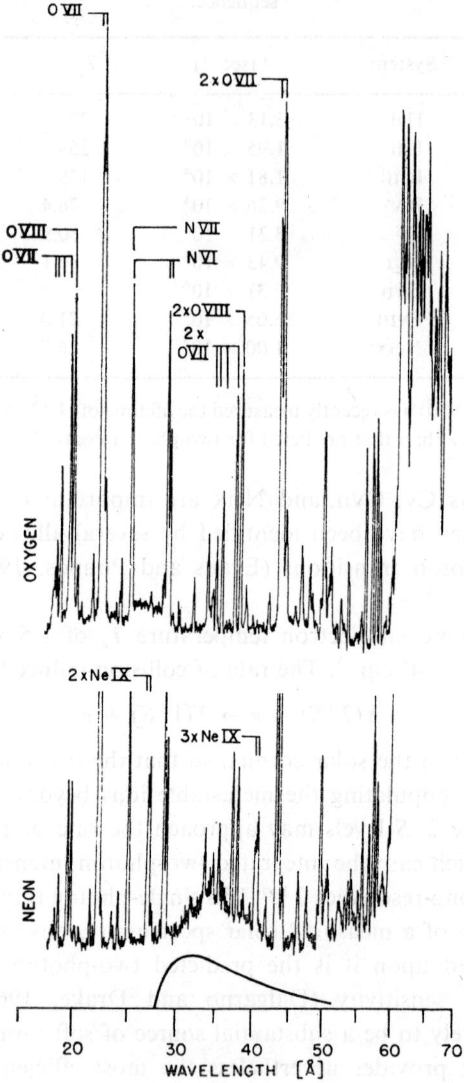

FIGURE 1-4. Densitometer tracings of spectra obtained with 6 percent (molecular) oxygen and 12 percent neon included in the deuterium filling gas. The VYNS-3 filter (~ 4000 Å thick) used is essentially opaque from 30 Å to the chlorine *K* edge at 62 Å, in which region the observed continuum radiation is of second-order origin. The lower curve represents the corresponding theoretical NeIX two-photon emission profile. No corrections have been made for variations in instrumental sensitivity with wavelength. From Elton, Palumbo, and Griem (1968) by courtesy of the Physical Review Letters.

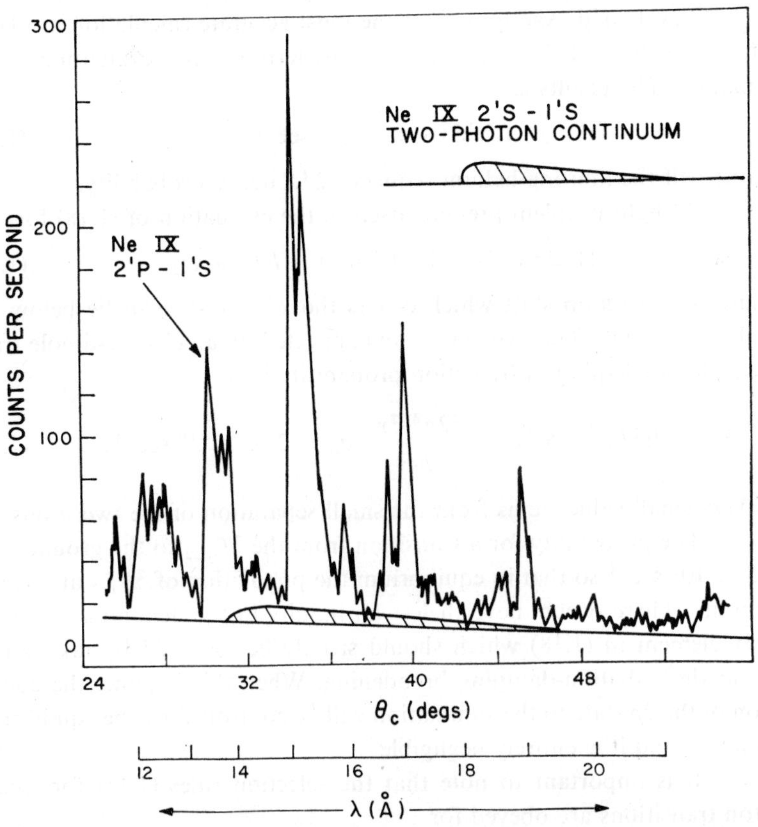

FIGURE 1-5. A reproduction of the solar spectrum of Evans and Pounds (1968) indicating the contribution from the $2\,^1S$–$1\,^1S$ two-photon decay of NeIX normalized to the $2\,^1P$–$1\,^1S$ transition of NeIX and modified by the instrumental sensitivity. From Dalgarno and Drake (1969) by courtesy of the Société Royale des Sciences de Liège.

is an important source of continuum emission in planetary nebulae (Spitzer and Greenstein, 1951), but the identification of it has not been firmly established.‡ That $He^+(2\,S_{1/2})$ does decay by two-photon emission of the predicted character appears to have been confirmed by laboratory studies (Artura, Novick, and Tolk, 1969). The observation has additional importaance in that it carries implications about the existence of a nuclear or electronic dipole moment (Salpeter, 1958; Feinberg, 1958).

For the single-electron system, the solution $\chi_f(\omega)$ of (1.23) can be determined analytically (Gavrila, 1967; Vetchinkin and Kchristenko, 1967)

‡ The large reddening found for η Carinae may arise from the metastable two-photon decay (Pagel, 1969).

and (1.18) evaluated exactly. In fact the most accurate calculation available (Shapiro and Breit, 1959) was performed by term by term evaluation of the summation. The results is

$$A(2\,^2S) \sim 8.226 Z^6 \text{ sec}^{-1}, \tag{1.42}$$

just one-half the limiting helium sequence $2\,^1S$ decay probability.

A slight problem presents itself in the evaluation of (1.18) for

$$\text{H}\,(2\,^2S_{1/2}) \rightarrow \text{H}\,(1\,^2S_{1/2}) + h\nu_1 + h\nu_2 \tag{1.43}$$

because of the Lamb shift which causes the $2\,^2P_{1/2}$ state to lie below the initial $2\,^2S_{1/2}$ state. The two states are connected by an electric-dipole transition. The single-photon transition probability is

$$A(^2P_{1/2}\text{–}^2S_{1/2}) = \frac{32\pi^3 e^2 \nu^3}{hc^3} a_0^2 = 2 \times 10^{-10} \text{ sec}^{-1}. \tag{1.44}$$

This very small value stems from the small separation of the two levels.

The probability for a transition from the $^2P_{1,2}$ to the ground state is 6.3×10^8 sec^{-1} so that in equilibrium the population of $^2P_{1/2}$ atoms will be minute. There occurs nevertheless an infinity in the denominator of the matrix element in (1.18) which should strictly be removed by taking into account the radiation-damping broadening. When this is done, the contribution of the $2p$ state to the summation will be controlled by the equilibrium population and it is entirely negligible.

It is important to note that the selection rules (1.11) for single-photon transitions are obeyed for

$$\text{H}(2\,^2S_{1/2}) \rightarrow \text{H}(1\,^2S_{1/2}) + h\nu \tag{1.45}$$

by magnetic-dipole transitions $L = 1$, $\lambda = 0$. The operator Q_{LM} of (1.10) leaves the radial part of any wave function unchanged and (1.8) vanishes because of the orthogonality of the initial and final $2s$ and $1s$ wave functions.

This vanishing of the magnetic-dipole transition probability occurs in the Pauli approximation but not in the Dirac approximation.

Let us return to Eq. (1.2) and insert the magnetic-dipole component of (1.4), still assuming that $kr \ll 1$. Then in place of (1.8) wo obtain (Akhiezer and Berestetskii, 1965)

$$w_{LM}^{(0)} = \frac{2e^2}{[(2L+1)!!]^2}\,\omega^{2L+1} \left| \int \psi_f^* \boldsymbol{\alpha}\, Y_{LM}^{(0)} r^2 \psi_i \, d\mathbf{r} \right|^2. \tag{1.46}$$

Now $\boldsymbol{\alpha}$ is the vector operator

$$\boldsymbol{\alpha} = \begin{pmatrix} 0 & \sigma \\ \sigma & 0 \end{pmatrix}, \tag{1.47}$$

where $\boldsymbol{\sigma}$ are the Pauli spin matrices, and $\mathbf{Y}_{LM}^{(0)}$ is the vector

$$\mathbf{Y}_{LM}^{(0)} = \frac{1}{(2L+1)^{1/2}} \mathbf{L}\, Y_{LM}. \tag{1.48}$$

Then

$$\begin{aligned} \boldsymbol{\alpha} \cdot \mathbf{Y}_{1M}^{(0)*} = (-1)\frac{M+1}{\sqrt{3}} \{ & \alpha^{+}[(1-M)(2+M)/2]^{1/2}\, Y_{1,-M-1} \\ & + \alpha^{-}[(1+M)(2-M)/2]^{1/2}\, Y_{1,-M+1} \\ & - \alpha^{0} Y_{1-M}\}, \end{aligned} \tag{1.49}$$

where

$$\begin{aligned} \sigma^{+} &= \frac{1}{\sqrt{2}}(\sigma_x + i\sigma_y) = \sqrt{2}\begin{pmatrix} 0 & 1 \\ 0 & 0 \end{pmatrix}, \\ \sigma^{-} &= \frac{1}{\sqrt{2}}(\sigma_x - i\sigma_y) = \sqrt{2}\begin{pmatrix} 0 & 0 \\ 1 & 0 \end{pmatrix}, \\ \sigma^{0} &= \sigma_z = \begin{pmatrix} 1 & 0 \\ 0 & -1 \end{pmatrix}. \end{aligned} \tag{1.50}$$

The Dirac wave functions for the $S_{1/2}$ states of hydrogen may be written (Bethe and Salpeter, 1957)

$$\psi = \begin{bmatrix} u_1 \\ u_2 \\ u_3 \\ u_4 \end{bmatrix} = \begin{bmatrix} g(r)\,(M+1/2)^{1/2}\, X_{0,M-1/2}(\hat{\mathbf{r}}) \\ -g(r)\,(-M+1/2)^{1/2}\, Y_{0,M+1/2}(\hat{\mathbf{r}}) \\ -if(r)\left(\dfrac{M+3/2}{3}\right)^{1/2} Y_{1,M-1/2}(\hat{\mathbf{r}}) \\ -if(r)\left(\dfrac{M+3/2}{3}\right)^{1/2} Y_{1,M+1/2}(\hat{\mathbf{r}}) \end{bmatrix}. \tag{1.51}$$

The evaluation of (1.43) accordingly reduces to the radial integrals

$$\begin{aligned} I = & \int_0^\infty g_{1s}(r)\, r^3 f_{2s}(r)\, dr \\ & + \int_0^\infty g_{2s}(r)\, r^3 f_{1s}(r)\, dr. \end{aligned} \tag{1.52}$$

To a close approximation

$$f = dg/dr. \tag{1.53}$$

Integrating (1.52) by parts reduces it to a multiple of the radial integral

$$\int_0^\infty g_{1s}(r)\, r^2 g_{2s}(r)\, dr$$

which vanishes. The expression (1.53) is not exactly correct, however. Breit and Teller gave the order-of-magnitude estimate for I of $\alpha^3 a_0$ from which they derive for the magnetic-dipole single-photon emission probability:

$$A \sim \frac{\pi}{81} \alpha^9 \nu \sim 5 \times 10^{-6}\ \mathrm{sec}^{-1}, \tag{1.54}$$

α being the fine-structure constant. This is negligible compared to the two-photon decay probability.

In actual physical situations, the two-photon decay competes with collision processes such as

$$\begin{aligned} e + \mathrm{H}(2s) &\to e + \mathrm{H}(2p), \\ \mathrm{H}^+ + \mathrm{H}(2s) &\to \mathrm{H}^+ + \mathrm{H}(2p), \\ \mathrm{H} + \mathrm{H}(2s) &\to \mathrm{H} + \mathrm{H}(2p), \end{aligned} \tag{1.55}$$

followed by electric-dipole emission of Lyman-alpha radiation from the $2p$ level. A substantial enhancement of the two-photon continuum will occur in plasmas which are optically thick towards Lyman-alpha. The repeated absorptions of Lyman-alpha populate the $2p$ level which is transferred to the $2s$ level by the reverse of process (1.52). A recent study of the effects of optical thickness is that of Gerola and Panagia (1968). There is a possibility that in optically thick but low-density plasmas the two-photon decay of $\mathrm{H}(2p)$,

$$\mathrm{H}(2p) \to \mathrm{H}(1s) + h\nu_1 + h\nu_2, \tag{1.56}$$

might be an important decay mode. It has a low efficiency because of parity requirements. A value of $1.94 \times 10^{-5}\ \mathrm{sec}^{-1}$ has been computed by Kipper and Tiit (1958).

1.2.3. $2\,^3S_1$ *States of the Helium Sequence*

The $2\,^3S_1$ state of helium can decay by a two-photon emission

$$\mathrm{He}(2\,^3S_1) \to \mathrm{He}(1\,^1S_0) + h\nu_1 + h\nu_2, \tag{1.57}$$

but the probability of (1.54) is much smaller than that of the $2\,^1S_0$ decay because of the change in spin multiplicity. The process occurs with the emission of two electric-dipole photons through the admixture of triplet character in the intermediate $n\,^1P$ states that connect with the final $1\,^1S_0$

state and the admixture of singlet character in the intermediate $n\,^3P$ states that connect with the initial $2\,^3S_1$ state. As Breit and Teller (1965) pointed out, this effect reduces the two-photon decay probability by a factor of about 10^{-6}.

Accordingly, it had been assumed that metastable triplet helium decayed by two-photon emission with a probability of about 10^{-5} sec^{-1}. There is, however, a further significant difference between the $2\,^1S_0$ and $2\,^3S_1$ decay processes (Drake and Dalgarno, 1968; Bely, 1968). Because one unit of angular momentum must be transported by the two photons in (1.54), a $\Delta M_J = \pm 1$ transition always follows a $\Delta M_J = 0$ transition and $(\mathbf{e}_1 \times \mathbf{e}_2)$ emerges in place of $(\mathbf{e}_1 \cdot \mathbf{e}_2)$. The decay probability is still symmetric about the mean transition frequency $\nu_T/2$, but instead of the maximum at $\nu_1 = \nu_2$ that occurs for the $2\,^1S$ decay, the $2\,^3S$ probability vanishes at $\nu_1 = \nu_2$. The vanishing of the transition probability when $\nu_1 = \nu_2$ is a reflection of a fundamental limitation on the possible states for two photons (Landau, 1948; Jacob and Wick, 1959). For the same reason, the $2\,^3S_1$ state of positronium cannot be annihilated by a two-photon emission.

The formal expression for the two-photon decay probability is

$$\begin{aligned} A(\nu_1)\,d\nu_1 &= \frac{1024\pi^6 e^4}{3c^6} (|\mathbf{e}_1 \times \mathbf{e}_2|^2)_{\text{av}}\, \nu_1^3 \nu_2^3 \\ &\times \frac{1}{2} \Bigg| \int_{m',m''} (2\,^3S\| P \| m'\,^3P)\, \mathscr{E}_{m',m''} (m''\,^1P \| P \| 1\,^1S) \\ &\times \Bigg[\frac{1}{E_i - E_{m'} + h\nu_2} - \frac{1}{E_i - E_{m''} + h\nu_2} \\ &- \frac{1}{E_i - E_{m'} + h\nu_1} + \frac{1}{E_i - E_{m''} + h\nu_1} \Bigg] \Bigg|^2 d\nu_1 \end{aligned} \tag{1.58}$$

where $(|e_1 \times e_2|)^2_{\text{av}} = 2/3$ and

$$\mathscr{E}_{m'm''} = \frac{\langle m'\,^3P_1 | H_1 | m''\,^1P_1 \rangle}{E_{m'} - E_{m''}}, \tag{1.59}$$

H_1 being the spin–orbit interaction operator. The double infinite summation in (1.58) can be approximated by a double finite summation by using the variational Greens functions described earlier, but it is now necessary to determine the Greens functions for both the initial- and final-state symmetries.

Figure 1-6 shows the predicted spectrum for $He(2\,^3S)$ as a function of frequency and Fig. 1-7 shows the spectrum as a function of wavelength. Table 1-2 lists the calculated decay rates (Drake, Victor, and Dalgarno, 1969). The values are in harmony with semi-empirical estimates (Bely and Faucher, 1969).

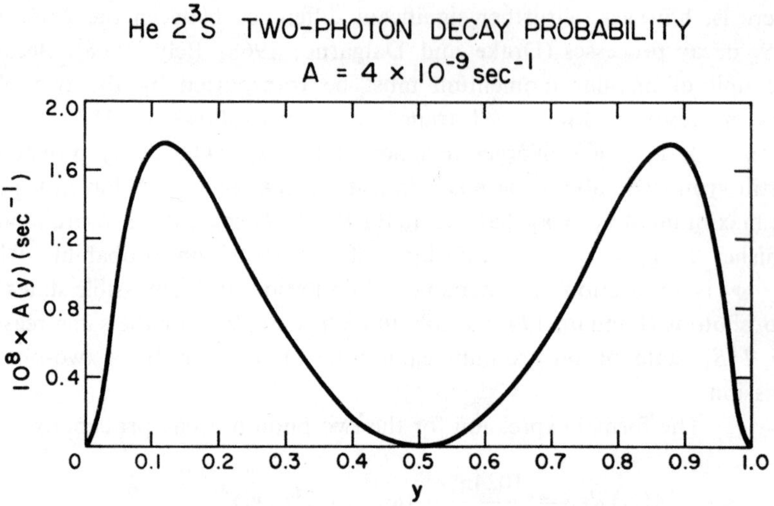

FIGURE 1-6. Photon energy distribution for the $2\,^3S$–$1\,^1S$ two-photon decay of HeI; y is the fraction of the energy transported by one of the two photons and $A = 4.02 \times 10^{+0}$ sec^{-1}. From Drake, Victor, and Dalgarno (1969) by courtesy of the Physical Review.

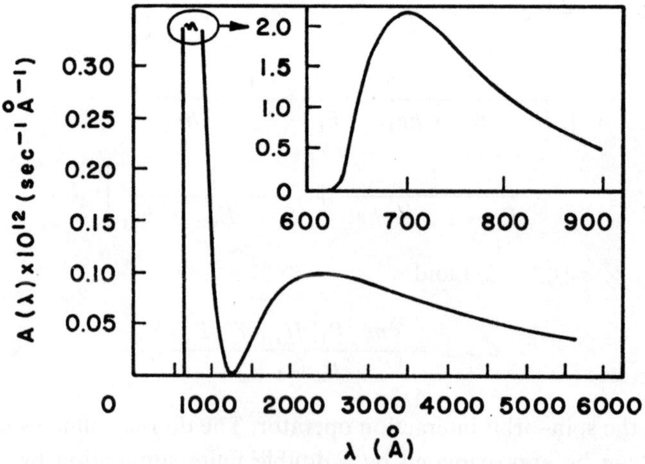

FIGURE 1-7. Photon energy distribution for the $2\,^3S$–$1\,^1S$ two-photon decay of HeI as a function of wavelength. From Drake, Victor, and Dalgarno (1969) by courtesy of the Physical Review.

TABLE 1-2. Two-photon decay probabilities of the $2\,^3S$ states of the helium sequence.

System	A (sec^{+1})	λ_{max} (Å)
HeI	4.02×10^{-0}	699, 2400
LiII	1.50×10^{-6}	228, 660
BeIII	6.36×10^{-5}	112, 445
BIV	1.01×10^{-3}	66.4, 278
CV	8.93×10^{-3}	44.4, 189
NVI	5.44×10^{-2}	30.9, 137
OVII	2.54×10^{-1}	23.1, 105
FVIII	9.73×10^{-1}	17.9, 82
NeIX	3.20	14.2, 67

Because of the dependence of spin–orbit interactions on the nuclear charge, the decay rates increase rapidly with increasing Z. Initially they increase as Z^{10} but for nuclear charges of the order of 25, the singlet–triplet mixing saturates and the $2\,^3S$ decay rates increase as Z^6 just as in the $2\,^1S$ case.

The two-photon decay rates for the $2\,^3S_1$ states are very small. For helium the rate is 4.0×10^{-9} sec^{-1}, so slow that it raises the possibility that a single photon process may be more rapid. The selection rules for

$$\mathrm{He}(2\,^3S_1) \rightarrow \mathrm{He}(1\,^1S_0) + h\nu \tag{1.60}$$

are satisfied by magnetic-dipole radiation $\lambda = 0, L = 1$.

Breit and Teller (1940) discussed the rate of (1.60) that arises because of the admixture of singlet and triplet states. The $2\,^3S_1$ state has small components of $^1P_1$, $^3P_1$, and $^3D_1$ character and the $1\,^1S_0$ has a small component of $^3S_0$ character. Because of parity, these components belong to highly excited *pp'* or *dd'* configurations. Breit and Teller estimate a value of 10^{-5} for the weights of the admixtures. A magnetic-dipole transition can occur between the $^3P_1$ component of $^3S_1$ and the $^3S_0$component of $^1S_0$ and Breit and Teller give an approximate estimate of 10^{-15} sec^{-1} for the $2\,^3S_1$–$1\,^1S_0$ single-photon transition probability. Griem (1969) similarly estimates for the helium sequence

$$A \sim 2 \times 10^{-16}(Z-1)^{10}\left[\frac{\nu}{(Z-1)^2\,\nu_0}\right]^3 \mathrm{sec}^{-1}, \tag{1.61}$$

where ν_0 is the frequency of the hydrogen Lyman-alpha line 2.47×10^{15} cm^{-1}. For large Z

$$A \sim 2 \times 10^{-16}(Z-1)^{10}. \tag{1.62}$$

G. W. F. Drake (1969) has recently carried out precise calculations of the admixtures and he computes

$$\begin{aligned} A(\text{He I}) &= 1.6 \times 10^{-13}\ \text{sec}^{-1}, \\ A(\text{O VII}) &= 4.4 \times 10^{-6}\ \text{sec}^{-1}. \end{aligned} \tag{1.63}$$

Although these values are substantially larger than earlier estimates, they are much less than those for two-photon decay.

Recently Gabriel and Jordan (1969a, b) have identified the $2\,^3S$–$1\,^1S$ lines of Cv, OvII, NeIX NaX, MgXI, and SiXIII in the coronal emission spectrum.† For OvII, they conclude that the single-photon decay rate must be at least of the order of 10 sec^{-1} in order that the line appear in the solar corona. If it were much slower, the $2\,^3S$ level would be depopulated by electron impact excitation,

$$e + \text{O VII}(2\,^3S) \rightarrow e + \text{O VII}(2\,^3P). \tag{1.64}$$

The conclusion, incidentally, provides a semiquantitative confirmation of the revized rates of two-photon decay. For OvII, the predicted two-photon decay rate of 0.25 sec^{-1} is in fact smaller than the derived minimum single-photon decay rate of 10 sec^{-1}.

Although it is possible in principle through plasma interactions to induce virtual $n\,^3P$ components in the $2\,^3S$ state which can then radiate by a single-photon electric-dipole transition, the coronal densities are insufficient in practice and the observations demand apparently a spontaneous single-photon radiative transition. The selection rules require that it be a magnetic-dipole transition.

Griem (1969) has pointed out that the relativistic mechanism discussed by Breit and Teller for $H(2\,^2S_{1/2})$ magnetic-dipole decay gives decay rates for the $2\,^3S_1$ levels that are faster than the single-photon decay induced by spin–orbit interactions and that are faster than the two-photon decay rates. After modifying (1.54) to take account of the different statistical weights and generalizing it to other nuclear charges, Griem obtains the approximate formula‡

$$A \sim 3.7 \times 10^{-6}(Z-1)^9\,(Z - 1/2)\left[\frac{\nu}{(Z-1)^2\,\nu_0}\right]^3 \text{sec}^{-1}. \tag{1.65}$$

---

† The single photon decays of the $2\,^3S_1$ states of SiXIII, SXV, and ArXVII have now been detected in the laboratory (Schmieder and Marrus, 1970).

‡ Drake (1970) gives the asymptotic formula $A \sim 4.41 \times 10^9 \lambda^{-5}$ sec^{-1} where λ is the wavelength in Å.

Expression (1.65) assumes that the $1s$ orbital can be described as hydrogenic with charge $Z - 1/2$ and the $2s$ orbital can be described as hydrogenic with charge $Z - 1$.

The formula can be derived from (1.46) on writing Ψ_i and Ψ_f as antisymmetrised products of hydrogenic Dirac wave functions. That a singlet–triplet magnetic-dipole transition occurs without an explicit spin–orbit interaction stems from the fact that in the relativistic description, the spin projection is not a rigorous quantum number.

The accuracy of (1.65) may be adequate for highly stripped ions where a hydrogenic description is appropriate. The values are large enough to be consistent with the coronal observations.

For neutral helium, (1.65) gives an emission probability of 4×10^{-5} sec^{-1}. The accuracy of this estimate is not high and a calculation of the radiative lifetime of the metastable $2\,^3S_1$ helium atom that takes proper account of electron correlation is needed.‡

TABLE 1-3. Single-photon decay rates $2\,^3S_1$–$1\,^1S_0$.[a]

System	A (sec^{-1})
CV	4.86×10
OVII	1.04×10^3
NeIX	1.09×10^4
MgXI	7.24×10^4
SiXIII	3.56×10^5

[a] From Drake (1971).

The value 4×10^{-5} sec^{-1} for the single-photon decay probability is close to the value 2×10^{-5} sec^{-1} for the two-photon decay probability that has been adopted in earlier discussions of the helium abundance (Mathis, 1957). Fortuitously then, these early discussions may still be appropriate.‡

The general conclusion of these studies was that an important mechanism for depopulating the $2\,^3S_1$ state of helium in planetary nebulae must exist that has not yet been identified (Osterbrock, 1964; Robbins, 1968). The studies work from measurements of the intensity of $\lambda 10880$ which is

‡ Drake (1971) has now performed such a calculation. His value for He is 1.272×10^{-4} sec^{-1} and a new discussion of helium in nebulas is warranted. Some of Drake's values for high Z are reproduced in Table 1-3.

produced by spontaneous electric-dipole radiation in the transition

$$\mathrm{He}(2\,^3P_{2,1,0}) \rightarrow \mathrm{He}(2\,^3S_1) + h\nu. \tag{1.66}$$

It is assumed that the $2\,^3P$ level is populated by radiative capture-cascade processes and by electron impact excitation,

$$e + \mathrm{He}(2\,^3S) \rightarrow e + \mathrm{He}(2\,^3\mathrm{P}). \tag{1.67}$$

The intensity of λ10830 is compared with the intensity of λ5876 due to

$$\mathrm{He}(3\,^3D) \rightarrow \mathrm{He}(2\,^3P) + h\nu. \tag{1.68}$$

It is assumed that $\mathrm{He}(3\,^3D)$ is populated by radiative capture-cascade processes,

$$\mathrm{He}^+ + e \rightarrow \mathrm{He}' + h\nu. \tag{1.69}$$

The $\mathrm{He}(2\,^3S)$ depopulation mechanisms that were adopted consist of spontaneous two-photon decay (now to be replaced by spontaneous one-photon decay), collisional transitions to singlet states,

$$e + \mathrm{He}(2\,^3S) \rightarrow e + \mathrm{He}(m\,^1L), \tag{1.70}$$

and photoionization by stellar continuous radiation and by Lyman-alpha radiation produced in the nebula (O'Dell, 1965; Capriotti, 1967),

$$h\nu + \mathrm{He}(2\,^3S) \rightarrow \mathrm{He}^+ + e. \tag{1.71}$$

The photoionization cross sections for $\mathrm{He}(2\,^3S)$ [and for $\mathrm{He}(2\,^1S)$] have been calculated by Huang (1948) using a simple correlated representation of Ψ_i and a single-channel approximation to Ψ_f in the electric-dipole matrix element. The final wave function Ψ_f describes the scattering of electrons by He^+ and it can be derived by a variety of approximate methods. It would be of interest to calculate the photoionization cross section using a final-state continuum wave function which contains an adequate representation of the resonance $(2s\,2p)\,^3P$ state. The photoionization cross section for

$$\mathrm{He}(1s\,2s)\,^3S + h\nu \rightarrow \mathrm{He}(2s\,2p)\,^3P$$

$$\rightarrow \mathrm{He}^+\,(1s)\,^2S + e \tag{1.72}$$

will be extremely large in the vicinity of the resonance state at 38.35 eV or 322 Å above the $2\,^3S$ state. Unlike the photoionization processes (Burke and McVicar, 1965)

$$\mathrm{He}(1s^2)\,^1S + h\nu \rightarrow \mathrm{He}(2s\,2p)\,^1P \rightarrow \mathrm{He}^+ + e \tag{1.73}$$

and (Macek, 1967)

$$\mathrm{H}^-(1s^2)\,{}^1S + h\nu \rightarrow \mathrm{H}^-(2s\,2p)\,{}^1P \rightarrow \mathrm{H} + e, \tag{1,74}$$

(1.72) occurs through a single-electron jump.

Robbins (1968) has suggested that photoionization through the series of resonances

$$\mathrm{He}(1s\,2s)\,{}^3S + h\nu \rightarrow \mathrm{He}(2s\,np)\,{}^3P \rightarrow \mathrm{He}^+ + e \tag{1.75}$$

may contribute importantly to the depopulation of $2\,{}^3S$ atoms in some nebulae.

Let us turn now to the coronal emissions of the highly stripped helium-like ions and inquire into the other excited states in which the valence electron has principal quantum number $n = 2$, the $(1s\,2p)\,{}^1P$ and $(1s\,2p)\,{}^3P$ states. The $(1s\,2p)\,{}^1P$ states are populated by electron impact of ground-state ions together with some contribution from radiative recombination. They decay by allowed electric-dipole transitions primarily to the ground $(1s^2)\,{}^1S$ state but also to the metastable $(1s\,2s)\,{}^1S$ state. The transition probabilities have been calculated to high accuracy (cf. Wiese, Smith, and Glennon, 1966). It is interesting to note that the major inaccuracy in the $2\,{}^1P$–$2\,{}^1S$ transition probabilities for high Z rests in the uncertainties in the transition frequencies and not in the electric-dipole matrix elements. Thus there are two wavelengths in the literature for the Cv $2\,{}^1P$–$2\,{}^1S$ line—one of 3541 Å (Prasad and El-Menshawy, 1968) and one of 3526.7 Å (Boland, Irons, and McWhirter, 1968). A theoretical decision requires an accurate calculation of the relativistic and radiative corrections to the energy eigenvalues.‡

The decay of the $2\,{}^3P$ states is more intersting. The $2\,{}^3P_1$ state can decay through an allowed electric-dipole transition to the $2\,{}^3S_0$ state. An electric-dipole transition to the $1\,{}^1S_0$ state, though not strictly forbidden by the selection rules, involves a change in spin multiplicity. We have remarked earlier that spin–orbit interaction effects increase rapidly with increasing nuclear charge and in any case the $2\,{}^3P$–$1\,{}^1S$ lines are strong features of laboratory and coronal emission spectra.

Recalling our previous measure of spin–orbit interaction (1.59), we may write for the electric-dipole matrix element

$$\langle 1\,{}^1S_0|\,\mathbf{r}\,|2\,{}^3P_1\rangle = \mathop{\mathbf{S}}_{m''=2} \langle 1\,{}^1S_0|\,\mathbf{r}\,|m''\,{}^1P_1\rangle\, \mathscr{E}_{2m''}, \tag{1.76}$$

‡ Such a calculation has now been carried out (Accad, Pekeris, and Schiff, 1969). The theoretical wavelength is 3525.7 Å.

so that the calculation can be completed once the Green's function of ${}^1P$ symmetry has been constructed.

The resulting transition probabilities (Drake and Dalgarno, 1969) for the sequence are listed in Table 1-4. The values are in reasonable accord with semiempirical results of Elton (1967) who uses a method of analyzing experimental fine-structure data that is essentially equivalent to curtailing the summation in (1.76) at $m'' = 2$.

TABLE 1-4. Transition probabilities ($\sec^{-1}$).

System	$2\,{}^3P_1$–$1\,{}^1S_0$	$2\,{}^3P_1$–$2\,{}^3S_1$	$2\,{}^3P_2$–$1\,{}^1S_0$
HeI	1.80×10^2	1.02×10^7	3.27×10^{-1}
LiII	1.81×10^4	2.88×10^7	3.50×10
BeIII	4.01×10^5	3.42×10^7	6.17×10^2
BIV	4.23×10^6	4.55×10^7	5.01×10^3
CV	2.84×10^7	5.65×10^7	2.62×10^4
NVI	1.40×10^8	6.78×10^7	1.03×10^5
OVII	5.53×10^8	7.94×10^7	3.31×10^5
FVIII	1.85×10^0	9.15×10^7	9.16×10^5
NeIX	5.43×10^0	9.80×10^7	2.26×10^6

The allowed electric-dipole transition probabilities of the $2\,{}^3P$–$2\,{}^3S$ transition are also listed in Table 1-4. For ions heavier than BIV, the spin-forbidden $2\,{}^3P$–$1\,{}^1S$ decay is actually more probable than the allowed $2\,{}^3P$–$2\,{}^3S$ decay.

The electric dipole transition to the $1\,{}^1S_0$ level depopulates only the $2\,{}^3P_1$ level. The $2\,{}^3P_2$ and $2\,{}^3P_0$ levels can be depopulated by allowed electric-dipole transitions to the $2\,{}^3S_1$ level and the $2\,{}^3P_2$ level can also be depopulated by magnetic-quadrupole transitions to the ground $1\,{}^1S_0$ level (Mizushima, 1964; Garstang, 1967). The magnetic-quadrupole transition probabilities have been calculated using highly accurate variational wave functions and also by a perturbation analysis by Drake (1969). Drake's results are included in Table 1-4.

For large values of the nuclear charge, the $2\,{}^3P_2$–$1\,{}^1S_0$ transition occurs more frequently than the $2\,{}^3P_2$–$2\,{}^3S_1$ transition but remains slow compared to the $2\,{}^3P_1$–$1\,{}^1S_0$ transition. It may be possible in a low-density plasma to observe the selective depopulation of the $2\,{}^3P_1$ state of helium-like ions with $Z > 6$. These differences between the radiative depopulation rates of the fine-structure levels of the $2\,{}^3P$ states of the helium-like coronal ions may be significant in determinating the population distribution. Collisions tend to offset the departure from statistical equilibrium. Proton-induced transitions

$$H^+ + m(2\,^3P_J) \rightarrow H^+ + m(2\,^3P_{J'}) \tag{1.77}$$

are inhibited by the Coulomb repulsion and electron processes

$$e + m(2\,^3P_J) \rightarrow e + m(2\,^3P_{J'}) \tag{1.78}$$

may be more significant.

2. COLLISION PROCESSES

I wish now to study some of the atomic particle collision processes that are important to astrophysics. To place the discussion in an astrophysical context, consider the hot model of the expanding universe. According to it, there occurs after some 10^5 years an "atomic physics" epoch. The atomic processes that take place at this evolutional stage must establish the physical conditions that lead to the formation of galaxies. At the beginning of the "atomic physics" stage, matter consisted of protons H^+, electrons e, and photons ν with possibly some helium. I intend in this second half of the lecture course to examine some of the atomic processes that occurred and to discuss how their efficiencies may be estimated.

2.1. Radiative Recombination

Hydrogen atoms can be formed by the *radiative recombination* of protons and electrons,

$$H^+ + e \rightarrow H(n) + h\nu, \tag{2.1}$$

and destroyed by the inverse process of *photoionization*

$$H(n) + h\nu \rightarrow H^+ + e. \tag{2.2}$$

The hydrogen atoms may be in an excited state and they can radiate discretely by allowed transitions,

$$H(n) \rightarrow H(n') + h\nu. \tag{2.3}$$

The recombination spectrum consists of discrete line emissions and a continuum. The atoms which cascade into the $2s$ level can also give rise to a continuum through the two-photon decay process discussed earlier.

The calculation of the rates of (2.1) and (2.2) reduces to the evaluation of electric-dipole matrix elements. Extensive tabulations are available (Burgess, 1964; Green, Rush, and Chandler, 1957; Menzel, 1968). Recombination preferentially populates low-lying levels. If the thermal energy of the electron is small compared to the ionization energy, the recombination coefficient for level n behaves with n and with electron temperature T_e as

$$\alpha_n(T_e) \sim \frac{1}{n} T_e^{-1/2}. \tag{2.4}$$

The total recombination coefficient is approximately

$$\alpha(T_e) \sim 5 \times 10^{-13} \left(\frac{1000}{T_e}\right)^{0.7}. \tag{2.5}$$

The cold electrons recombine more readily and recombination is itself a mechanism for increasing T_e.

Transitions can also occur through electron impact:

$$e + \mathrm{H}(n) \rightarrow e + \mathrm{H}(n'), \tag{2.6}$$

$$e + \mathrm{H}(n) \rightarrow e + \mathrm{H}^+ + e. \tag{2.7}$$

In some physical circumstances, it may be necessary to distinguish between the angular-momentum sublevels (nl) and if the electron density n_e is large, *three-body recombination* (Bates, Kingston, and McWhirter, 1962),

$$e + e + \mathrm{H}^+ \rightarrow e + \mathrm{H}, \tag{2.8}$$

is important.

Because of its importance in nebulae, numerous detailed studies of the recombination of protons and electrons have been performed (cf. Seaton, 1968; Sejnowski and Hjellming, 1969). Of special interest are the observations of lines in the radio spectrum originating in levels with large values n of the order of 100. To improve the completeness with which these observations can be interpreted more accurate cross sections are required not only for (2.5) and (2.6) (Saraph, 1964; Pluta and McDowell, 1966) but also for the analogous process

$$\mathrm{H}^+ + \mathrm{H}(nl) \rightarrow \mathrm{H}^+ + \mathrm{H}(n'l'), \tag{2.9}$$

$$\mathrm{H} + \mathrm{H}(nl) \rightarrow \mathrm{H} + \mathrm{H}(n'l'). \tag{2.10}$$

Process (2.10) may be important in HI regions which, lying along the line of sight, can modify the observed radiofrequency emission of HII regions (Dupree and Goldberg, 1969).

2.2. Radiative Attachment

The hydrogen atoms, formed by recombination, can attach an electron in a *radiative attachment* process to form negative hydrogen ions,

$$e + \mathrm{H} \rightarrow \mathrm{H}^- + h\nu, \tag{2.11}$$

giving rise to a continuum emission. Because of its importance to the solar opacity at long wavelengths, the inverse of (2.9), *photodetachment*,

$$h\nu + \mathrm{H}^- \rightarrow e + \mathrm{H}, \tag{2.12}$$

has been the subject of extensive investigation both experimentally and theoretically (cf. Branscomb, 1962; Doughty, Fraser, and McEachran, 1966; Macek, 1967).

The predicted radiative attachment rates are shown in Fig. 2-1 as a function of electron temperature T_e.

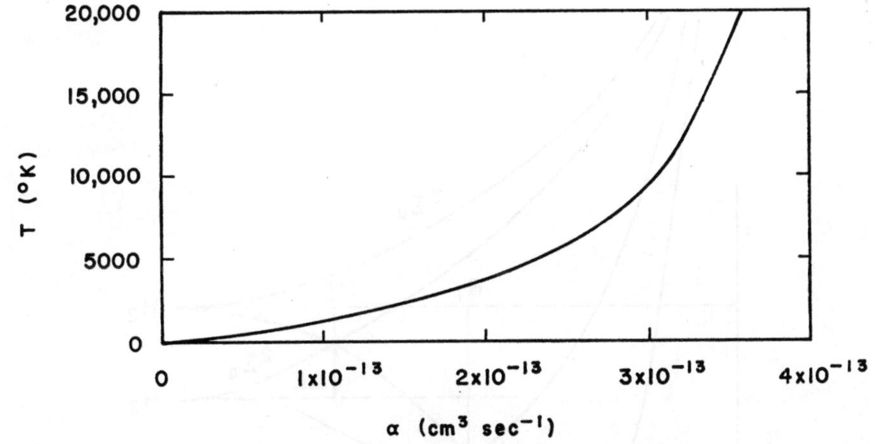

FIGURE 2-1. The rate coefficient for the radiative attachment of electrons to hydrogen atoms. From Dalgarno and Kingston (1963) by courtesy of The Observatory.

2.3. Associative Detachment

A critical aspect of the discussions of galaxy formation is the production of the hydrogen molecule H_2 since its existence gives rise to a cooling mechanism which converts kinetic energy into radiation. With the formation of H^-, H_2 can be produced by the process of *associative detachment*,

$$\mathrm{H} + \mathrm{H}^- \rightarrow H_2 + e. \tag{2.13}$$

Associative detachment can be regarded as proceeding through the formation of a temporary quasimolecular ion which can decay by an autodetachment process with spontaneous emission of an electron (Bates and Massey, 1954). Consider the slow approach of H and H^-, so slow that an average can be taken over the electronic motion—the Born–Oppenheimer approximation of molecular spectroscopy. The averaged electronic motion generates an effective force field in which the nuclei move. Because of symmetry there are two such interaction-potential curves corresponding to the $^2\Sigma_u$ and $^2\Sigma_g$ molecular states. Similarly two molecular states, the $^1\Sigma_g$ and $^3\Sigma_u$ states, are produced by the slow approach of two ground-state hydrogen atoms.

At infinite separation R of the two nuclei, the $^2\Sigma_u$ and $^2\Sigma_g$ potentials coincide at an energy 0.75 eV—the electron affinity of H^-—below the common limit of the $^1\Sigma_g$ and $^3\Sigma_u$ states of H_2, as Fig. 2-2 illustrates. Now He^-

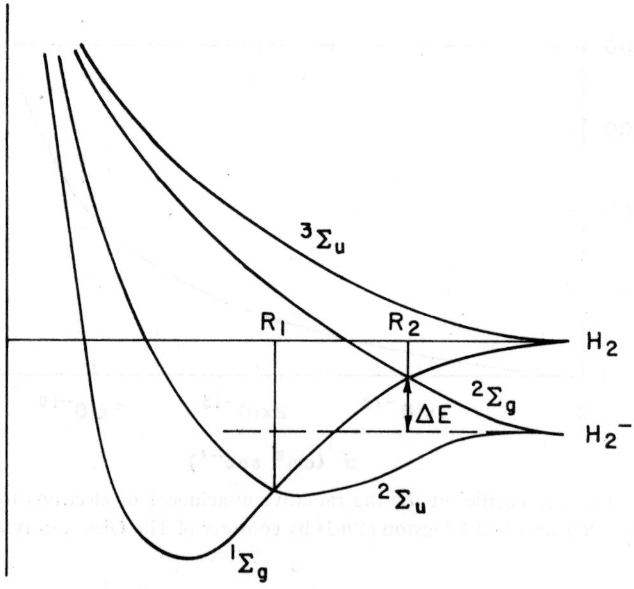

FIGURE 2-2. Schematic representation of the real parts of the $^2\Sigma_u$ and $^2\Sigma_g$ interaction potentials of H^{2+}, showing their intersections with the $^1\Sigma_g$ and $^3\Sigma_u$ potentials of H_2.

does not exist in a doublet spin state so that the $^2\Sigma_u$ and $^2\Sigma_g$ potential-energy curves of H_2^- must at some finite values of R intersect at least one of the $^1\Sigma_g$ or $^3\Sigma_u$ potential curves of H_2. Inside the points of intersection, a spontaneous radiationless transition can occur with the emission of an electron. The concept of an interatomic potential can be retained by allowing the potential to acquire an imaginary component so that the complex $(H_2^-)^*$ can decay. Thus, if we write the interaction potential as

$$E(R) = W(R) - \tfrac{1}{2}i\Gamma(R), \tag{2.14}$$

the wave function at time t can be written

$$\psi(t) = \psi(0)\exp[-i\{W - \tfrac{1}{2}i\Gamma\}t/\hbar] \tag{2.15}$$

and its probability density decays as

$$|\psi(t)|^2 = |\psi(0)|^2 \exp(-\Gamma t/\hbar). \tag{2.16}$$

Γ is the width of the resonating $^2\Sigma_u$ or $^2\Sigma_g$ state or complex and $\hbar/\Gamma$ is its lifetime.

The $^2\Sigma_u$ state will appear as a resonance in the scattering of electrons by molecular hydrogen in its ground state. The existence of the two resonance states also leads to peaks in the cross sections for vibrational excitation (see Fig. 2-3, Bardsley and Mandl, 1968)

$$e + H_2(v = 0) \rightarrow e + H_2(v) \tag{2.17}$$

and for *dissociative attachment* (see Fig. 2-4)

$$e + H_2 \rightarrow H + H^- \tag{2.18}$$

(Schulz and Asundi, 1967).

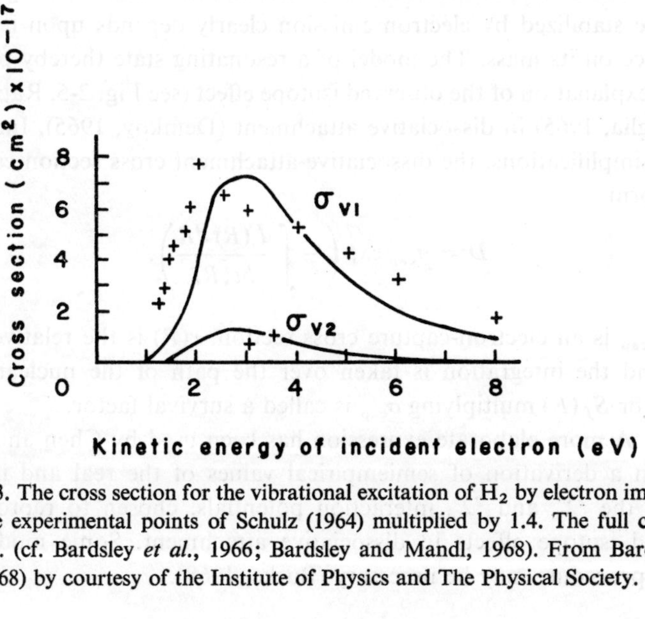

FIGURE 2-3. The cross section for the vibrational excitation of H_2 by electron impact. The crosses are experimental points of Schulz (1964) multiplied by 1.4. The full curves are theoretical. (cf. Bardsley *et al.*, 1966; Bardsley and Mandl, 1968). From Bardsley and Mandl (1968) by courtesy of the Institute of Physics and The Physical Society.

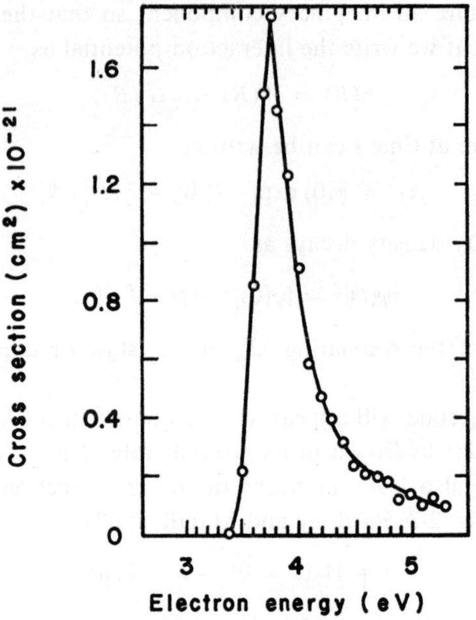

FIGURE 2-4. The cross section for dissociative attachment in collisions of electrons and H_2. From Schulz and Asundi (1957) by courtesy of the Physical Review.

The time that the quasi-molecular ion remains in the region where it can be stabilized by electron emission clearly depends upon its velocity and hence on its mass. The model of a resonating state thereby provides a natural explanation of the observed isotope effect (see Fig. 2-5, Rapp, Sharp, and Briglia, 1965) in dissociative attachment (Demkov, 1965). Indeed with certain simplifications, the dissociative-attachment cross section can be put in the form

$$D = \sigma_{cap} \exp\left(- \int \frac{\Gamma(R)\, dR}{hv(R)} \right), \qquad (2.19)$$

where σ_{cap} is an electron-capture cross section, $v(R)$ is the relative velocity at R, and the integration is taken over the path of the nuclear motion. The factor $S_f(E)$ multiplying σ_{cap} is called a survival factor.

A more elaborate expression has been used by Chen and Peacher (1968) in a derivation of semiempirical values of the real and imaginary parts of the $^2\Sigma_u$ and $^2\Sigma_g$ interaction potentials, chosen to reproduce the measured isotope effects in dissociative attachment. Some modifications of their procedure may be necessary (Burke, 1968).

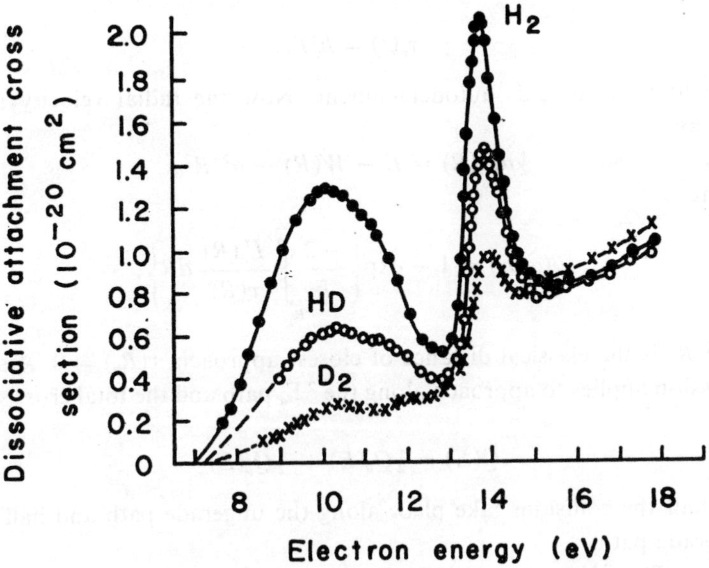

FIGURE 2-5. Cross sections for dissociative attachment in collisions of elctrons with H_2, HD, and D_2. From Rapp, Sharp, and Briglia (1956) by courtesy of the Physical Review Letters.

The theoretical prediction of the real and imaginary parts of the potential-energy curves presents a severe problem. The customary Rayleigh–Ritz variational principle, commonly used in molecular-structure calculations fails inside the stabilization regions. Suitably modified it can (if used with care) yield reliable real parts (Eliezer, Taylor, and Williams, 1967) but only partial success has attended efforts to predict the imaginary parts. Approximate theoretical values are available, however, for the $^2\Sigma_u$ and $^2\Sigma_g$ states of H_2^- (Bardsley, Herzenberg, and Mandl, 1966).

If we regard the nuclei as moving classically in the force fields derived from the real parts of the interaction potentials, we may write for the cross section for associative detachment from approach along the $^2\Sigma_u$ path

$$Q(E) = 2\pi \int_0^\infty \beta(\varrho, E)\, d\varrho, \tag{2.20}$$

where ϱ is the impact parameter, E is the energy of relative motion, and β_u is the probability that detachment has occured during the approach and separation of H and H^-. The detachment probability may be expressed

$$\beta = 1 - \exp(- \int dt/\tau_u), \tag{2.21}$$

where

$$\tau_u(t) = h/T_u \tag{2.22}$$

is the lifetime towards autodetachment. Now the radial velocity $v(R)$ is given by

$$\tfrac{1}{2}\mu v^2(R) = E - W(R) - \varrho^2/R^2, \tag{2.23}$$

so that

$$\beta(\varrho, E) = 1 - \exp\left\{\frac{-2}{\hbar}\int_{R_u}^{\infty} \frac{\Gamma_u(R)}{v(R)}\,dR\right\}, \tag{2.24}$$

where R_u is the classical distance of closest approach, $v(R_u) = 0$. A similar expression applies to approach along the ${}^2\Sigma_g$ path and the total cross section is

$$Q(E) = \tfrac{1}{2}Q_u(E) + \tfrac{1}{2}Q_g(E) \tag{2.25}$$

since half the collisions take place along the ungerade path and half along the gerade path.

The ${}^2\Sigma_g$ curve of H_2^- crosses the ${}^1\Sigma_g$ curve of H_2 at an energy ΔE above the energy at infinite separation so that no particles can penetrate into the region $R < R_2$ (cf. Fig. 2-2) unless their energy exceeds ΔE. Thus the cross section Q_g (E) is effectively zero for $E \leqq \Delta E$. In contrast the ${}^2\Sigma_u$ curve of H_2^- crosses the ${}^1\Sigma_g$ curve of H_2 at an energy lying below the energy at infinite separation and in the limit of low velocities all collisions which surmount the centrifugal barrier enter the stabilization region where they undergo detachment. At large internuclear distances, the interaction is dominated by the polarization term

$$W(R) \sim -\tfrac{1}{2}\alpha e^2/R^{-4}, \tag{2.26}$$

where α is the polarizability of atomic hydrogen.

The critical or orbiting impact parameter ϱ_0 below which the particle spirals into the origin is given by

$$\left(1 - \frac{W}{E} - \frac{\varrho^2}{R_u^2}\right) = \frac{\partial}{\partial R}\left(1 - \frac{W}{E} - \frac{\varrho^2}{R}\right)\bigg|_{R_u} = 0. \tag{2.27}$$

Using (2.26) we obtain

$$\varrho_0 = \left(\frac{2}{v}\right)^{1/2}\left(\frac{\alpha e^2}{\mu}\right)^{1/2} \tag{2.28}$$

and

$$Q = \frac{2\pi}{v}(\alpha e^2/\mu)^{1/2}, \tag{2.29}$$

where v is the initial velocity. Expression (2.27) leads to a constant rate coefficient

$$vQ = 2\pi(\alpha e^2/\mu)^{1/2} \text{ cm}^3 \text{ sec}^{-1}. \tag{2.30}$$

This is the famous Langevin formula which has been widely used in the analysis of ion–molecule reaction rates. For H–H$^-$

$$k = \tfrac{1}{2} vQ_u = 1.33 \times 10^{-9} \text{ cm}^3 \text{ sec}^{-1}. \tag{2.31}$$

The results of an explicit evaluation of (2.20) are given in Table 2-1. The value measured in the laboratory at 300 °K is 1.3×10^{-9} cm^3 sec^{-1} (Schmeltekopf, Fehsenfeld, and Ferguson, 1967).

TABLE 2-1. Rate coefficients for associative detachment, $H + H^- \rightarrow H_2 + e$, in cm^3 sec^{-1}.[a]

T(°K)	$^2\Sigma_u$	$^2\Sigma_g$	Sum $\times 10^9$
100	1.7×10^{-9}	...	1.7
300	1.95×10^{-9}	...	1.9_5
500	1.9×10^{-9}	...	1.9
1000	1.8×10^{-9}	5.8×10^{-14}	1.8
2000	1.5×10^{-9}	8.4×10^{-12}	1.5
4000	1.3×10^{-9}	1.6×10^{-10}	1.5
8000	1.3×10^{-9}	7.6×10^{-10}	2.1
16 000	1.3×10^{-9}	1.9×10^{-9}	3.2
32 000	1.5×10^{-9}	3.8×10^{-9}	5.3

[a] From Browne and Dalgarno (1969).

At high temperatures some fractions of the detachments leave H_2 in a dissociating state

$$H + H^- \rightarrow H + H + e. \tag{2.32}$$

Further work is necessary to predict the magnitude of (2.32).

It is clear from the nature of the associative–detachment reaction (2.13) and especially from its reverse—dissociative attachment—that the resulting molecules will be vibrationally developed and explicit confirmation of this fact has been given by Chen and Peacher (1968). Chen and Peacher have used their semiempirical potentials to calculate cross sections for specific vibrational levels,

$$H + H^- \rightarrow H_2(v) + e. \tag{2.33}$$

The vibrational development of H_2 may be an important aspect of the physics at early times. The influence of H_2^*, produced by (2.31), on the physics of stellar atmospheres has been explored (Lambert and Pagel, 1968).

2.4. Mutual Neutralization

The positive and negative hydrogen ions can destroy each other by a process of *mutual neutralization*,

$$H^+ + H^- \rightarrow H + H, \tag{2.34}$$

the efficiency of which is much enhanced at thermal energies by the Coulomb attraction. Suppose we allow H^+ and H^- to approach adiabatically. The initial interaction potential-energy curve $V_i(R)$ will be dominated at large separations R of the nuclei by the term

$$u_i = -e^2/R. \tag{2.35}$$

The endothermic channels have one hydrogen atom in the ground state and the other in a state of principal quantum number $n = 1, 2, 3,$ or 4:

$$H(1s) + H(n = 1, 2, 3, 4). \tag{2.36}$$

The potential curves arising from the approach of H(1s) and H(n) all decrease at least as fast as R^{-3} with increasing R. Hence in the absence of interaction between the initial and final states, there would occur crossings of the initial potential-energy curve and the possible final potential-energy curves. In practice adiabatic molecular states of the same symmetry cannot cross. Figure 2-6 illustrates the effect of the avoided crossing.

Because of the Frank–Condon principle, transitions are probable only in the regions around the crossing points and because of the essentially classical nature of the nuclear motion a particular crossing can only be traversed if it exceeds the classical distance of closest approach. The probability of crossing from one adiabatic potential curve to another is often computed using the Landau–Zener formula.

The Landau–Zener formula represents an approximate solution of a two-state description of the collision. Let $\psi_i(\mathbf{r}|R)$ and $\psi_f(\mathbf{r}|R)$ be the exact adiabatic eigenfunctions of the quasimolecule formed by the colliding systems as they approach-in initial and final states of the reacting particles. Thenif H_R is the Hamiltonian for a fixed nuclear separation R,

$$\langle \psi_i(\mathbf{r}|\mathbf{R})| \, H \, |\psi_f(\mathbf{r}|\mathbf{R})\rangle_R = 0. \tag{2.37}$$

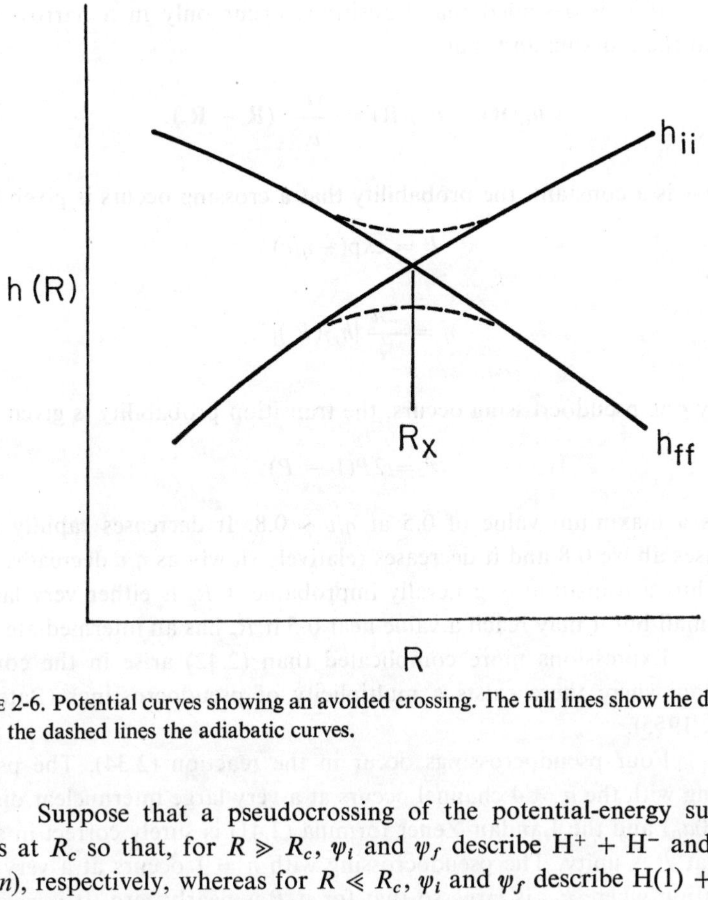

FIGURE 2-6. Potential curves showing an avoided crossing. The full lines show the diabatic curves, the dashed lines the adiabatic curves.

Suppose that a pseudocrossing of the potential-energy surfaces occurs at R_c so that, for $R \gg R_c$, ψ_i and ψ_f describe $H^+ + H^-$ and H(1) + H(n), respectively, whereas for $R \ll R_c$, ψ_i and ψ_f describe H(1) + H(n) and $H^+ + H^-$, respectively. Following Bates (1962), introduce wave functions $\phi_i(\mathbf{r}|\mathbf{R})$ and $\phi_f(\mathbf{r}|\mathbf{R})$ and define

$$h_{st}(\mathbf{R}) = \langle \phi_s(\mathbf{r}|\mathbf{R})|\, H \,|\phi_t(\mathbf{r}|\mathbf{R})\rangle. \tag{2.38}$$

The diabatic wave functions ϕ_i and ϕ_f are chosen so that h_{ii} and h_{ff} intersect at some separation R_c.

The specification of the diabatic wave functions ϕ_s and of the associated potential energy curves $h_{ss}(R)$ and transition operators $h_{st}(R)$ is not unique and different choices will be appropriate to different reactions (cf. O'Malley, 1968; Smith, 1969). The Landau–Zener formula is based upon a two-state expansion of the total wave function for the colliding system with the choice $\phi_i(\mathbf{r}|\mathbf{R}) = \psi_i(\mathbf{r}|\infty)$ and $\phi_f(\mathbf{r}|\mathbf{R}) = \psi_f(\mathbf{r}|\infty)$, $\psi_i(\mathbf{r}|\infty)$ and $\psi_f(\mathbf{r}|\infty)$ being products of atomic wave functions.

If it is assumed that transitions occur only in a narrow region around the crossing and that

$$h_{ii}(\mathbf{R}) - h_{ff}(\mathbf{R}) = \frac{\alpha \mathbf{v}}{v} \cdot (\mathbf{R} - \mathbf{R}_c), \tag{2.39}$$

where α is a constant, the probability that a crossing occurs is given by

$$P = \exp(-\eta/v), \tag{2.40}$$

where

$$\eta = \frac{2\pi}{\alpha} |h_{if}(R_c)|^2. \tag{2.41}$$

If only one pseudocrossing occurs, the transition probability is given by

$$\mathscr{P} = 2P(1 - P). \tag{2.42}$$

$\mathscr{P}$ has a maximum value of 0.5 at $\eta/v \sim 0.8$. It decreases rapidly as η/v increases above 0.8 and it decreases relatively slowly as η/v decreases below 0.8. Thus a transition is generally improbable if R_c is either very large or very small but it may reach a value near 0.5 if R_c has an intermediate value.

Expressions more complicated than (2.42) arise in the common situation where there exists a multiplicity of pseudocrossings (Bates and Lewis, 1955).

Four pseudocrossings occur in the reaction (2.34). The pseudocrossing with the $n = 4$ channel occurs at a very large internuclear distance ($\approx 200a_0$) and the Landau–Zener formula (2.41) is surely correct in asserting that P is unity. The pseudocrossing with $n = 1$ occurs at a very small separation where h_{if} is large so that for it P is nearly zero. It is necessary to take account only of the $n = 2$ and $n = 3$ state pseudocrossings.

The crossing point corresponding to the description of ϕ_i as an H^- wave function and of ϕ_f as a product of an H(1s) wave function and an H(n) wave function is near $35.6a_0$ for $n = 3$ and near $11.1a_0$ for $n = 2$. The separation h_{if} (R_c) is 9.27×10^{-3} eV for $n = 3$ and 5.26×10^{-1} eV for $n = 2$ (Bates and Lewis, 1955).

These estimates do not accurately reflect the interatomic potentials outside the crossing region where the adiabatic potentials are physically meaningful and an attempt has been made to derive alternative estimates from a semiempirical analysis of the computed adiabatic noncrossing interaction potentials (Victor, Webb, Dalgarno, and Browne 1969). The semiempirical crossing points are $25.0a_0$ and $12.1a_0$ and the semiempirical separations are 1.50×10^{-2} eV and 1.2 eV.

With the revized estimates, the contribution from the inner region leading to the $n = 2$ final channel is negligible at thermal velocities.

The first set of parameters leads to a rate coefficient for (2.32)

$$k = 7.2\left(\frac{1000}{T}\right)^{1/2} 10^{-8}\ \text{cm}^3\ \text{sec}^{-1} \tag{2.43}$$

and the second set to

$$k = 9.0\left(\frac{1000}{T}\right)^{1/2} 10^{-8}\ \text{cm}^3\ \text{sec}^{-1}. \tag{2.44}$$

The large values of k and the $T^{-1/2}$ dependence of k are consequences of the collecting action of the Coulomb attraction in the incident channel.

The limits within which the Landau–Zener formula (2.40) is valid are seldom met in practice (cf. Bates, 1962). A more elaborate multistate impact-parameter description, based upon approximate adiabatic eigenfunctions and interaction potentials (Victor *et al.*, 1969) gives values at thermal energies that are very different from the Landau–Zener results.

Recent experimental data (Rundel, Aitken, and Harrison, 1969; Moseley, Aberth, and Peterson, 1970; Gaily and Harrison, 1970) are not in harmony with either the multistate or the Landau–Zener calculations. It may be that interactions between molecular states of different angular-momentum projections play a critical role. The experiments of Moseley *et al.* suggest a rate coefficient of

$$k = 2\left(\frac{1000}{T}\right)^{1/2} 10^{-7}\ \text{cm}^3\ \text{sec}^{-1}. \tag{2.45}$$

2.5. Radiative Association

Molecular ions can be formed by *radiative association*,

$$\text{H} + \text{H}^+ \rightarrow \text{H}_2^+ + h\nu, \tag{2.46}$$

which gives rise to a continuum emission. The process, being radiative, will be much less rapid than associative detachment or mutual neutralization. Thus the collision duration at thermal velocities is of the order of 10^{-13} sec and an allowed transition may occur at a rate of the order of 10^9 sec^{-1} for visible radiation. A stabilizing photon is emitted therefore in one of every 10^4 collisions. The frequency of collisions is given by the Langevin formula (2.29) as about 10^{-9} cm^3 sec^{-1} so that the rate coefficient for radiative association will rarely exceed 10^{-13} cm^3 sec^{-1}.

A model of radiative association can be constructed analogous to that for associative detachment with the difference that stabilization of the quasimolecule (H_2^+) can occur at all separations due to the emission of a photon. Thus $\Gamma(R)/h$ is replaced in (2.24) by the radiative transition probability $A(R)$.

Since $A(R)$ is small, we may expand the exponential and employ the approximation

$$\beta(p, E) = -2\int_{R_0} \frac{A(R)}{v(R)}\, dR. \tag{2.47}$$

Expression (2.47) was used by Bates (1951) to calculate the rate of (2.46) through a transition from the $2p\sigma_u$ to $1s\sigma_g$ state of H_2^+. The rate coefficients are reproduced in Table 2-2.

TABLE 2-2. Rate coefficients for radiative association, $H + H^+ \rightarrow H_2^+ + h\nu$.

T (°K)	k (cm³ sec⁻¹)
500	1.3×10^{-18}
1000	5.2×10^{-18}
2000	1.9×10^{-17}
4000	6.2×10^{-17}
8000	1.7×10^{-16}
16000	3.5×10^{-16}
32000	5.6×10^{-16}
64000	6.0×10^{-16}

Substantial vibrational development of the product H_2^+ ions is to be expected. It is demonstrated by some calculations (Buckingham, Reid, and Spence, 1952) in which the classical description of the nuclear motion is replaced by a quantal description of the process as a free-bound transition between the initial continuum state and the final vibrational states. The calculations involve a summation over the rotational angular-momentum quantum numbers in the initial and final states.

The H_2^+ ions can be destroyed by the inverse process of *photo-dissociation*

$$H_2^+(v) + h\nu \rightarrow H + H^+. \tag{2.48}$$

Cross sections for (2.36) have been tabulated by Dunn (1968). The calculations involve the same electric-dipole matrix elements as did those for (2.46).

The cross sections show that the vibrational development of the H_2^+ ions produced by radiative association can markedly enhance the destruction rate by photodissociation. The vibrationally excited H_2^+ cannot decay radiatively by electric-dipole transitions.

2.6. Charge Transfer

The H_2^+ ions can be converted into H_2 molecules by *charge transfer*,

$$H_2^+ + H \rightarrow H_2 + H^+, \tag{2.49}$$

and the sequence of (2.46) and (2.49) has been advocated by Saslaw and Zipoy (1967) as a mechanism for the early production of H_2. It appears that (2.13) is more efficient (Hirasawa, Aizu, and Taketani, 1969). McDowell (1961) had earlier noted that the associative detachment of H and H^- is a source of H_2 in HI regions.

There is no adequate prescription for calculating charge-transfer cross sections except in cases of exact symmetrical resonance such as

$$H^+ + H \rightarrow H + H^+, \tag{2.50}$$

$$H^- + H \rightarrow H + H^-. \tag{2.51}$$

Resonance charge-transfer reactions can be important cooling mechanisms in physical circumstances where the ion and neutral particles have different temperatures and the symmetry effects markedly reduce the ionic mobilities and diffusion coefficients (cf. Dalgarno, 1961).

From laboratory studies, it appears that many charge-transfer reactions involving molecules proceed at every collision and in cases where no explicit measurements exist it is usual to adopt a value for the cross section less than but of the order of the Langevin cross section (2.29). Thus Saslaw and Zipoy (1967) adopt a cross section of 10^{-15} cm^2 for (2.49).

The H_2^+ ions can also be removed by mutual neutralization,

$$H_2^+ + H^- \rightarrow H_2 + H, \tag{2.52}$$

$$H_2^+ + H^- \rightarrow H + H + H. \tag{2.53}$$

There are no calculations or measurements but some general considerations by Bates and Boyd (1956) based upon the Landau–Zener formula and the observation that the internal vibrational modes generate a large number of pseudocrossings suggest rate coefficients of the order of 10^{-7} cm^3 sec^{-1} or larger.

2.7. Ion–Molecule Reactions

The H_2^+ and the H_2 can undergo an *ion–molecule reaction*,

$$H_2^+ + H_2 \rightarrow H_3^+ + H, \tag{2.54}$$

producing H_3^+. The molecular ion H_2^+ has an interesting radio-frequency spectrum (cf. Somerville, 1968; Luke, 1969) but (2.54) and (2.49) may be effective in limiting the abundance of H_2^+ in the interstellar medium.

The Langevin formula predicts a rate coefficient of about 2×10^{-9} cm³ sec⁻¹ if reaction proceeds on every collision and it appears that this is the case. Laboratory data yield a rate coefficient of 10^{-9} cm³ sec⁻¹ at 300 °K.

2.8. Dissociative Recombination

Dissociative recombination is the process

$$H_2^+ + e \rightarrow H + H. \tag{2.55}$$

Dissociative recombination is generally the most efficient recombination process in low-density plasmas. The reverse process is *associative ionization*,

$$H + H \rightarrow H_2^+ + e. \tag{2.56}$$

Like dissociative attachment, *dissociative recombination* (and associative ionization) can be regarded as proceding through the formation of a temporary resonance state or molecular complex which stabilizes by dissociation (Bates, 1950; Bates and Dalgarno, 1962; Bardsley, 1968). Thus

$$H_2^+ + e \rightarrow (H_2)^* \rightarrow H + H. \tag{2.57}$$

In order for the process to be efficient, the interaction potential of the initial state of H_2^+ must cross that of the resonance state near to the equlibrium separation of H_2^+. The required configuration of real interaction potentials is illustrated in Fig. 2-7. Analogous to the cross section (2.19) for dissociative attachment, the cross section can be written as a product of a capture cross section $\sigma_{\mathrm{cap}}(E)$ for formation of the resonance state and a survival factor $S_f(E)$. If the electron energy is ε, the capture cross section is proportional to the probability $f(\varepsilon)$ of finding the nuclei in a small region around R_c. Bates (1950) adopted for the survival factor $S_f(E)$

$$S_f(E) = \frac{\tau_a(\varepsilon)}{\tau_a(\varepsilon) + \tau_p(\varepsilon)}, \tag{2.58}$$

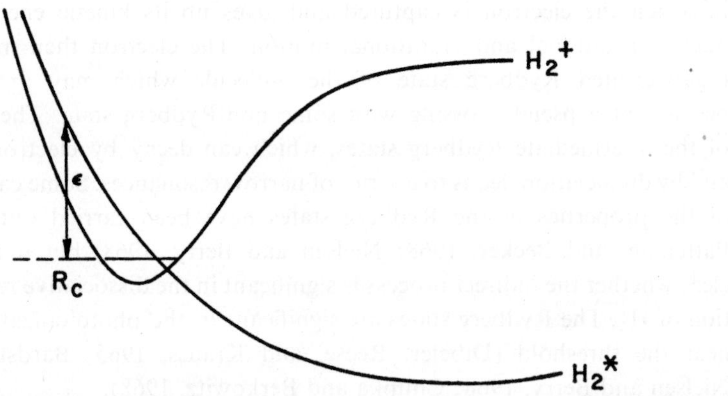

FIGURE 2-7. Interaction potentials for dissociative recombination. The electron energy is ε.

where $\tau_a(\varepsilon)$ is the lifetime of H_2^+ (at R_c) towards autoionization (electron emission) and $\tau_p(\varepsilon)$ is the time taken for the atoms to move apart to distances where the molecule is stable against autoionization and must dissociate. Bardsley (1968) has derived an alternative form for $S_f(E)$ which is less easy to interpret physically.

If the autoionization lifetime is long compared to the time for the stabilizing dissociation process, the recombination coefficient for a particular vibrational level varies approximately with electron temperature T_e as $T_e^{-1/2}$ (Bardsley, 1968). The recombination coefficient may well be sensitive to the vibrational distribution of the molecular ion. The behaviour with vibrational temperature is difficult to predict because it depends upon the detailed nature of the interaction potentials.

The resonant states that contribute substantially to the dissociative recombination of H_2^+ are those for which both electrons are in $2s\sigma_g$, $2p\sigma_u$, or $2p\pi_u$ orbitals. Some preliminary studies have been carried out of the molecular states

$$(2p\sigma_u)(2p\pi_u)\,^3\Pi_g, \quad (2p\sigma_u)(2s\sigma_g)\,^3\Sigma_u, \quad (2p\sigma_u)^2\,^1\Sigma_g^+$$

(Bauer and Wu, 1956; Wilkins, 1966; Dubrovsky, Ob'edkov, and Janev, 1967) and rate coefficients of the order of several times 10^{-8} cm^3 sec^{-1} have been obtained. The accuracy of these predictions is low. The major uncertainty is probably the calculation of the interaction potentials of the resonating states.

Bardsley (1968) has pointed out that in addition to the direct process for dissociative recombination there may occur an indirect process in which the electron is captured and gives up its kinetic energy to the nuclear rotational and vibrational motion. The electron then moves in a highly-excited Rydberg state of the molecule which may predissociate because of a pseudocrossing with some non-Rydberg state. The existence of the intermediate Rydberg states, which can decay by electron emision and by dissociation, leads to a series of narrow resonances. Some calculations of the properties of the Rydberg states have been carried out (Russek, Patterson, and Becker, 1968; Nielsen and Berry, 1968) but it is not yet clear whether the indirect process is significant in the dissociative recombination of H_2^+. The Rydberg states are significant in the photoionization of H_2 near the threshold (Dibeler, Reese, and Krauss, 1965; Bardsley, 1968; Nielsen and Berry, 1968; Chupka and Berkowitz, 1968).

Experimentally Persson and Brown (1955) have concluded that in H_2 the coefficient is less than 3×10^{-8} cm^3 sec^{-1}, a conclusion in harmony with the observations of Popov and Afaneseva (1960). The positive ions in the H_2 afterglow have not been identified. They may not be H_2^+. Because of (2.54), the ions may be (and probably are) H_3^+. The dissociative recombination of H_3^+ may lead back to H_2,

$$H_3^+ + e \rightarrow H_2 + H. \tag{2.59}$$

2.9. Photodissociation

The destruction of H_2^+ by *photodissociation*,

$$H_2^+(1s\sigma_g) + h\nu \rightarrow H_2^+(2p\sigma_u) \rightarrow H^+ + H, \tag{2.60}$$

was noted earlier in Sec. 2.5. The photodissociation proceeds efficiently through an electric-dipole transition, producing a proton and a hydrogen atom in its ground state.

The analogous process for H_2,

$$H_2(X\,{}^1\Sigma_g^+) + h\nu \rightarrow H_2(b\,{}^3\Sigma_u^+) \rightarrow H + H, \tag{2.61}$$

is very slow because of the change in spin multiplicity. The process has the same characteristics as the absorption,

$$He(1\,{}^1S_0) + h\nu \rightarrow He(2\,{}^3P), \tag{2.62}$$

to which it tends in the limit of vanishing nuclear separations. Our earlier discussion of the reverse of (2.62) tells us that (2.61) can proceed through an electric-dipole transition with spin–orbit mixing and by a magnetic-quadrupole transition, and it is to be expected from Table 1-4 that the electric-dipole transition will be much more probable.

The oscillator strength of (2.58) has been estimated by Kahn (1955) as 10^{-5} and by Goldu and Salpeter (1963) as 10^{-8}. An explicit calculation has been carried out by Bottcher and Browne (1969) who evaluate the spin–orbit interaction between the $b\,^3\Sigma_u^+$ and $C\,^1\Pi_u$ states. They obtain a lifetime of 1 sec for the $b\,^3\Sigma_u^+$ state corresponding to an oscillator strength of the order of 10^{-11}. The process is unlikely to compete with other destruction mechanisms.

Because (2.58) is slow so is the reverse process of radiative association,

$$H(1s) + H(1s) \rightarrow H_2 + h\nu. \tag{2.63}$$

A rate coefficient of 4×10^{-27} cm³ sec⁻¹ at 100 °K has been estimated by Malville (1964). The value is probably an overestimate.

Molecular hydrogen can be photodissociated directly by absorption of radiation into the continua of the electronically excited singlet states. The strongest transitions are the electric-dipole transitions to the $B\,^1\Sigma_g^+$ and $C\,^1\Pi_u$ states which dissociate into H(1*s*) and H(2*p*). Their common threshold is 846 Å if all the H_2 is in the ground vibrational level.

We have noted already that H_2 is produced by associative detachment in a highly vibrating state and the vibrational development can also be enhanced through fluorescence by discrete absorption and emission,

$$H_2(X\,^1\Sigma_g^+, 0) + h\nu \rightarrow H_2(B\,^1\Sigma_u^+, v'), \tag{2.64}$$

followed by

$$H_2(B\,^1\Sigma_u^+, v') \rightarrow H_2(X\,^1\Sigma_g^+, v'') + h\nu. \tag{2.65}$$

Similar processes can occur through other excited levels.

Photodissociation of vibrationally excited $H_2(X\,^1\Sigma_g^+, v'')$ not only has a threshold at longer wavelengths but the cross sections may be much larger also. It is clear that depending upon the physical environment a substantial increase in the destruction rate of H_2 can result from vibrational excitations. The vibrationally excited levels can decay only slowly through electric-quadrupole transitions. Of critical importance are *vibrational deactivation processes*,

$$H + H_2(v_2) \rightarrow H + H_2(v_2), \tag{2.66}$$

and similar reactions caused by electrons and protons, and *vibrational interchange* processes,

$$H_2(v_1) + H_2(v_2) \rightarrow H_2(v_3) + H_2(v_4). \tag{2.67}$$

The photodissociation cross sections are readily evaluated once the electric-dipole transition moment has been determined. If $\psi_i(\mathbf{r}/\mathbf{R})$ and $\psi_f(\mathbf{r}|\mathbf{R})$ are the adiabatic wave functions of the initial and final states, the transition moment is given by

$$D(R) = e\langle\psi_i(\mathbf{r}|\mathbf{R})|\,\mathbf{r}\,|\psi_f(\mathbf{r}|\mathbf{R})\rangle \tag{2.68}$$

and the photodissociation cross section is obtained from the matrix element

$$m = \langle\chi_i(R)|\,D(R)\,|\chi_f(R)\rangle, \tag{2.69}$$

where χ_i is the initial discrete rotation-vibration nuclear eigenfunction and χ_f is the final continuum nuclear eigenfunction. The functions χ_i and χ_f are determined by solving differential equations for motion in the adiabatic interaction potentials.

Various calculations of $D(R)$ have been carried out (cf. Browne, 1969). $D(R)$ can also be obtained from analysis of measurements of the energy losses at fixed angles of fast electrons scattered by the atoms or molecules of the gas (cf. Geiger and Schmoransky, 1969). The experiments can be interpreted using the Born approximation, according to which the scattering cross section at zero angle is related directly to the electric-dipole moment (cf. Lassettre, 1969). Thus the Born approximation reduces the scattering matrix element to that of the operator $\exp(i\mathbf{K}\cdot\mathbf{r})$ where $\mathbf{K}$ is the momentum change. When K is small, $\exp(i\mathbf{K}\cdot\mathbf{r})$ can be replaced by $i\mathbf{K}\cdot\mathbf{r}$. $D(R)$ can also be obtained more directly from absorption measurements (cf. Hesser, Brooks, and Lawrence, 1968).

The predicted photodissociation cross sections of $H_2(X\,^1\Sigma_g^+, v'' = 0)$ are shown in Fig. 2-8 (Dalgarno and Allison, 1969). It is interesting to observe that the major contribution to the direct photodissociation is the weaker transition

$$H_2(X\,^1\Sigma_g^+) + h\nu \rightarrow H_2(B'\,^1\Sigma_u^+) \rightarrow H(1s) + H(2s). \tag{2.70}$$

This is a consequence of the adiabatic potentials which control the overlap of the initial and final vibrational wave functions. Most of the oscillator strengths of the stronger $B\,^1\Sigma_u^+$ and $C\,^1\Pi_u$ transitions are taken up by transitions to discrete vibrational levels.

The absorption spectrum of H_2 at wavelengths shorter than 845 Å is much more complicated than that presented by the three molecular states considered in Fig. 2-8. Many other states contribute significantly, some of which predissociate and some of which autoionize (or pre-ionize). There is a great deal of resonance structure. Actually the absorption to the $C\,^1\Pi_u$ state has structural features near the spectral head, arising from

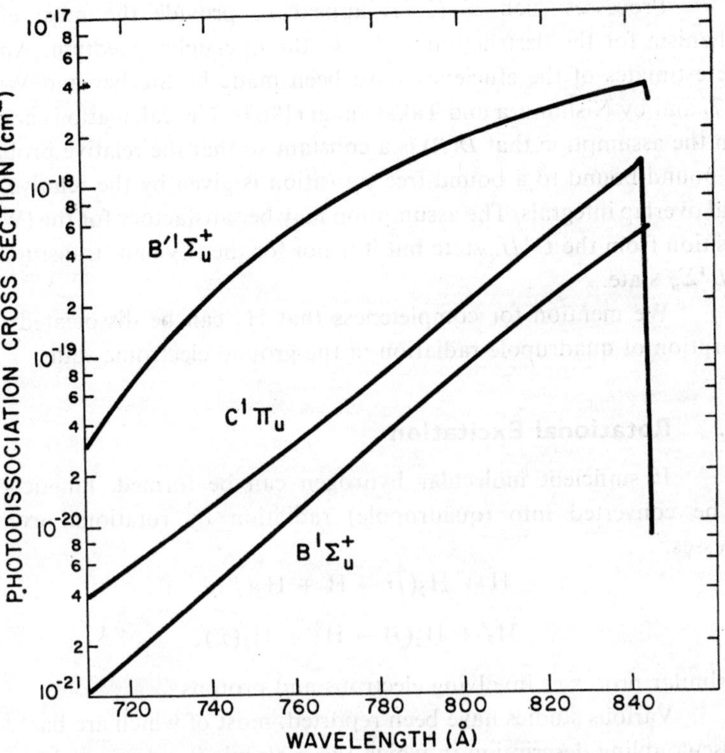

FIGURE 2-8. The photodissociation cross sections of H_2 into the *B*, *B′*, and *C* states. There is structure close to the dissociation limit which is not shown.

shape resonances trapped by the repulsive barrier that exists in the adiabatic interaction potential. The photoionization cross section

$$H_2 + h\nu \rightarrow H_2^+ + e \tag{2.71}$$

has peaks near the threshold from the autoionization of Rydberg states (Chupka and Berkowitz, 1968).

Even in the absence of vibrational excitation, H_2 can still be destroyed by radiation longer than 846 Å because some fraction of the emissions from the electronically excited states will terminate in continuum vibrational levels of the ground electronic state. Thus

$$H_2(B\,^1\Sigma_u^+, v') \rightarrow H_2(X\,^1\Sigma_g^+, k'') + h\nu, \tag{2.72}$$

where k'' is the wave number of the nuclear motion.

Processes such as (2.72) appear to provide the most efficient mechanism for the destruction of H_2 in the interstellar medium. Approximate estimates of the efficiency have been made by Stecher and Williams (1967) and by Nishimura and Takayanagi (1969). The calculations are based upon the assumption that $D(R)$ is a constant so that the relative probability of a bound-bound to a bound-free transition is given by the relative vibrational overlap integrals. The assumption may be satisfactory for the (Werner) transition from the $C\,^1\Pi_u$ state but it is not for the (Lyman) transition from the $B\,^1\Sigma_u^+$ state.

We mention for completeness that H_2 can be dissociated by the absorption of quadrupole radiation in the ground electronic state.

2.10. Rotational Excitation

If sufficient molecular hydrogen can be formed, kinetic energy can be converted into (quadrupole) radiation by rotational excitation processes,

$$\mathrm{H} + \mathrm{H}_2(j) \rightarrow \mathrm{H} + \mathrm{H}_2(j'), \tag{2.73}$$

$$\mathrm{H}_2 + \mathrm{H}_2(j) \rightarrow \mathrm{H}_2 + \mathrm{H}_2(j'), \tag{2.74}$$

and similar processes involving electrons and protons.

Various studies have been reported, most of which are based upon a close-coupling description in which the molecule is represented as a rigid rotator and the interaction between the colliding systems is expanded in spherical harmonics

$$V(\mathbf{R}) = V_0(\mathbf{R}) + V_2(\mathbf{R})\,P_2(\cos\theta), \tag{2.75}$$

where θ is the angle between the rotator axis and the direction of initial motion (cf. Arthurs and Dalgarno, 1960). The orientation-dependent term couples the different rotational levels.

In a partial-wave analysis, the total scattering wave function is expressed as sum of products of radial waves and angular eigenfunctions, each of which is associated with a total angular momentum J, compounded from the angular momentum j of the rotating molecule and the angular momentum l of the relative motion. The scattering problem reduces to the solution of a set of coupled second-order differential equations, coupled by the potential $\langle jlJ|\, V_2 \,|j'l'J\rangle$ such that $\mathbf{j} + \mathbf{l} = \mathbf{j}' + \mathbf{l}' = \mathbf{J}$ (cf. Takayanagi, 1965).

For H–H_2 and H_2–H_2 collisions, the most refined calculations are those of Allison and Dalgarno (1967) who solved numerically the close-coupling equations. They retained only open channels. It would be interesting to explore the narrow rotational resonances that the inclusion of closed channels can generate.

The results of Allison and Dalgarno (1967) for

$$H + H_2(j = 0) \rightarrow H + H_2(j = 2)$$

and (2.76)

$$H + H_2(j = 0) \rightarrow H + H_2(j = 4)$$

are reproduced in Table 2-3. Much more extensive calculations have been carried out by Takayanagi and Nishimura (1960) and Nishimura (1968).

TABLE 2-3. Rate coefficients for rotational excitation of H_2, $H + H_2(j) \rightarrow H + H_2(j')$, in $cm^3\ sec^{-1}$

T(°K)	j–j' 0–2	j–j' 0–4
20	4×10^{-23}	...
100	1.3×10^{-13}	1.6×10^{-20}
200	3.1×10^{-12}	2.1×10^{-16}
500	3.9×10^{-11}	1.7×10^{-13}
1000	1.4×10^{-10}	3.0×10^{-12}

The uncertainties in all these calculations are high—perhaps an order of magnitude or more even for the $j = 0$–2 transition. If the orientation-dependent term of the potential is of the order of λ, it is obvious that for small λ the cross section is of the order λ^2. Present methods for predicting intermolecular forces are inadequate even for the simple case of a pair of interacting atoms.

The rotational excitation of molecular hydorgen is a basic process in the cooling of the interstellar clouds (cf. Field, Rather, Aannestad, and Orszag, 1968; Nishimura, 1968).

2.11. Penning Ionization

The inclusion of helium amongst the reactants probably has little effect on the overall development of the system. Of possible importance are the metastable atoms which can be destroyed in *Penning ionization* reactions with the production of electrons.

An example of a Penning ionization process is

$$He^* + H \rightarrow He + H^+ + e, \quad (2.77)$$

where He* is an excited helium atom. Similar processes can occur in the collision of two excited hydrogen atoms:

$$H(n) + H(n') \rightarrow H + H^+ + e, \tag{2.78}$$

$$H(n) + H(n') \rightarrow H_2^+ + e. \tag{2.79}$$

A semiclassical description can be used for the nuclear motion. At any separation R there is a finite probability that the highly excited molecule (HeH)* will spontaneously autoionize in a radiationless transition with the emission of an electron.

The theoretical difficulties encountered in calculating the complex interaction potential of the initial state are severe. In the case when He* and He are connected by an allowed transition, a perturbation description leads at large separations to the formula

$$\Gamma(R) = \frac{2\pi}{R^6} |\boldsymbol{\mu}_A \cdot \boldsymbol{\mu}_B - 3(\boldsymbol{\mu}_A \cdot \mathbf{R})(\boldsymbol{\mu}_B \cdot \mathbf{R})/R^2|^2 \tag{2.80}$$

for the autoionizing width at separation R, where μ_A is the dipole moment of the transition from He* to He and μ_B is the dipole moment of the ionizing transition from H to $H^+ + e$. If a straight line path is taken for the relative motion of the particles, the cross section is

$$\sigma = 13.9\left(\frac{\mu_A^2 \mu_B^2}{\hbar v}\right)^{2/5} \tag{2.81}$$

(Watanabe and Katsuura, 1967).

Deviations from the straight line path may have significant effects. The orbiting model that leads to the Langevin formula for ion–molecule reactions has been used by Bates, Bell, and Kingston (1967) and by Bell, Kingston, and Dalgarno (1968). The applicability of these simple models is open to question but there are adequate experimental data to support the general conclusion that the process can be rapid. The rate coefficients of (2.74) may be of the order of 10^{-9} cm^3 sec^{-1}.

The subject has been opened to more detailed theoretical study by recent developments by which the energy distribution of the electron can be measured (cf. Cermak and Herman, 1968).

REFERENCES

ACCAD, Y., C. PEKERIS, and B. SCHIFF (1969), Phys. Rev. **183**, 78.

AKHIEZER, A. I., and V. B. BERESTETSKII (1965), *Quantum Electrodynamics* (John Wiley, New York).

ALLISON, A. C., and A. DALGARNO (1967), Proc. Phys. Soc. (London) **90**, 609.

ARTHURS, A. M., and A. DALGARNO (1960), Proc. Roy. Soc. (London) ***A*156**, 551.

ARTURA, C. J., R. NOVICK, and N. TOLK (1969), Astrophys. J. Letters **157**, L181.
BARDSLEY, J. N., A. HERZENBERG, and F. MANDL (1966), Proc. Phys. Soc. (London) **89**, 305 and 321.
BARDSLEY, J. N. (1968), Proc. Phys. Soc. (London) **89**, 305 and 321.
BARDSLEY, J. N., and F. MANDL (1968), Rept. Progr. Phys. **31**, 471.
BATES, D. R. (1950), Phys. Rev. **77**, 718; **78**, 492.
BATES, D. R. (1951), Monthly Notices Roy. Astron. Soc. **111**, 303.
BATES, D. R. (1952), *ibid.* **112**, 40.
BATES, D. R. (1962), in *Atomic and Molecular Processes* (Academic Press, New York).
BATES, D. R., and H. S. W. MASSEY (1954), Phil. Mag. **45**, 111.
BATES, D. R., and J. T. LEWIS (1955), Proc. Phys. Soc. (London) **A68**, 173.
BATES, D. R., and T. J. M. BOYD (1956), Proc. Phys. Soc. (London) **A69**, 910.
BATES, D. R., and A. DALGARNO (1962), in *Atomic and Molecular Processes* (Academic Press, New York).
BATES, D. R., A. E. KINGSTON, and R. W. C. MCWHIRTER (1962), Proc. Roy. Soc. (London) **A267**, 297.
BATES, D. R., K. L. BELL, and A. E. KINGSTON (1967), Proc. Phys. Soc. (London) **91**, 288.
BAUER, E., and T. Y. WU (1956), Can. J. Phys. **34**, 1436.
BELL, K. L., A. DALGARNO, and A. E. KINGSTON (1968), J. Phys. B (Proc. Phys. Soc.) **1**, 18.
BELY, O. (1968), J. Phys. B (Proc. Phys. Soc.) **1**, 718.
BELY, O., and P. FAUCHER (1969), Astron. Astrophys. **1**, 37.
BETHE, H., and E. E. SALPETER (1957), *Quantum Mechanics of One- and Two-Electron Systems* (Academic Press, New York).
BOLAND, B. C., F. E. IRONS, and R. W. MCWHIRTER (1968), J. Phys. B (Proc. Phys. Soc.) **1**, 1180.
BOTTCHER, C., and J. C. BROWNE (1969), private communication.
BRANSCOMB, L. M. (1962), in *Atomic and Molecηlar Processes*, edited by D. R. Bates (Academic Press, New York).
BREIT, G., and E. TELLER (1940), Astrophys. J. **91**, 215.
BROWNE, J. C. (1969), Astrophys. J. **156**, 397.
BROWNE, J. C., and A. DALGARNO (1969), J. Phys. B (Proc. Phys. Soc.) **2**, 885.
BUCKINGHAM, R. A., S. REID, and R. SPENCE (1952), Monthly Notices Roy. Astron. Soc. **112**, 382.
BURGESS, A. (1964), Monthly Notices Roy. Astron. Soc. **69**, 1.
BURKE, P. G., and D. D. MCVICAR (1865), Proc. Phys. Soc. (London) **86**, 989.
BURKE, P. G. (1968), J. Phys. B (Proc. Phys. Soc.) **1**, 586.
CAPRIOTTI, E. R. (1967), Astrophys. J. **150**, 95.
CASIMIR, H. B. G., and D. POLDER (1948), Phys. Rev. **73**, 360
CERMAK, V , and Z HERMAN (1968), Chem Phys Letters **2**, 359
CHAN, Y M , and A DALGARNO (1965), Proc Phys Soc (London) **85**, 227.
CHEN, J. Y. C., and J. L. PEACHER (1968), Phys. Rev. **167**, 30.
CHUPKA, W. A., and J. BERKOWITZ (1968), J. Chem. Phys. **48**, 5726.
DALGARNO, A. (1961), Ann. Geophys. **17**, 16.
DALGARNO, A., and A. C. ALLISON (1969), J. Geophys. Res. **74**, 4178.
DALGARNO, A., and G. W. F. DRAKE (1969), Mém. Soc. Roy. Sci. Liège **17**, 69.
DALGARNO, A., and S. T. EPSTEIN (1969), J. Chem. Phys. **50**, 2837.

DEMKOV, Y. N. (1965), Phys. Letters **15**, 235.
DIBELER, V. H., R. M. REESE, and M. KRAUSS (1965), J. Chem. Phys. **42**, 2045.
DOUGHTY, N. A., P. A. FRASER, and R. P. MCEACHRAN (1966), Monthly Notices Roy. Astron. Soc. **192**, 255.
DRAKE, G. W. F., and A. DALGARNO (1968), Astrophys. J. Letters **152**, L121.
DRAKE, G. W. F. (1969), Astrophys. J. **158**, 1199.
DRAKE, G. W. F. (1969), unpublished.
DRAKE, G. W. F. (1972), Phys. Rev., **A3**, 908.
DRAKE, G. W. F., and A. DALGARNO (1969), Astrophys. J. **157**, 459.
DRAKE, G. W. F., G. A. VICTOR, and A. DALGARNO (1969), Phys. Rev. **180**, 25.
DUBORVSKY, G. V , V. D. OB'EDKOV, and R. K. JANEV (1967), *Fifth International Conference on Atomic Collisions, Abstracts* (Nauka, Leningrad).
DUNN, G. H. (1968), Phys. Rev. **172**, 1.
DUPREE, A., and L. GOLDBERG (1969), Astrophys. J. Letters **158**, L49.
ELIEZER, I., H. S. TAYLOR, and J. K. WILLIAMS (1967), J. Chem. Phys. **47**, 2165.
ELTON, R. C. (1967), Astrophys. J. **148**, 573.
ELTON, R. C., L. J. PALUMBO, and H. R. GRIEM (1968), Phys. Rev. Letters **20**, 783.
EVANS, K., and K. A. POUNDS (1968), Astrophys. J. **152**, 319.
FEINBERG, G. (1958), Phys. Rev. **112**, 1637.
FELD, G. B., J. D. G. RATHER, P. A. AANNESTAD, and S. A. ORSZAG (1968), Astrophys. J. **151**, 953.
GABRIEL, A. H., and C. JORDAN (1969a), Nature **221**, 947.
GABRIEL, A. H., and C. JORDAN (1969b), Monthly Notices Roy. Astron. Soc. **145**, 241.
GAILY, T. D., and H. F. A. HARRISON (1970), J. Phys. B (Atomic Molecular Phys.) **3**, L25.
GARSTANG, R. H. (1967), Astrophys. J. **148**, 579.
GAVRILA, M. (1967), Phys. Rev. **163**, 147.
GEIGER, J., and H. SCHMORANSKY (1969), J. Mol. Spectr. **32**, 39.
GEROLA, H., and N. PANAGIA (1968), Astrophys. Space Sci. **2**, 285.
GOULD, R. J., and E. E. SALPETER (1963), Astrophys. J. **138**, 393.
GREEN, L. C., P. P. RUSH, and C. D. CHANDLER (1957), Astrophys. J. Suppl. **3**, 37.
GRIEM, H. R. (1969), Astrophys. J. Letters **156**, L109; **161**, 155 (1970) [Erratum].
HESSER, J. E., N. H. BROOKS, and G. M. LAWRENCE (1968), J. Chem. Phys. **49**, 5488.
HIRASAWA, T., K. AIZU, and M. TAKETANI (1969), Progr. Theoret. Phys. (Kyoto) **43**, No. 3.
HUANG, S. S. (1948), Astrophys. J. **108**, 354.
JACOB, M., and G. C. WICK (1959), Ann. Phys. (New York) **7**, 404.
KAHN, F. D. (1955), Mém. Soc. Roy. Sci. Liège **15**, 393.
KIPPER, A. YA., and V. M. TIIT (1958), Vopr. Kosmogonii, Akad. Nauk SSSR **6**, 99.
LAMBERT, D. L., and B. E. J. PAGEL (1968), Monthly Notices Roy. Astron. Soc. **141**, 299.
LANDAU, L. (1948), Dokl. Akad. Nauk SSSR **60**, 207.
LASSETTRE, E. N. (1969), Can. J. Chem. **47**, 1733.
LEVINSON, I. B., and A. A. NIKITIN (1965), *Handbook for Theoretical Computation of Line Intensities in Atomic Spectra* (Davey Press, New York).
LUKE, S. K. (1969), Astrophys. J. **156**, 761.
MACEK, J. (1967), Proc. Phys. Soc. (London) **92**, 265.
MALVILLE, J. M. (1964), Astrophys. J. **139**, 198.
MATHIS, J. S. (1957), Astrophys. J. **125**, 328.
MCDOWELL, M. R. C. (1961), Observatory **81**, 240.

MENZEL, D. H. (1968), Astrophys. J. Suppl. **18**, 221.
MIZUSHIMA, M. (1964), Phys. Rev. **134**, A883.
MOSELEY, J., W. ABERTH, and J. R. PETERSON (1970), Phys. Rev. Letters **24**, 435.
NIELSEN, S. E., and R. S. BERRY (1968), Chem. Phys. Letters **2**, 503.
NISHIMURA, S. (1968), Publ. Astron. Soc. Japan **20**, 39.
NISHIMURA, S., and L. TAKAYANAGI (1969), Publ. Astron. Soc. Japan **21**, 111.
O'DELL, C. R. (1965), Astrophys. J. **142**, 1093.
O'MALLEY, T. F. (1967), Phys. Rev. **162**, 98.
OSTERBROOK, D. E (1964), Ann. Rev. Astron. Astrophys. **2**, 95.
PAGEL, B. E. J. (1969), Nature **221**, 325.
PEARL, A. S. (1970), Phys. Rev. Letters **24**, 703.
PERSSON, K. B., and S. C. BROWN (1955), Phys. Rev. **100**, 729.
PLUTA, K. M., and M. R. C. MCDOWELL (1966), Proc. Phys. Soc. (London) **89**, 299.
POPOV, N. A., and E. A. AFANESEVA (1960)G Soviet Phys.—Tech. Phys. **4**, 764.
PRASAD, A. N., and M. F. EL-MENSHAWY (1968), J. Phys. B (Proc. Phys. Soc.) **1**, 47.
RAPP, S., T. E. SHARP, and D. D. BRIGLIA (1965), Phys. Rev. Letters **14**, 533.
ROBBINS, R. R. (1968), Astrophys. J. Letters **151**, L35.
RUNDEL, R. D., K. L. AITKEN, and M. F. A. HARRISON (1969), J. Phys. B **2**, 954.
RUSSEK, A., M. R. PATTERSON, and R. L. BECKER (1968), Phys. Rev. **167**, 17.
SALPETER, E. E. (1958), Phys. Rev. **112**, 1645.
SARAPH, H. E. (1964), Proc. Phys. Soc. (London) **83**, 763.
SASLAW, W. C., and D. ZIPOY (1967), Nature **216**, 976.
SCHMELTEKOPF, A. L., F. C. FEHSENFELD, and E. E. FERGUSON (1967), Astrophys. J. Letters **148**, L155.
SCHMIEDER, R. W., and MARRUS, R. (1970) Phys. Rev. Letters **25**, 1245.
SCHULZ, G. J. (1964), Phys. Rev. **135**, A988.
SCHULZ, G. J., and R. K. ASUNDI (1967), Phys. Rev. **158**, 25.
SEATON, M. J. (1968), Advan. Atom. Mod. Phys. **4**, 331.
SEJNOWSKI, T. J., and R. M. HJELLMING (1969), Astrophys. J. **156**, 915.
SHAPIRO, J., and G. BREIT (1959), Phys. Rev. **113**, 179.
SITZ, P., and R. YARIS (1968), J. Chem. Phys. **49**, 3546.
SMITH, F. T. (1969), Phys. Rev. **179**, 111.
SOMERVILLE, W. B. (1968), Monthly Notices Roy. Astron. Soc. **139**, 163.
SPITZER, L., and J. L. GREENSTEIN (1951), Astrophys. J. **114**, 407.
STAUFFER, A. D., and M. R. C. MCDOWELL (1966), Proc. Phys. Soc. (London) **89**, 289.
STECHER, T. P., and D. A. WILLIAMS (1967), Astrophys. J. Letters **149**, L29.
TAKAYANAGI, K. (1965), Advan. Atom. Mol. Phys. **1**, 149.
TAKAYANAGI, K., and S. NISHIMURA (1960), Publ. Astron. Soc. Japan **12**, 77.
VETCHINKIN, S. I., and S. V. KCHRISTENKO (1967), Chem. Phys. Letters **1**, 437.
VICTOR, G. A., G. WEBB, A. DALGARNO, and J. C. BROWNE (1969), in preparation.
WATANABE, T., and K. KATSUURA (1967), J. Chem. Phys. **47**, 800.
WIDING, K. G., and G. D. SANDLIN (1968), Astrophys. J. **152**, 345.
WIESE, W. L. ,M. W. SMITH, and B. M. GLENNON (1966), *Atomic Transition Probabilities* (National Bureau of Standards, Washington, D.C.), Vol. I.
WILKINS, R. L. (1966), J. Chem. Phys. **44**, 1884.